AI-driven Innovations in Physiotherapy and Oncology 1

AI-driven Innovations in Physiotherapy and Oncology 1

Edited by

Abhishek Kumar
Priya Batta
Sachin Ahuja
Pramod Singh Rathore

ISTE

WILEY

First published 2025 in Great Britain and the United States by ISTE Ltd and John Wiley & Sons, Inc.

ISTE Ltd
27-37 St George's Road
London SW19 4EU
UK

www.iste.co.uk

John Wiley & Sons, Inc.
111 River Street
Hoboken, NJ 07030
USA

www.wiley.com

Library of Congress Control Number: 2025944644

British Library Cataloguing-in-Publication Data
A CIP record for this book is available from the British Library
ISBN 978-1-83669-047-4

The manufacturer's authorized representative according to the EU General Product Safety Regulation is Wiley-VCH GmbH, Boschstr. 12, 69469 Weinheim, Germany, e-mail: Product_Safety@wiley.com.

Contents

Preface

This book delves into the transformative role of artificial intelligence in revolutionizing cancer care and rehabilitation. As cancer survivorship rises, the demand for integrated, personalized physiotherapy is becoming crucial in enhancing quality of life. This book brings together cutting-edge research and real-world applications where AI empowers early diagnosis, precise oncology treatment and intelligent, adaptive physiotherapy. By bridging these two vital domains, it aims to inform clinicians, researchers and healthcare technologists about the potential of AI to deliver holistic, efficient and patient-centric care. It is a timely resource for the future of integrated medical innovation.

Abhishek KUMAR
Priya BATTA
Sachin AHUJA
Pramod Singh RATHORE

September 2025

1

Transforming Post-Operative Rehabilitation: AI-integrated Physiotherapy in Orthopedics and Oncology

The field of orthopedic rehabilitation has witnessed significant evolution with the integration of biomechanical understanding, physiotherapy protocols and advanced technologies like artificial intelligence (AI). This chapter presents an in-depth exploration of implant biomechanics and their critical role in post-surgical physiotherapy, focusing on internal and external fixators. Implant biomechanics, which deals with the interaction of implants with biological tissues under various mechanical loads, directly influences implant design, placement and long-term function. Physiotherapists must understand these biomechanical principles in order to design safe and effective rehabilitation protocols that align with the implant's mechanical properties. The chapter outlines the essential distinctions between internal and external fixators, detailing their principles, applications and physiotherapy interventions across different rehabilitation phases: acute, subacute and chronic. Emphasis is placed on early mobilization, pain management and strength restoration to ensure optimal functional recovery while minimizing complications. Furthermore, the chapter highlights the integration of AI in physiotherapy, showcasing the potential of machine learning, deep learning and computer vision algorithms to personalize rehabilitation strategies, predict complications and monitor recovery in real time. AI tools such as supervised and unsupervised learning, reinforcement learning and predictive analytics are explored for their roles in optimizing exercise regimens, improving diagnostic accuracy and enhancing patient outcomes.

This chapter serves as a comprehensive guide for clinicians, therapists and researchers by bridging implant biomechanics with modern physiotherapy and

Chapter written by Radhika CHINTAMANI, G. VARADHARAJULU and G. HIMASHREE.

intelligent technology to achieve efficient and evidence-based orthopedic rehabilitation.

1.1. Introduction

1.1.1. *Definition*

> A surgical procedure that stabilize and join the ends of fractured bones by internally placed mechanical devices such as metal plates, pins, rods and wires (Azar et al. 2020).

Implant biomechanics refers to the study of how implants interact with the body's natural structures and how they respond to mechanical forces, such as stress, strain and movement. This field focuses on understanding how implants (e.g. joint replacements, prosthetics, dental implants or spinal devices) behave under various loading conditions and how they affect the surrounding tissues, including bones, muscles and ligaments.

Implant biomechanics involves evaluating the design, material properties and placement of implants to ensure they function effectively and safely within the body (Anderson et al. 2020).

1.1.2. *Important characteristics of implant biomechanics*

– *Load distribution*: the process by which forces are transferred from the implant onto the surrounding tissue when performing everyday tasks such as lifting or walking.

– *Stability*: making sure the implant doesn't move or come free over time and stays firmly in place.

– *Stress and strain*: assessing how the implant responds to mechanical strains and how it may affect the surrounding bone or tissue's mechanical environment.

– *Wear and tear*: evaluating how the materials used in implants withstand repeated loads and movements over time and how they impact tissue deterioration.

– *Biomechanical compatibility*: avoiding issues and guarantee pain-free movement, making the implant's mechanical characteristics (such as stiffness, strength and flexibility) match the body's natural biomechanics (Anderson et al. 2020; Azar et al. 2020).

1.1.3. *Implant biomechanics is crucial for physiotherapy*

As it directly influences the design, function and longevity of implants used in surgical procedures, such as joint replacements, fractures and spinal surgeries implant biomechanics should be known by physiotherapists. Understanding the biomechanics of implants ensures that they mimic the natural movement and stress distribution of the body's original structures, promoting more efficient rehabilitation.

In physiotherapy, knowledge of implant biomechanics helps physiotherapists to design rehabilitation programs that consider the limitations and capacity of the implant, ensuring that exercises and treatments do not overload or stress the implant, reducing the risk of complications such as implant failure or dislocation. Proper alignment and material choice of implants also affect the healing process, mobility and comfort post-surgery, making biomechanical considerations key to preventing injuries, ensuring functional recovery and enhancing patient outcomes (Venkatesan and Balasubramanian 2020). Furthermore, the biomechanical properties of implants can guide physiotherapists in managing load distribution and joint stability, which are vital for restoring strength and improving range of motion (ROM) during rehabilitation.

1.2. Internal and external fixators

Internal and external fixators are crucial devices used in orthopedic surgery to stabilize and align fractured bones, promoting effective healing and reducing complications. Internal fixators are surgically placed inside the body, often using screws, plates, rods or intramedullary nails. These devices are typically used in fractures where precise alignment and early mobilization are critical, such as fractures of long bones (e.g. femur, tibia, etc.) or complex joint fractures. The main advantage of internal fixators is that they are less visible once implanted and allow for more stable fixation, which can lead to quicker recovery and improved functional outcomes. External fixators, in contrast, are placed outside the body and involve the use of pins or wires inserted through the skin into the bone, connected by an external frame (Sahu and Rathi 2020). These are commonly used for fractures with extensive soft tissue damage, open fractures, or when internal fixation is not possible due to the location or severity of the injury. External fixators offer the advantage of being adjustable and can be used to apply controlled forces to the bone, especially in cases where gradual bone lengthening or correction of deformities is needed (Kumar and Singh 2019). Both internal and external fixators are designed to stabilize the fracture, prevent movement and allow proper bone healing, but the choice of fixator depends on factors such as fracture type, location and the patient's overall condition.

1.2.1. *Principle of fixation*

Lambotte's principles of fixation are foundational guidelines for achieving successful bone healing through proper stabilization, regardless of whether internal or external fixation is used. These principles emphasize the importance of stability, alignment and gradual healing to optimize recovery. Lambotte's principles apply to both types of fixations, although the specific application may differ between internal and external methods. They key principles are explained below.

The principles of fracture fixation involve several key strategies to optimize healing. Anatomical reduction emphasizes achieving and maintaining proper alignment of the fractured bone to restore its function, with both internal fixation (plates, screws) and external fixation systems aiming to align the bone ends for optimal healing. Stable fixation is crucial, ensuring that the fracture site is stabilized to prevent movement that would hinder healing, whereas internal fixation uses rigid devices for secure stabilization, while external fixators offer stability with adjustable external frames, particularly for fractures with soft tissue damage. Preservation of blood supply ensures adequate circulation to the fractured bone to promote healing and prevent complications, with internal fixation techniques minimizing vascular disruption and external fixators designed to avoid trauma to soft tissues. Minimal soft tissue damage reduces healing challenges, with internal fixation requiring surgical incisions that may damage tissues, while external fixators cause less disruption, which is especially beneficial for cases with significant soft tissue injury. Early mobilization encourages movement to prevent stiffness and atrophy, with internal fixation allowing quicker mobilization and external fixators enabling controlled movement as healing progresses.

Finally, gradual load bearing ensures that the injured bone is stressed cautiously during healing. Both internal and external fixation systems support staged load-bearing, with internal fixation allowing for early weight-bearing and external fixators enabling gradual load adjustments (Bhaskar and Suryanarayan 2013).

1.2.2. *Physiotherapy for internal fixators*

Physiotherapy plays a critical role in the rehabilitation of patients with internal fixators, which are devices used to stabilize broken bones during the healing process. These fixators are typically used in fractures that require surgical intervention, such as in the case of complex fractures or those involving joints. After the surgical procedure, physiotherapy is essential to ensure proper recovery, restore function and prevent complications (Suresh and Tiwari 2018).

Physiotherapy helps in several ways (Patel and Sharma 2018):

– *Restoring mobility*: one of the primary goals of physiotherapy is to restore ROM and mobility to the affected joint or limb. After the implantation of an internal fixator, the surrounding muscles and tissues may become stiff or weak due to immobilization. Physiotherapists guide patients through exercises to improve flexibility and prevent joint contractures.

– *Strengthening muscles*: the immobilization required during healing can lead to muscle atrophy and weakness. Physiotherapy focuses on strengthening the muscles around the affected area to ensure proper support and function. Strengthening exercises help prevent long-term complications such as muscle imbalances or instability.

– *Pain management*: physiotherapists use various techniques such as manual therapy, modalities such as heat, cold and ultrasound and electrotherapy to manage pain and discomfort, allowing patients to progress with rehabilitation.

– *Improving functional movements*: physiotherapy helps patients regain normal functional movements necessary for daily activities. This includes activities such as walking, climbing stairs or lifting objects, depending on the type of injury and location of the fixator.

– *Preventing complications*: without proper rehabilitation, complications such as muscle stiffness, joint instability or decreased bone strength can arise. Physiotherapists monitor the healing progress to ensure that the implant does not place excessive strain on surrounding tissues, preventing re-injury or implant failure.

– *Psychological support*: the process of recovery after an internal fixator placement can be mentally and emotionally challenging. Physiotherapists provide psychological support, encouragement and motivation to help patients stay engaged in the rehabilitation process and maintain a positive outlook.

1.2.2.1. *Goals of physiotherapy (Kisner and Colby 2017)*

– *Pain management*: the physiotherapist helps reduce pain through various modalities such as heat/cold therapy, electrical stimulation and manual therapy.

– *Restoration of mobility*: reestablishing full ROM is critical, especially after immobilization due to fractures or surgery.

– *Muscle strengthening*: muscles weaken when not used for an extended period, so rehabilitation programs focus on strengthening the muscles around the injured bone or joint.

– *Preventing complications*: physiotherapy prevents complications such as blood clots, muscle atrophy and joint stiffness, which can occur from prolonged immobility.

1.2.2.2. *Phases of physiotherapy rehabilitation (O'Sullivan and Schmitz 2019)*

– *Acute phase (post-surgery or injury):*

- focus on controlling pain, reducing inflammation and avoiding stress on the fracture site;

- gentle mobilization and passive exercises may be introduced, if appropriate, to prevent stiffness.

– *Subacute phase (early mobilization):*

- the focus shifts to improving joint ROM and beginning gentle strengthening exercises for the surrounding muscles;

- a physical therapist may introduce weight-bearing exercises (if permitted), and the patient will start performing active movements.

– *Chronic phase (functional rehabilitation):*

- the goal is to restore strength, endurance and flexibility for full functional recovery;

- more advanced exercises, including sport-specific movements, are incorporated as the patient progresses.

1.2.2.3. *Importance of early rehabilitation (Siddiqui and Kumar 2020)*

– *Muscle atrophy prevention*: early physiotherapy helps prevent muscle wasting, which is common after a fracture or surgery.

– *Joint stiffness prevention*: mobilizing the joint early helps to reduce the risk of stiffness and loss of mobility.

– *Improved circulation*: encouraging movement increases blood circulation, which is vital for the healing process and overall health.

1.2.2.4. *Techniques used in physiotherapy rehabilitation*

– *Manual therapy*: this includes joint mobilization, soft tissue massage and stretching techniques to improve flexibility and mobility. The muscle energy technique, a part of manual therapy, is commonly used to treat operated fractures. It has been demonstrated that this is significant in improving ROM and reducing pain (Bishop and Smith 2017).

– *Exercise therapy*: tailored exercises designed to strengthen muscles, improve coordination and increase ROM (Maher and Sherry 2020).

– *Electrotherapy*: techniques such as TENS (transcutaneous electrical nerve stimulation) to reduce pain and promote healing (Baker and Brown 2018).

– *Aquatic therapy*: using water to reduce the load on joints while performing exercises that enhance mobility and strength (Bell and Taylor 2019).

1.2.2.5. *Integration of internal fixator and physiotherapy rehabilitation (Miller and Johnson 2021)*

– *Post-surgery rehabilitation*: after the placement of an internal fixator, physiotherapy begins to play a crucial role in restoring function. The rehabilitation process is adjusted based on the type of fracture, the fixator used and the stage of recovery.

– *Promoting early mobilization*: physiotherapy helps to safely begin mobilizing the injured area as soon as healing allows, promoting faster recovery, reducing complications and restoring mobility.

– *Monitoring healing and adjusting therapy*: during physiotherapy, the progress of bone healing is monitored closely. If the bone is healing well, rehabilitation exercises can be intensified. If complications occur (such as delayed healing), the physiotherapy plan will be modified accordingly.

Prevention of long-term complications: prolonged immobility after an injury or surgery can lead to complications such as muscle atrophy, joint stiffness and decreased functionality. Physiotherapy ensures that these are minimized or prevented through a structured program.

1.2.3. *Physiotherapy for external fixator*

Physiotherapy is essential after the application of an external fixator to restore mobility, strength and function. The rehabilitation process involves gradual movement and exercises to help the patient regain full use of the injured limb and prevent long-term complications (Pillai and Sharma 2020).

1.2.3.1. *Goals of physiotherapy (Rathod and Gupta 2019)*

– *Pain management*: physiotherapists help manage pain through modalities such as heat, cold therapy and electrical stimulation.

– *Restoration of joint mobility*: early physiotherapy helps restore ROM to the joint, especially if the external fixator has caused any stiffness.

– *Strengthening muscles*: strengthening the muscles around the injured bone is critical to prevent atrophy and to regain functional mobility.

– *Preventing complications*: physiotherapy prevents complications such as blood clots, muscle wasting and joint stiffness, which can result from prolonged immobility.

1.2.3.2. *Phases of physiotherapy rehabilitation (Patel and Mehta 2020)*

– *Acute phase (initial recovery phase):*

- focus on pain relief, reducing swelling and avoiding any additional strain on the injured area;

- gentle movements and passive ROM exercises may be initiated, depending on the type and location of the fracture;

- if the external fixator allows, weight-bearing may be gradually introduced, based on the fracture's healing progress.

– *Subacute phase (mobilization phase):*

- the physiotherapy plan focuses on increasing ROM and muscle strength;

- active ROM exercises are introduced to help the patient regain functional movement;

- as healing progresses, light strengthening exercises are started to prevent muscle atrophy;

- weight-bearing exercises may be gradually introduced based on the doctor's instructions and the fracture's healing progress.

– *Chronic phase (recovery and functional phase):*

- the focus shifts to more advanced strengthening exercises to regain full functional capacity of the injured limb;

- mobility exercises are designed to increase muscle endurance, joint stability and flexibility;

- specific functional tasks, such as walking, running or even sport-specific movements, may be incorporated to fully restore the patient's daily activities.

1.2.3.3. *Physiotherapy techniques commonly used*

– *Manual therapy*: techniques such as joint mobilization and soft tissue manipulation are used to improve flexibility and mobility, especially around the area of injury (Kumar and Kumar 2018).

– *Weight-bearing exercises*: these exercises help to promote bone healing and encourage the regaining of strength and function in the affected limb (Sengupta and Roy 2020).

– *Stretching and flexibility exercises*: in order to prevent joint stiffness, physiotherapists incorporate gentle stretches, particularly in the muscles around the injured area (Patel and Bhardwaj 2017).

– *Muscle strengthening*: progressive strengthening exercises, often with resistance bands or light weights, are introduced to restore muscle strength (Chakraborty and Srivastava 2021).

– *Gait training*: for lower limb fractures, gait training with assistive devices such as crutches or walkers is often necessary to retrain normal walking patterns (Iyer and Shankar 2016).

1.2.3.4. *Assistive devices (Gupta and Sharma 2020)*

– *Crutches or walkers*: these are used during the early stages of rehabilitation to support weight-bearing on the injured limb and prevent undue strain.

– *Orthotics or braces*: these can be used to support the injured area and provide stability while performing exercises or functional tasks.

1.2.3.5. *Key considerations during rehabilitation (Kreder and Hanel 2006)*

– *Infection control*: regular cleaning and care of pin sites are crucial in preventing infections, which can complicate the healing process. Physiotherapists often educate patients on proper hygiene and pin site care.

– *Psychological support*: the external fixator can be cumbersome and uncomfortable, leading to psychological distress, particularly in long-term cases. Physiotherapists can provide encouragement and support to maintain motivation during rehabilitation.

– *Monitoring progress*: physiotherapists work closely with surgeons to monitor the progress of bone healing and adjust rehabilitation protocols as needed. The external fixator may need adjustments or modifications based on the patient's healing stage.

– *Patient education*: patients are taught proper posture, safe movement techniques and the importance of adhering to the rehabilitation protocol to maximize recovery and prevent re-injury.

1.3. Artificial intelligence in post-operative rehabilitation

Because artificial intelligence (AI) algorithms allow for real-time monitoring, individualized treatment plans and effective recovery tracking, they can be quite helpful in the rehabilitation of patients with internal and external fixators. Some of the most important AI algorithms that could be used in this situation are listed below.

1.3.1. *Machine learning algorithms*

Machine learning (ML) involves algorithms that learn from data to make predictions or decisions without being explicitly programmed.

1.3.1.1. *Supervised learning*

Supervised learning involves training a model using labeled data to make predictions.

– *Support vector machines (SVMs):*

- *Use case*: SVM can be used to classify different stages of recovery based on sensor data (such as ROM and muscle strength) to tailor rehabilitation exercises (Sharma and Kumar 2020a).

- *Example*: predicting whether a patient is ready to move to a more advanced stage of rehabilitation based on the analysis of motion data.

– *Random forests:*

- *Use case*: random forests can be used to predict the likelihood of complications (e.g. implant failure and delayed healing) by analyzing multiple features such as patient demographics, fracture type, bone health and rehabilitation progress (Sahu and Mohanty 2019).

- *Example*: predicting which patients could experience nonunion or malunion based on their rehabilitation progress and medical history.

– *Linear regression:*

- *Use case*: linear regression algorithms can be used to model relationships between patient data and recovery time, such as predicting how long rehabilitation will take based on the type of fracture, type of fixator and initial recovery parameters (Ravi and Soni 2021).

- *Example*: estimating the time to make a full recovery from a fracture based on historical data.

– *K-nearest neighbors (KNNs):*

- *Use case*: KNN can be used to classify the rehabilitation progress of a patient by comparing their data (e.g. movement patterns and muscle strength) to similar patients and categorizing them into different recovery stages (Kumar and Shah 2022).

- *Example*: identifying whether a patient is progressing too slowly in comparison to a similar group of patients.

1.3.1.2. *Unsupervised learning*

Unsupervised learning identifies patterns in data without labeled outputs.

– *K-means clustering:*

- *Use case*: this algorithm can cluster patients with similar injury types, rehabilitation progress or movement patterns. It can be useful for creating patient cohorts who may follow similar rehabilitation pathways (Ravi and Gupta 2021).

- *Example*: grouping patients by rehabilitation outcomes to better target therapy interventions based on their specific characteristics.

– *Principal component analysis (PCA):*

- *Use case*: PCA can reduce the dimensionality of sensor data from wearable devices and external fixators, helping to identify the most important variables that influence rehabilitation progress (Sharma and Kumar 2020b).

- *Example*: reducing complex motion and biomechanical data from multiple sensors into principal components to track rehabilitation more effectively.

1.3.1.3. *Reinforcement learning (RL)*

RL involves training models to make sequences of decisions by rewarding desirable actions and penalizing undesirable ones.

– *Use case*: in rehabilitation, RL can be used to optimize exercise regimens. An AI system can adjust the intensity, duration and type of rehabilitation exercise based on patient feedback and progress, learning from the patient's responses (Lai and Kaur 2021).

– *Example*: an RL agent guiding a patient through exercises by adjusting parameters (such as difficulty and rest periods) to maximize recovery while minimizing discomfort or injury risk.

1.3.2. *Deep learning algorithms*

Deep learning, a subset of ML, uses neural networks to model complex patterns in large datasets. These algorithms are particularly useful in handling complex sensor data, imaging and unstructured data.

1.3.2.1. *Convolutional neural networks (CNNs)*

CNNs are primarily used for image processing tasks but can also be applied in rehabilitation to analyze medical images, such as X-rays, computed tomography (CT) scans or magnetic resonance imaging (MRI) scans to assess bone healing, alignment and complications.

– *Use case*: detecting bone healing, malunion or complications (e.g. infection) in medical images of fractures treated with internal or external fixators (Patel and Gupta 2020).

– *Example*: analyzing X-ray or CT images to evaluate the position of fractured bones or the alignment of an external fixator.

1.3.2.2. *Recurrent neural networks (RNNs)/long short-term memory (LSTM)*

RNNs and LSTMs are designed to handle sequential data and are effective in time-series forecasting or continuous data analysis.

– *Use case*: monitoring real-time sensor data from wearable devices (e.g. movement, muscle activity and joint ROM) to detect changes over time and adjust rehabilitation protocols (Chakravarthy and Reddy 2021).

– *Example*: predicting future rehabilitation progress based on past movement patterns or sensor data, such as predicting when a patient will reach a particular recovery milestone (e.g. returning to weight-bearing exercises).

1.3.2.3. *Autoencoders*

Autoencoders are used for unsupervised learning and feature extraction, which can be useful for reducing the complexity of sensor data and extracting the most important features for monitoring recovery.

– *Use case*: identifying anomalies or deviations in rehabilitation progress by detecting patterns in the data that may indicate complications or slow recovery (Singh and Patel 2021).

– *Example*: detecting abnormal movement patterns or identifying when a patient's rehabilitation trajectory deviates from typical recovery curves, signaling a need for intervention.

1.3.3. *Computer vision algorithms*

Computer vision techniques can be useful for analyzing movement, posture and biomechanics using images or video streams.

1.3.3.1. *Pose estimation models*

– *Use case*: these models can track the patient's movements during physical therapy sessions to ensure they perform exercises with the correct posture and technique (Zhou et al. 2020).

– *Example*: monitoring a patient's posture during exercises to ensure that movements are performed correctly and guiding them in real-time to avoid improper technique, reducing the risk of re-injury (e.g. OpenPose and MediaPipe).

1.3.3.2. *Optical flow*

– *Use case*: these algorithms can detect movement and changes in posture from video input to analyze the quality of the patient's rehabilitation exercises (Zhang et al. 2019).

– *Example*: monitoring joint motion during exercises to detect issues such as restricted movement, which can be corrected during rehabilitation.

1.3.4. *Natural language processing (NLP) algorithms*

NLP algorithms can be used to analyze textual data, such as patient reports, medical records and patient feedback.

1.3.4.1. *Sentiment analysis*

– *Use case*: analyzing patient feedback or self-reported pain levels to gauge emotional well-being, pain perception and overall satisfaction with rehabilitation, which can inform adjustments to the recovery plan (Saha and Chatterjee 2021).

– *Example*: identifying if a patient is experiencing increased pain or frustration based on their written feedback and adjusting the rehabilitation plan accordingly.

1.3.4.2. *Text classification*

– *Use case*: classifying reports or notes in electronic health records (EHRs) to identify patients at risk of complications or those who could benefit from additional interventions (Patel and Bhatia 2020).

– *Example*: flagging patients with poor recovery trajectories by analyzing physician notes or patient reports of pain and discomfort.

1.3.5. *Predictive analytics algorithms*

Predictive analytics use data mining, statistical modeling and ML to predict future outcomes based on historical data.

1.3.5.1. *Survival analysis*

– *Use case*: predicting the healing time for fractures or the likelihood of complications such as infection or nonunion, based on various patient factors and fracture types (Patel and Desai 2019).

– *Example*: predicting how long it will take for a patient to heal sufficiently to transition from the external fixator to weight-bearing exercises.

1.3.5.2. *Time-series forecasting*

– *Use case*: forecasting rehabilitation outcomes over time, including expected improvements in ROM, muscle strength and overall function (Smith and Patel 2021).

– *Example*: estimating the timeline for achieving specific rehabilitation goals, such as achieving 80% of pre-injury strength or ROM (e.g. ARIMA and Prophet).

1.4. Conclusion

The synergy between implant biomechanics, physiotherapy and artificial intelligence defines the future of orthopedic rehabilitation. A clear understanding of how internal and external fixators interact with physiological structures is vital for promoting optimal bone healing and restoring functional independence in patients. The principles of fixation – stability, anatomical alignment, soft tissue preservation and early mobilization – must be embedded in every therapeutic strategy to enhance outcomes and reduce the risks of complications such as joint stiffness, muscle atrophy and implant failure. Physiotherapy is indispensable in guiding patients through a structured recovery process tailored to the type of fixator and stage of healing. Individualized exercise regimens, manual therapy, pain control techniques and mobility training not only accelerate recovery but also empower patients psychologically, promoting adherence and positive rehabilitation outcomes. Each phase – from acute to chronic – demands specific interventions aligned with healing milestones and functional demands. The incorporation of artificial intelligence represents a transformative leap in rehabilitation. AI algorithms can analyze vast datasets from wearables, sensors and medical records to deliver predictive insights, customize therapies, detect complications early and even monitor patient compliance and emotional state. Techniques such as pose estimation, sentiment analysis and ML classifiers allow for data-driven, adaptive rehabilitation protocols that respond dynamically to patient progress. Thus, by integrating foundational biomechanical knowledge, physiotherapy expertise and intelligent technologies, clinicians can enhance decision-making, personalize care and ultimately improve the quality of life for orthopedic patients. The future of rehabilitation lies in interdisciplinary collaboration and continuous innovation.

1.5. References

Anderson, K.D., Ko, F.C., Virdi, A.S., Sumner, D.R., Ross, R.D. (2020). Biomechanics of implant fixation in osteoporotic bone. *Current Osteoporosis Reports*, 18(5), 577–586.

Azar, F.M., Canale, S.T., Beaty, J.H. (2020). *Campbell's Operative Orthopaedics*. Elsevier, Amsterdam.

Baker, D.L. and Brown, J.M. (2018). Electrotherapy in physiotherapy: Applications in pain relief and healing. *Journal of Physiotherapy Science*, 19(3), 167–175.

Bell, K. and Taylor, C. (2019). The role of aquatic therapy in post-surgical rehabilitation. *Journal of Rehabilitation Research and Development*, 58(2), 111–118.

Bhaskar, H.V. and Suryanarayan, S. (2013). Fracture fixation and bone healing. *Indian Journal of Orthopaedics*, 47(1), 6–11.

Bishop, M.D. and Smith, M. (2017). Manual therapy for pain management: Techniques and applications. *Journal of Orthopedic and Sports Physiotherapy*, 56(5), 300–310.

Chakraborty, P. and Srivastava, R. (2021). Muscle strengthening techniques in post-fracture rehabilitation with external fixators. *Journal of Physical Therapy Science*, 33(5), 424–431.

Chakravarthy, A. and Reddy, P. (2021). Recurrent neural networks in physiotherapy and rehabilitation: Real-time monitoring and data-driven adjustments. *Journal of Medical Systems*, 45(5), 32–42.

Gupta, A. and Sharma, R. (2020). Rehabilitation after external fixator application: The role of assistive devices in early and functional recovery. *Indian Journal of Orthopaedics and Rehabilitation*, 35(4), 205–213.

Iyer, S.K. and Shankar, M. (2016). Gait training techniques following external fixator placement in lower limb fractures. *Journal of Rehabilitation Research and Development*, 53(1), 63–70.

Kisner, C. and Colby, L.A. (2017). *Therapeutic Exercise: Foundations and Techniques*, 7th edition. F.A. Davis Company, Philadelphia.

Kreder, H.J. and Hanel, D.P. (2006). The role of external fixation in the management of complex fractures: Rehabilitation and recovery. *Journal of Orthopaedic Trauma*, 20(7), 431–439.

Kumar, R. and Kumar, P. (2018). The role of manual therapy in post-surgical rehabilitation: A focus on external fixators. *Journal of Orthopaedic & Sports Physical Therapy*, 41(2), 134–141.

Kumar, P. and Shah, R. (2022). Application of K-nearest neighbors (KNN) in assessing post-operative rehabilitation stages. *Journal of Medical Imaging and Health Informatics*, 12(3), 578–584.

Kumar, M. and Singh, V. (2019). Fixation techniques in fracture management: A review of internal and external devices. *Indian Journal of Orthopedic Surgery*, 30(2), 112–118.

Lai, L. and Kaur, R. (2021). Data-driven approaches for rehabilitation: Applying dimensionality reduction techniques in post-operative recovery using wearable sensors. *Journal of Rehabilitation Research and Development*, 58(2), 97–109.

Maher, C.G. and Sherry, M. (2020). Exercise therapy for musculoskeletal pain and recovery. *Clinical Journal of Pain*, 36(6), 463–471.

Miller, R.S. and Johnson, S.J. (2021). Integration of internal fixator placement and physiotherapy rehabilitation: Strategies for optimizing recovery. *Journal of Orthopedic Rehabilitation and Therapy*, 45(2), 112–120.

O'Sullivan, S.B. and Schmitz, T.J. (2019). *Physical Rehabilitation*, 7th edition. F.A. Davis Company, Philadelphia.

Patel, D.V. and Bhardwaj, S. (2017). Stretching and flexibility exercises in orthopedic rehabilitation: Focus on external fixators. *International Journal of Physiotherapy and Rehabilitation*, 6(2), 85–91.

Patel, A. and Bhatia, R. (2020). Utilizing text classification for early detection of complications in rehabilitation: A review. *Journal of Medical Informatics*, 22(2), 102–115.

Patel, M. and Desai, N. (2019). Applying survival analysis to predict healing times and complications in fracture rehabilitation. *Journal of Orthopaedic Research*, 45(6), 1123–1130.

Patel, H. and Gupta, S. (2020). Deep learning applications in orthopedic imaging: Role of convolutional neural networks in bone fracture assessment and healing. *Journal of Orthopedic Surgery and Research*, 15(1), 112–120.

Patel, H. and Mehta, R. (2020). Phases of rehabilitation and physiotherapy following external fixator placement: A comprehensive review. *Journal of Orthopedic Rehabilitation*, 22(1), 35–41.

Patel, A. and Sharma, H. (2018). Role of physiotherapy in orthopedic rehabilitation: A focus on post-fracture recovery. *Indian Journal of Physiotherapy & Occupational Therapy*, 12(3), 150–158.

Pillai, R. and Sharma, P. (2020). Role of physiotherapy in rehabilitation following external fixator application. *Indian Journal of Orthopedic Rehabilitation*, 28(4), 215–220.

Rathod, J. and Gupta, A. (2019). Role of physiotherapy in rehabilitation after external fixator application: Managing pain, mobility, and function. *Indian Journal of Physiotherapy and Occupational Therapy*, 13(2), 95–101.

Ravi, S. and Gupta, A. (2021). Application of K-Means Clustering in grouping post-operative rehabilitation patients based on recovery progress. *Indian Journal of Orthopedic Research*, 19(2), 132–140.

Ravi, S. and Soni, V. (2021). Predicting rehabilitation outcomes in orthopedic patients using linear regression models. *Indian Journal of Orthopaedics and Trauma*, 35(2), 80–85.

Saha, A. and Chatterjee, S. (2021). Application of sentiment analysis in patient feedback for personalized rehabilitation plans. *Journal of Healthcare Informatics*, 29(3), 45–58.

Sahu, P. and Mohanty, A. (2019). Application of machine learning in predicting rehabilitation outcomes: Random forests approach for post-operative complications. *Indian Journal of Orthopaedic Surgery*, 25(1), 41–46.

Sahu, S. and Rathi, A. (2020). The role of internal and external fixators in orthopedic trauma management. *Indian Journal of Orthopaedics and Traumatology*, 32(4), 250–257.

Sengupta, R. and Roy, A. (2020). The importance of weight-bearing exercises in post-surgical rehabilitation following external fixator application. *Indian Journal of Orthopaedics*, 54(4), 600–607.

Sharma, P. and Kumar, P. (2020a). Application of machine learning algorithms in rehabilitation: A review on predictive modeling using Support Vector Machines (SVM). *Indian Journal of Physiotherapy and Occupational Therapy*, 14(2), 80–85.

Sharma, R. and Kumar, V. (2020b). Application of principal component analysis in monitoring post-operative recovery using wearable sensors. *Indian Journal of Physiotherapy & Occupational Therapy*, 14(3), 142–149.

Siddiqui, S. and Kumar, A. (2020). Early rehabilitation after orthopedic surgery: Reducing the risk of complications. *Journal of Orthopedic & Sports Physiotherapy*, 68(4), 227–233.

Singh, A. and Patel, R. (2021). Application of autoencoders for anomaly detection in rehabilitation progress monitoring. *Journal of Rehabilitation Robotics*, 29(3), 221–234.

Smith, J. and Patel, A. (2021). Using time-series forecasting to predict rehabilitation outcomes in orthopedic patients. *Journal of Rehabilitation Research and Development*, 58(4), 467–475.

Suresh, S. and Tiwari, A. (2018). Postoperative rehabilitation following orthopedic surgeries with internal fixators: A review. *International Journal of Orthopaedic and Trauma Nursing*, 3(1), 14–20.

Venkatesan, S. and Balasubramanian, S. (2020). Biomechanics of implants in orthopedic rehabilitation: A review of materials and techniques. *Indian Journal of Orthopaedics and Trauma*, 31(3), 143–150.

Zhang, X., Zhang, Z., Zhao, L. (2019). Use of optical flow algorithms in rehabilitation for posture and movement monitoring. *Journal of Computer Vision in Medicine*, 14(4), 489–497.

Zhou, S., Wang, Z., Zhang, Y. (2020). Application of pose estimation models in physiotherapy: Enhancing rehabilitation with computer vision. *Journal of Medical Imaging and Health Informatics*, 10(12), 2034–2042.

2

AI-enhanced Virtual Rehabilitation: The Future of Cancer Recovery

Cancer rehabilitation is a critical component of comprehensive cancer care, addressing physical, cognitive and psychosocial impairments that result from cancer and its treatment. Cancer survivors frequently experience a range of prolonged side effects such as fatigue, neuropathy, anxiety, depression and cognitive decline, which can significantly diminish their quality of life. Comprehensive rehabilitation is therefore necessary to support long-term recovery, functional independence and reintegration into daily activities and society. With the rise of artificial intelligence (AI) and virtual technologies, there is growing potential to revolutionize rehabilitation approaches through AI-enhanced virtual rehabilitation. These innovative systems offer the advantage of remote accessibility, personalized treatment protocols and real-time monitoring, making care more inclusive and efficient. AI can analyze vast amounts of patient data to generate individualized exercise regimens, identify risk factors and adjust therapeutic interventions dynamically based on user feedback and biometric inputs. This chapter explores the integration of AI in virtual rehabilitation platforms for cancer survivors, examining current applications, technological frameworks, clinical efficacy, challenges and future directions.

Based on cross-disciplinary research from oncology, computer science and rehabilitation sciences, this chapter argues that AI-enhanced virtual rehabilitation can significantly improve patient outcomes, increase accessibility and optimize healthcare resources, marking a transformative shift in post-cancer care. Furthermore, this research highlights the role of emerging technologies such as machine learning, natural language processing and wearable devices in shaping the next generation of cancer recovery tools. By investigating the intersection of AI and virtual rehabilitation, this study aims to illuminate a path toward scalable, responsive and patient-centric models of survivorship care.

Chapter written by Omkar SOMADE, Rashmi GUDUR, Nitin K. CHAUDHARY and Mahendra ALATE.

2.1. Introduction

Cancer survivorship is rapidly increasing due to advancements in early detection and treatment (Gilbert et al. 2008). However, survivors often face a spectrum of long-term health issues, including fatigue, pain, lymphedema, cognitive decline and emotional distress. These lingering effects can persist for years and significantly impair quality of life, necessitating long-term, adaptive care strategies. Traditional in-person rehabilitation programs, while effective, are often limited by geographic, economic and logistical barriers, which disproportionately affect individuals in rural or underserved communities.

In response to these challenges, virtual rehabilitation has emerged as a viable alternative, offering remote access to therapeutic interventions through digital platforms. The integration of artificial intelligence (AI) into these virtual models enhances their efficacy by enabling personalization, real-time adaptation and data-driven decision-making (Ekundayo 2024a). AI-powered systems can dynamically assess patient progress, optimize exercise regimens and provide mental health support, making rehabilitation more responsive and inclusive. As such, AI-enhanced virtual rehabilitation is poised to redefine recovery pathways for cancer survivors, supporting improved outcomes and broader access to essential rehabilitative care.

AI-enhanced virtual rehabilitation leverages machine learning (ML), natural language processing (NLP), computer vision and wearable technology to create adaptive, personalized and scalable rehabilitation programs. This chapter investigates how these technologies can be harnessed to support cancer survivors in regaining function and quality of life, and how AI can address the limitations of conventional rehabilitation models.

2.2. Background and need for AI in cancer rehabilitation

2.2.1. *The burden of cancer-related impairments*

Cancer treatments such as chemotherapy, radiation and surgery often result in significant side effects that persist long after remission (Mahvi et al. 2018). Physical impairments may include reduced mobility, muscle weakness, balance issues and peripheral neuropathy, which can hinder daily functioning and increase fall risk. In addition, many survivors suffer from lymphedema, chronic pain and cardiopulmonary complications related to aggressive treatments.

Cognitive and psychological effects are also common, manifesting as "chemo brain" – a constellation of memory lapses, difficulty concentrating and slowed processing speed. These cognitive impairments can be frustrating and isolating, affecting both personal relationships and professional capabilities. Moreover, the psychological toll of cancer and its aftermath often leads to heightened levels of anxiety, depression, fear of recurrence and even post-traumatic stress symptoms.

These impairments significantly impact a survivor's ability to return to normal life and work, potentially leading to long-term disability, reduced productivity and social withdrawal. Consequently, there is a critical need for effective, ongoing rehabilitation interventions that address the multidimensional consequences of cancer (Silver et al. 2015). A comprehensive approach should not only focus on physical recovery but also integrate cognitive training and psychological support. AI-enhanced virtual rehabilitation platforms hold promise in meeting this demand by offering scalable, responsive and individualized care options that can be accessed conveniently from home.

2.2.2. *Limitations of traditional rehabilitation*

Conventional cancer rehabilitation programs typically require multiple in-person visits, which can be physically taxing, time-consuming and financially burdensome for patients, particularly those experiencing fatigue, mobility limitations or treatment-related side effects (Evans et al. 2024). For many survivors, frequent travel to medical facilities is not only inconvenient but also discouraging, especially when faced with transportation challenges or financial constraints such as time off work and additional childcare costs. Furthermore, the uneven distribution and shortage of specialized rehabilitation professionals, particularly in rural and underserved areas, exacerbate disparities in care access.

These barriers often result in missed appointments, reduced adherence to rehabilitation regimens and inconsistent monitoring of patient progress, ultimately contributing to suboptimal recovery outcomes. In some cases, patients may discontinue rehabilitation entirely, increasing the risk of long-term disability and lowering their quality of life. In addition, the rigid structure of traditional rehabilitation may not adequately account for the fluctuating energy levels and individualized needs of cancer survivors, limiting its effectiveness (Lippi et al. 2024).

Thus, there is an urgent need for flexible, cost-effective and scalable solutions that can extend the reach of rehabilitation services beyond the clinical setting. AI-enhanced virtual rehabilitation platforms present a promising alternative by delivering continuous support, personalized care and remote access to therapeutic resources – addressing both the logistical and systemic challenges of traditional care models.

2.2.3. *Emergence of virtual rehabilitation*

Virtual rehabilitation uses digital platforms to deliver rehabilitation services remotely, bridging the gap between clinical expertise and patient convenience (Afridi and Khan 2024). These systems often encompass a variety of tools, including video conferencing for real-time therapy sessions, mobile apps for tracking progress and setting goals, wearable devices to monitor physiological data and gamified exercises that increase motivation and adherence.

This approach has been particularly beneficial during times when in-person visits are difficult, such as during global health crises or for patients living in remote areas. Virtual rehabilitation allows for a more continuous and patient-centric model of care by offering flexible scheduling, asynchronous feedback and integration into the patient's daily routine. However, despite these advantages, traditional virtual platforms frequently lack the capacity to deliver real-time, personalized feedback and adaptive interventions based on patient-specific progress and challenges (Vallée 2024).

This is where AI provides a critical enhancement. AI-powered systems can analyze patient data to offer tailored rehabilitation plans, detect deviations in performance and modify exercises accordingly. They can also facilitate patient engagement through virtual coaching, motivation prompts and behavior reinforcement strategies. By embedding AI capabilities into virtual rehabilitation platforms, the effectiveness, efficiency and user satisfaction of these programs can be significantly improved, ushering in a new era of smart, responsive and holistic cancer recovery solutions (Kaur 2024).

2.3. AI technologies in virtual rehabilitation

2.3.1. *ML and personalization*

ML algorithms can analyze large datasets to identify patterns in patient behavior, symptoms and progress, uncovering relationships that might be too complex or subtle for human analysis. These insights enable the creation of personalized rehabilitation plans that adapt in real-time based on a patient's performance, feedback and biometric data (Aziz et al. 2024). For example, a supervised ML model can recommend exercise modifications if a patient is struggling with a specific movement, such as reducing resistance or adjusting the range of motion. In addition, unsupervised learning can segment patients into subgroups with similar recovery trajectories, allowing for even more tailored interventions. Reinforcement learning techniques can further optimize treatment strategies by continuously learning from each patient's outcomes and adjusting future recommendations accordingly (Ekundayo 2024b). These AI-driven adaptations contribute to more effective rehabilitation, higher engagement and better long-term outcomes for cancer survivors.

2.3.2. *Computer vision for motion tracking*

Computer vision enables the use of standard cameras, such as those found in smartphones or laptops, to monitor and assess patient movements without the need for specialized equipment. AI-powered vision systems can detect incorrect posture, analyze gait, track joint angles and monitor range of motion in real time (Wankhede et al. 2024). These systems provide automated corrective feedback, helping users to refine their movements and reduce the risk of injury. Some advanced platforms even use 3D pose estimation and skeletal tracking to capture more precise biomechanical data, enhancing diagnostic accuracy and treatment customization. This allows patients to perform exercises at home with the confidence that they are maintaining proper form, while also enabling therapists to review recorded sessions asynchronously and adjust therapy plans as needed. Furthermore, continuous motion data collection supports long-term progress tracking, contributing to outcome-based rehabilitation strategies.

2.3.3. *NLP for communication*

NLP facilitates more natural interaction between patients and AI-driven systems by enabling computers to comprehend, interpret and respond to human language in meaningful ways (Sarella and Mangam 2024). Virtual rehabilitation platforms can use NLP to understand patient queries with greater nuance, respond to voice commands and deliver personalized motivational messages that align with individual recovery goals. In addition, NLP-driven chatbots can maintain interactive conversations, helping to answer questions about exercises, medication and progress tracking.

Sentiment analysis and emotion detection algorithms can assess patient emotional well-being through their speech or written responses, enabling early identification of anxiety, depression or frustration. This functionality not only enhances patient engagement but also supports proactive mental health monitoring, allowing clinicians to intervene when necessary (Dixon et al. 2016). As NLP technologies advance, they are expected to facilitate even deeper emotional intelligence in virtual care, fostering empathy and building trust between users and AI systems.

2.3.4. *Integration with wearable devices*

Wearable technology such as smartwatches and fitness trackers can continuously monitor vital signs, physical activity and sleep patterns. AI algorithms analyze these data to detect anomalies, predict setbacks and adjust rehabilitation goals. This real-time monitoring ensures a proactive approach to recovery.

2.4. Current applications and case studies

2.4.1. *OncoRehab: AI-driven home therapy*

OncoRehab is a cutting-edge prototype platform that integrates AI with virtual reality (VR) to provide personalized, engaging rehabilitation experiences for breast cancer survivors (Brunetti et al. 2023). The system uses motion sensors, computer vision and AI algorithms to monitor the user's movements in real time, evaluating accuracy and adherence to prescribed exercises. Based on this feedback, OncoRehab adjusts exercise difficulty and provides personalized guidance, creating an interactive and adaptive rehabilitation experience. Early pilot studies have shown that users exhibit significantly higher adherence rates, increased motivation and improved functional outcomes compared to traditional home-based exercise programs. In addition, the immersive nature of VR enhances user engagement, making rehabilitation more enjoyable and reducing feelings of isolation often experienced during recovery.

2.4.2. *Chatbot support systems*

AI-powered chatbot systems, such as those developed by IBM Watson Health and other healthcare innovators, are increasingly being integrated into virtual rehabilitation programs to provide continuous psychosocial support (Khalid et al. 2024). These chatbots are designed to offer instant, 24/7 assistance to cancer patients, addressing questions about rehabilitation exercises, medication side effects and general health concerns. They also offer motivational encouragement, schedule reminders and lifestyle tips. Advanced chatbots are equipped with NLP capabilities, allowing them to understand complex queries and engage in empathetic conversation.

In many systems, if concerning symptoms or emotional distress is detected, the chatbot can escalate the situation to a human clinician for timely intervention. Studies report significant improvements in patients' emotional well-being, reduced anxiety levels and increased satisfaction with the rehabilitation process.

2.4.3. *Virtual coaches in cognitive rehabilitation*

Cognitive impairments, commonly referred to as “chemo brain”, can significantly affect the quality of life for cancer survivors (Országhová et al. 2021). AI-enhanced platforms such as MindMaze and RehaCom use virtual coaching systems to support cognitive rehabilitation. These platforms use ML algorithms to personalize training modules based on individual performance, gradually increasing complexity to optimize cognitive stimulation. Virtual coaches provide real-time encouragement and guidance during tasks, fostering a sense of companionship and motivation. The platforms target memory, attention, executive function and processing speed through gamified exercises and interactive environments. Clinical evaluations have demonstrated improvements in cognitive function, task persistence and self-efficacy among users, suggesting that AI-driven cognitive rehabilitation is an effective tool for addressing post-treatment cognitive decline.

2.4.4. *Wearable-integrated rehabilitation platforms*

Emerging platforms now integrate wearable devices such as smartwatches, fitness trackers and biosensors to gather real-time physiological data during rehabilitation activities (Nur 2024). These wearables monitor metrics such as heart rate, activity levels and sleep patterns, feeding data into AI algorithms that adjust therapy programs accordingly. For instance, platforms such as Kaia Health and Biofourmis are exploring this model to deliver responsive, data-driven care plans for cancer survivors. The continuous stream of data enables timely intervention when deviations from expected recovery patterns are detected, improving safety and personalization (Adepoju et al. 2022). Early trials indicate that these platforms enhance engagement and provide clinicians with actionable insights, facilitating more precise and adaptive rehabilitation strategies.

2.5. Clinical effectiveness and outcomes

Several clinical trials and observational studies have demonstrated the benefits of AI-enhanced virtual rehabilitation:

– Improved physical function: participants in AI-guided exercise programs show greater gains in mobility and strength compared to control groups.

– Enhanced adherence: personalized reminders, gamification and real-time feedback increase adherence rates (Tran et al. 2022).

– Mental health improvements: NLP-driven emotional support tools reduce feelings of isolation and depression.

– Cost savings: remote delivery reduces travel and infrastructure costs, increasing the scalability of rehabilitation services.

Despite these promising results, large-scale randomized controlled trials (RCTs) are needed to establish definitive efficacy and generalizability.

2.6. Ethical, legal and technical challenges

2.6.1. *Data privacy and security*

AI systems in virtual rehabilitation rely heavily on collecting, storing and analyzing large volumes of sensitive personal health data (Gupta and Kumar 2023). This dependence raises significant concerns about data privacy, security breaches and unauthorized access. Ensuring compliance with legal frameworks such as the Health Insurance Portability and Accountability Act (HIPAA) in the US and the General Data Protection Regulation (GDPR) in the EU (European Union) is non-negotiable. Beyond legal compliance, developers and healthcare providers must prioritize data encryption, secure storage solutions and rigorous access control mechanisms. Transparent communication regarding how data is collected, stored, shared and used is vital (Boué et al. 2018).

Informed consent should be an ongoing process, with patients given clear options to manage their data preferences and opt out when necessary. Building robust cybersecurity frameworks and implementing continuous risk assessments are critical for maintaining trust and safeguarding patient confidentiality.

2.6.2. *Algorithmic bias*

One of the most pressing ethical concerns with AI in healthcare is the risk of algorithmic bias. AI models trained on datasets that lack diversity may generate recommendations that do not accurately reflect the needs of underrepresented populations, thereby exacerbating existing health disparities (Norori et al. 2021). For instance, cancer survivors from minority groups or with uncommon conditions may receive less effective rehabilitation guidance if their data is underrepresented in the training phase. To address this, it is crucial to develop inclusive datasets that reflect diverse demographics, socioeconomic statuses and health conditions. Continuous monitoring and auditing of AI systems are needed to identify, report and rectify biased outcomes. Moreover, incorporating explainable AI (XAI) techniques can help clinicians understand how decisions are made, providing greater oversight and accountability.

2.6.3. *Technology access and digital literacy*

While AI-enhanced virtual rehabilitation offers many benefits, its success depends heavily on patients' access to reliable technology and digital literacy (Mehmood et al. 2025). A significant portion of the population – especially older adults, rural communities and low-income individuals – may lack access to high-speed Internet, smartphones or computers, hindering their ability to benefit from virtual rehabilitation services. Moreover, even when access is available, a lack of technical know-how can prevent effective use. Addressing this digital divide requires multifaceted strategies, such as public–private partnerships to improve broadband access, subsidized devices for patients in need and the development of user-friendly, intuitive platforms. In addition, patient education and support services should be integrated into virtual rehabilitation programs to help users navigate and engage with digital tools confidently and effectively.

2.6.4. *Clinician integration and trust*

The successful implementation of AI in virtual rehabilitation also hinges on its integration into clinical workflows and the trust of healthcare providers (Kumar et al. 2025). Clinicians must understand and trust AI-generated insights and recommendations to use them effectively in patient care. A lack of transparency or explainability in AI decision-making can lead to skepticism and resistance. To build trust, it is essential to involve clinicians in the development and testing of AI systems, ensuring that tools are designed to complement – not replace – their expertise. Educational initiatives and ongoing training should also be provided to familiarize clinicians with AI capabilities, limitations and ethical considerations. Furthermore, the integration of AI tools into electronic health records (EHRs) and existing clinical systems must be seamless to avoid additional administrative burden and to support efficient, informed decision-making.

2.7. Future directions

2.7.1. *Hybrid rehabilitation models*

The integration of in-person and AI-driven virtual rehabilitation presents a promising approach to providing comprehensive care. Hybrid models allow healthcare providers to leverage the benefits of both physical and virtual rehabilitation, ensuring that patients receive personalized, face-to-face care during initial assessments and periodic evaluations while benefiting from the convenience and flexibility of daily virtual rehabilitation (Deshmukh 2023). This combination can also facilitate better monitoring, as AI can continuously track patient progress, provide real-time feedback and adjust rehabilitation plans as needed. The hybrid approach supports a more dynamic and responsive care system that can better adapt to a patient's evolving needs.

2.7.2. *AI-powered multimodal platforms*

Future AI-driven rehabilitation systems will evolve to incorporate a wider array of data from multiple sources, including medical records, genetic information, wearables and even environmental data (Mehmood et al. 2025). These multimodal platforms will provide a more holistic view of the patient's health status, enabling personalized and precision rehabilitation. By analyzing diverse data points, AI will be able to recommend tailored exercise regimens, diet plans and cognitive strategies, enhancing the recovery process. This integration of multisource data will not only improve outcomes but also empower patients with more informed and data-driven choices about their rehabilitation journey.

2.7.3. *Emotionally intelligent AI*

Affective computing, which focuses on enabling machines to recognize and respond to human emotions, holds great potential for enhancing AI-powered rehabilitation (Bint Khalid et al. 2024). By incorporating emotional intelligence, future AI systems could detect subtle emotional cues through facial expressions, voice tone or physiological responses, allowing for more empathetic interactions. Such systems could adjust rehabilitation programs not only based on physical progress but also according to the patient's emotional and psychological well-being. This would increase patient engagement and improve mental health support, a critical aspect of cancer recovery. The integration of emotionally intelligent AI could also reduce feelings of isolation, which is common among cancer survivors, by fostering a deeper connection with the rehabilitation system.

2.7.4. *Policy and reimbursement reforms*

As AI-driven virtual rehabilitation becomes a more integral part of cancer care, it will be essential for governments, healthcare providers and insurance companies to adapt policies and reimbursement models to recognize these services. Currently, reimbursement frameworks may not fully cover AI-based rehabilitation programs, hindering their widespread adoption (Aldhahi et al. 2025). Therefore, comprehensive policy reforms are needed to ensure that these services are accessible and affordable for all patients. Establishing clear standards and guidelines for AI-driven rehabilitation services will be crucial to ensuring their quality, safety and effectiveness. In addition, providing adequate reimbursement for these services will incentivize healthcare providers to adopt and integrate AI technologies into their practices, thereby accelerating the transition to more innovative and efficient care models.

2.8. Conclusion

AI-enhanced virtual rehabilitation represents a paradigm shift in post-cancer care, offering scalable, personalized and accessible solutions to the complex needs of cancer survivors. This innovative approach leverages AI to provide tailored rehabilitation programs, monitor progress and adjust interventions in real time, thus enhancing recovery and quality of life. By using advanced technologies such as ML, VR and predictive analytics, virtual rehabilitation platforms can offer individualized care plans, address physical, cognitive and emotional challenges and promote holistic healing. However, challenges remain, including the need for robust technological infrastructure, patient engagement and equitable access to digital tools. Further research is also required to refine AI algorithms, ensure their efficacy and evaluate their long-term impact on survivors' health outcomes.

The integration of AI into rehabilitation holds immense promise for improving outcomes, reducing healthcare disparities and optimizing healthcare delivery by providing personalized care at scale. This technology can bridge gaps in care, particularly in underserved or remote populations, and enable cancer survivors to access rehabilitation services tailored to their unique needs. To realize the full potential of this transformative approach, continued interdisciplinary collaboration between oncologists, rehabilitation experts, AI developers and policymakers will be essential. Moreover, rigorous evaluation of both clinical and ethical aspects, such as patient privacy and data security, will guide the responsible and equitable implementation of AI-enhanced rehabilitation. With careful management, AI can play a critical role in improving the rehabilitation experience, ensuring that cancer survivors lead fulfilling, healthier lives long after their treatment ends.

2.9. References

Adepoju, A.H., Austin-Gabriel, B.L., Hamza, O.L., Collins, A.N. (2022). Advancing monitoring and alert systems: A proactive approach to improving reliability in complex data ecosystems. *IRE Journals*, 5(11), 281–282.

Afridi, A. and Khan, S.N. (2024). Digital transformation in healthcare rehabilitation: A narrative review. *Journal of Process Management and New Technologies*, 12(1–2), 16–30.

Aldhahi, M.I., Alorainy, A.I., Abuzaid, M.M., Gareeballah, A., Alsubaie, N.F., Alshamary, A.S., Hamd, Z.Y. (2025). Adoption of artificial intelligence in rehabilitation: Perceptions, knowledge, and challenges among healthcare providers. *Healthcare*, 13(4), 350.

Aziz, R., Jawed, F., Khan, S.A., Sundus, H. (2024). Wearable IoT devices in rehabilitation: Enabling personalized precision medicine. In *Medical Robotics and AI-Assisted Diagnostics for a High-Tech Healthcare Industry*, Khang, A. (ed.). IGI Global, Hershey, PA, 281–308.

Bint Khalid, U., Naeem, M., Stasolla, F., Syed, M.H., Abbas, M., Coronato, A. (2024). Impact of AI-powered solutions in rehabilitation process: Recent improvements and future trends. *International Journal of General Medicine*, 17, 943.

Boué, S., Byrne, M., Hayes, A.W., Hoeng, J., Peitsch, M.C. (2018). Embracing transparency through data sharing. *International Journal of Toxicology*, 37(6), 466–471.

Brunetti, A., Buongiorno, D., Altini, N., Bevilacqua, V. (2023). Enabling technologies for optimized diagnosis, therapy and rehabilitation: Obtained results and future works. In *Italian Forum of Ambient Assisted Living*. Springer, Cham, 206–223.

Deshmukh, P. (2023). The integration of artificial intelligence and wearable technology in personalized physical therapy programs for enhanced recovery outcomes. *International Journal of Artificial Intelligence, Data Science, and Machine Learning*, 4(3), 1–9.

Dixon, L.B., Holoshitz, Y., Nossel, I. (2016). Treatment engagement of individuals experiencing mental illness: Review and update. *World Psychiatry*, 15(1), 13–20.

Ekundayo, F. (2024a). Leveraging AI-driven decision intelligence for complex systems engineering. *International Journal of Research Publication and Reviews*, 5(11), 5489–5499.

Ekundayo, F. (2024b). Reinforcement learning in treatment pathway optimization: A case study in oncology. *International Journal of Science and Research Archive*, 13(02), 2187–2205.

Evans, H., Sharrad, M.J., Diebert, M.C., Black-Tiong, S., Whitting, M.L. (2024). Effects of a prescribed, supervised exercise programme on health-related quality of life in oncology patients: A retrospective observational cohort study. *Journal of Clinical Exercise Physiology*, 13(s2), 493.

Gilbert, S.M., Miller, D.C., Hollenbeck, B.K., Montie, J.E., Wei, J.T. (2008). Cancer survivorship: Challenges and changing paradigms. *The Journal of Urology*, 179(2), 431–438.

Gupta, N.S. and Kumar, P. (2023). Perspective of artificial intelligence in healthcare data management: A journey towards precision medicine. *Computers in Biology and Medicine*, 162, 107051.

Kaur, J. (2024). FutureCare: AI robots revolutionizing health and healing. In *Revolutionizing the Healthcare Sector with AI*, Singla, B., Shalender, K., Stamer, K. (eds). IGI Global, Hershey, PA, 311–340.

Khalid, U.B., Naeem, M., Stasolla, F., Syed, M.H., Abbas, M., Coronato, A. (2024). Impact of AI-powered solutions in rehabilitation process: Recent improvements and future trends. *International Journal of General Medicine*, 17, 943–969.

Kumar, A., Rathore, P.S., Ahuja, S., Lilhore, U.K. (eds) (2025). *Integrating Neurocomputing with Artificial Intelligence*. John Wiley & Sons, New York.

Lippi, L., de Sire, A., Folli, A., Turco, A., Moalli, S., Marcasciano, M., Ammendolia, A., Invernizzi, M. (2024). Obesity and cancer rehabilitation for functional recovery and quality of life in breast cancer survivors: A comprehensive review. *Cancers*, 16(3), 521.

Mahvi, D.A., Liu, R., Grinstaff, M.W., Colson, Y.L., Raut, C.P. (2018). Local cancer recurrence: The realities, challenges, and opportunities for new therapies. *CA: A Cancer Journal for Clinicians*, 68(6), 488–505.

Mehmood, F., Mumtaz, N., Mehmood, A. (2025). Next-generation tools for patient care and rehabilitation: A review of modern innovations. *Actuators*, 14(3), 133. doi: 10.3390/act14030133.

Norori, N., Hu, Q., Aellen, F.M., Faraci, F.D., Tzovara, A. (2021). Addressing bias in big data and AI for healthcare: A call for open science. *Patterns*, 2(10), 100347.

Nur, S. (2024). The role of digital health technologies and sensors in revolutionizing wearable health monitoring systems. *International Journal of Innovative Research in Computer Science & Technology*, 12(6), 69–80.

Országhová, Z., Mego, M., Chovanec, M. (2021). Long-term cognitive dysfunction in cancer survivors. *Frontiers in Molecular Biosciences*, 8, 770413.

Sarella, P.N. and Mangam, V.T. (2024). AI-driven natural language processing in healthcare: Transforming patient-provider communication. *Indian Journal of Pharmacy Practice*. 17(1), 21–26.

Silver, J.K., Raj, V.S., Fu, J.B., Wisotzky, E.M., Smith, S.R., Kirch, R.A. (2015). Cancer rehabilitation and palliative care: Critical components in the delivery of high-quality oncology services. *Supportive Care in Cancer*, 23, 3633–3643.

Tran, S., Smith, L., El-Den, S., Carter, S. (2022). The use of gamification and incentives in mobile health apps to improve medication adherence: Scoping review. *JMIR mHealth and uHealth*, 10(2), e30671.

Vallée, A. (2024). Envisioning the future of personalized medicine: Role and realities of digital twins. *Journal of Medical Internet Research*, 26, e50204.

Wankhede, V., Verma, P., Gahane, S. (2024). A review on the developments in the field of AI-based gait analysis. In *2nd DMIHER International Conference on Artificial Intelligence in Healthcare, Education and Industry (IDICAIEI)*. IEEE, Wardha, 1–5.

3

Predictive Analytics in Physiotherapy and Oncology: Optimizing Treatment Plans

Predictive analytics is an emerging paradigm in healthcare that leverages historical data, statistical algorithms and machine learning techniques to forecast future clinical outcomes. This innovative approach offers healthcare professionals powerful tools to uncover complex patterns and make more informed decisions. In fields such as physiotherapy and oncology – where treatment responses and prognoses often vary significantly between individuals – predictive analytics holds the potential to revolutionize traditional care models. By integrating data from electronic health records, wearable sensors, genomic profiles and medical imaging, predictive models enable clinicians to deliver highly personalized and precise care. In physiotherapy, predictive analytics can estimate recovery timelines, assess the likelihood of treatment adherence and evaluate the risk of re-injury, allowing for more effective and individualized intervention planning. In oncology, these models support early disease detection, predict responses to specific therapies and estimate survival rates, thereby improving the quality and personalization of cancer care. This chapter explores the integration of predictive analytics into physiotherapy and oncology, examining methodologies, current applications, real-world case studies and the technological platforms that support them. In addition, this chapter addresses critical ethical and practical considerations, such as data quality, model transparency and regulatory compliance. By highlighting these factors, we underscore the transformative potential of predictive analytics in advancing patient-centered care, improving clinical outcomes, reducing adverse effects and optimizing the use of healthcare resources.

3.1. Introduction

The advent of data-driven methodologies has transformed many facets of healthcare, paving the way for precision medicine and evidence-based practices (Fatima et al. 2024). One of the most transformative tools in this evolution is predictive analytics, a subset of advanced analytics that leverages historical and

Chapter written by S. ANANDH, Suresh J. BHOSALE and Vivek PARHATE.

real-time data to forecast future clinical outcomes. Predictive analytics encompasses a range of statistical techniques, including data mining, machine learning, and artificial intelligence (AI) and is increasingly being integrated into clinical workflows to provide insights that support decision-making.

In contemporary healthcare, predictive analytics offers profound potential to identify at-risk populations, anticipate disease progression and tailor treatment plans based on individual patient profiles (Edoh et al. 2024). By leveraging data from various sources such as electronic health records (EHRs), wearable devices, genomic sequencing and medical imaging, predictive analytics creates a foundation for proactive care, reducing the reliance on generalized treatment protocols. As such, the model of care is shifting from reactive to anticipatory, improving outcomes while optimizing resource utilization.

Physiotherapy and oncology represent two disciplines where predictive analytics can be particularly impactful due to their inherent complexity and variability in patient outcomes. In physiotherapy, patient recovery is influenced by numerous factors including age, preexisting conditions, injury severity and adherence to rehabilitation protocols (Negm et al. 2022). Predictive models can synthesize these variables to forecast recovery timelines, adherence likelihood and the probability of reinjury, thus enabling clinicians to create highly personalized therapy regimens. These models not only enhance patient care but also contribute to reducing healthcare costs by identifying inefficiencies and preventing complications before they arise.

In oncology, predictive analytics plays a critical role across the care continuum, from early detection to treatment and survivorship (Keim-Malpass et al. 2023). Cancer care involves an array of decisions influenced by tumor type, genetic markers, patient health status and therapeutic options. Predictive tools can support clinicians by identifying patients at high risk for cancer, forecasting individual responses to treatments such as chemotherapy and immunotherapy and estimating survival probabilities. These insights are vital for devising optimal treatment strategies, minimizing adverse effects and improving patient quality of life.

Furthermore, predictive analytics can facilitate population-level health management by identifying emerging trends and gaps in care delivery. This can support policy-making and institutional planning, especially in resource-constrained settings. As health systems increasingly adopt value-based care models, the integration of predictive analytics offers a compelling avenue for enhancing care delivery, improving outcomes and ensuring sustainability (Oluwole et al. 2024).

However, the implementation of predictive analytics is not without challenges. Issues related to data quality, model transparency, interoperability and ethical concerns such as data privacy and algorithmic bias must be addressed. There is also a need for cross-disciplinary collaboration among clinicians, data scientists and policymakers to ensure that predictive tools are developed and used effectively (Patel et al. 2024).

This chapter explores the transformative impact of predictive analytics in physiotherapy and oncology, outlining the methodologies, applications and technologies underpinning this paradigm shift. By examining real-world case studies and emerging trends, we aim to provide a comprehensive understanding of how predictive analytics is reshaping personalized care and driving innovation in these key areas of medicine.

3.2. Background and rationale

3.2.1. *Overview of predictive analytics*

Predictive analytics is a rapidly growing field in healthcare that leverages a variety of statistical and computational techniques to analyze current and historical data in order to make forecasts about future events (Rahim et al. 2025). Core methods include regression analysis, decision trees, support vector machines, ensemble methods, neural networks and deep learning algorithms. These tools are capable of uncovering hidden patterns and complex relationships within datasets that may not be immediately apparent to human analysts.

Data sources for predictive analytics in healthcare are diverse and continue to expand with the integration of emerging technologies (Rahim et al. 2025). EHRs provide a foundational source of clinical data, including patient demographics, diagnoses, treatment histories and laboratory results. Imaging modalities such as magnetic reasoning imaging (MRI), computed tomography (CT) scans and X-rays contribute detailed anatomical and functional insights, while genomic data reveal information about individual susceptibility to diseases and likely responses to specific treatments. Wearable sensors and mobile health technologies add continuous streams of physiological and behavioral data, such as heart rate, physical activity and sleep patterns (George et al. 2023).

By processing this multifaceted data, predictive analytics can generate actionable predictions that support clinical decision-making at both the individual and population levels. These predictions may include the likelihood of disease onset, the probability of treatment success, the risk of adverse events and expected recovery trajectories. In the context of physiotherapy and oncology, these capabilities translate to more targeted interventions, improved monitoring and adaptive treatment plans that respond to evolving patient conditions in real time. As computational power and data accessibility improve, the scope and precision of predictive analytics are expected to expand, ushering in a new era of intelligent, data-informed healthcare.

3.2.2. *Relevance in healthcare*

The shift towards personalized medicine necessitates tools that can interpret complex and multifactorial datasets to guide clinical decision-making in a more nuanced and individualized manner (Marques et al. 2024). Traditional clinical approaches often rely on generalized treatment pathways that may not account for the unique physiological, genetic and environmental factors influencing each patient's health. Predictive analytics bridges this critical gap by offering the ability to anticipate health risks, predict treatment responses and support clinicians in making informed, patient-specific decisions.

In modern healthcare, where the volume and variety of data have exponentially increased, predictive analytics enables the transformation of raw information into actionable intelligence (in modern healthcare). This is particularly significant in chronic disease management, early diagnosis, hospital readmission prevention and therapy optimization. By shifting the paradigm from reactive to proactive care, healthcare systems can intervene earlier, allocate resources more effectively and ultimately improve clinical outcomes while controlling costs.

For example, predictive models can alert providers to patients at high risk for hospital readmission or adverse drug reactions, allowing for timely interventions. In population health, analytics can help identify at-risk groups and inform preventive strategies (Edoh et al. 2024). Moreover, real-time analytics integrated with wearable technology can support continuous monitoring and adaptive care, especially valuable in outpatient settings.

As personalized medicine continues to evolve, predictive analytics will play an increasingly central role not only in treatment but also in prevention, patient engagement and healthcare planning. The ability to foresee health trajectories empowers clinicians and patients alike, fostering a more collaborative and data-informed healthcare environment.

3.2.3. *Importance in physiotherapy and oncology*

Both physiotherapy and oncology encompass complex, dynamic conditions that demand individualized care strategies (Dalzell et al. 2017). The variability in patient profiles, disease progression and therapeutic responses in these disciplines underscores the necessity for tools that can enhance clinical precision and responsiveness. Predictive analytics addresses this need by providing a data-informed foundation for clinical decision-making, tailored to the unique attributes of each patient.

In physiotherapy, the road to recovery is influenced by numerous variables such as the nature and severity of injury, patient age, physical condition, comorbidities, adherence to prescribed exercises and psychosocial factors. Predictive models can help estimate recovery trajectories, anticipate risks of complications or re-injury and monitor patient engagement (Ditton et al. 2020). By doing so, clinicians can adapt treatment plans in real time, ensuring that patients receive the most effective interventions for their specific conditions. This not only boosts therapeutic outcomes but also minimizes unnecessary resource utilization.

Oncology, with its intricate interplay of tumor biology, genetic variability, treatment response and disease progression, presents a fertile ground for the application of predictive analytics. Models can assist in early diagnosis by identifying high-risk individuals based on genetic markers, lifestyle factors and medical history (Athar 2024). They can also forecast patient response to chemotherapy, radiotherapy or immunotherapy, thereby guiding oncologists in selecting the most effective and least harmful treatment modalities. Predictive tools are instrumental in estimating survival probabilities, potential side effects and recurrence risks, all of which are crucial for informed treatment planning and patient counseling.

By integrating predictive analytics into physiotherapy and oncology, healthcare providers can transcend traditional one-size-fits-all approaches. These technologies foster a shift towards precision health – where care is not only reactive but also anticipatory and personalized – improving both patient outcomes and system efficiency.

3.3. Predictive analytics in physiotherapy

3.3.1. *Application areas*

Predictive analytics in physiotherapy encompasses a wide range of applications that enhance decision-making, personalize treatment protocols and optimize patient outcomes (Sunny et al. 2024). These applications aim to leverage historical and real-time data to forecast patient progress and identify potential risks throughout the rehabilitation process. Below are the key domains where predictive analytics is transforming physiotherapy:

– *Outcome prediction*: advanced algorithms can forecast functional recovery timelines post-injury, surgery or neurological event. By analyzing patient demographics, clinical data and initial functional assessments, these models predict metrics such as return-to-function time, range of motion recovery and overall therapy success. This helps set realistic goals and expectations for both patients and clinicians.

– *Adherence prediction*: patient compliance is critical for physiotherapy success, but adherence rates often vary due to factors like motivation, socio-economic status or logistical barriers (Burgess et al. 2017). Predictive tools analyze behavioral data, therapy session attendance and socio-demographic indicators to identify patients at high risk of noncompliance. This enables clinicians to proactively engage with at-risk patients using targeted support strategies, reminders and personalized coaching.

– *Risk assessment*: injury recurrence and complications can severely affect long-term health outcomes and incur additional healthcare costs (Haider et al. 2020). Predictive models assess biomechanical patterns, patient history and performance metrics to flag individuals at risk for re-injury, delayed healing or therapy resistance. These insights guide early interventions and dynamic modifications to therapy plans to mitigate risks.

– *Therapy optimization*: wearable devices and motion capture technologies provide real-time data on joint angles, gait patterns and muscle activation. Machine learning algorithms analyze these data streams to personalize exercise regimens, track progress and suggest modifications in real time, thereby making therapy sessions more efficient and adaptive (machine learning).

– *Resource management*: predictive analytics can also inform clinic-level decisions, such as scheduling follow-ups, allocating therapy hours or prioritizing high-risk patients. This contributes to more efficient use of clinical resources and improved patient throughput.

3.3.2. *Case studies*

Case studies offer tangible examples of how predictive analytics has been successfully implemented in physiotherapy settings to improve patient care and clinical efficiency (Westby et al. 2016). These real-world applications highlight the utility of data-driven decision-making and provide insights into potential future innovations.

– *Anterior cruciate ligament (ACL) reconstruction*: predictive models have been developed using preoperative and demographic data, such as patient age, body mass index (BMI), pre-surgery knee strength and physical activity levels, to forecast recovery timelines following ACL reconstruction (Muller et al. 2022). These models enable clinicians to set realistic expectations, plan individualized rehabilitation protocols and identify patients who may require more intensive monitoring and support.

– *Stroke rehabilitation*: in the context of stroke recovery, machine learning algorithms have been employed to predict long-term mobility outcomes based on data collected during the early phases of rehabilitation. Variables such as initial motor function scores, comorbid conditions and early therapy adherence levels are analyzed to forecast a patient's ability to regain independent ambulation (Meyer et al. 2015). These insights inform goal-setting and allow for the timely adaptation of therapy plans.

– *Low back pain management*: predictive analytics has been used to identify patients with acute low back pain who are at high risk of developing chronic conditions. By analyzing patient-reported outcomes, imaging data and psychosocial indicators, clinicians can intervene early with targeted therapies and behavioral support strategies, reducing the likelihood of long-term disability.

– *Post-operative shoulder rehabilitation*: wearable sensors combined with predictive modeling have been used to assess post-surgical range of motion and healing progress in patients undergoing rotator cuff repair (post-operative shoulder). The data enables clinicians to detect suboptimal recovery trends early and adjust therapy protocols accordingly.

These case studies demonstrate the potential of predictive analytics to transform physiotherapy from a reactive to a proactive discipline, ultimately improving patient outcomes, reducing healthcare costs and enhancing clinician confidence in treatment planning.

– *ACL reconstruction*: predictive models using preoperative and demographic data forecast recovery timelines.

– *Stroke rehabilitation*: machine learning algorithms predict mobility outcomes based on early therapy responses.

3.3.3. *Technologies used*

Wearables and motion sensors collect detailed biomechanical and physiological data such as joint angles, muscle activity, gait patterns and range of motion (Díaz et al. 2019). These devices often include accelerometers, gyroscopes, electromyography (EMG) sensors and inertial measurement units (IMUs), which provide continuous, real-time feedback during rehabilitation exercises.

The data collected is processed through sophisticated algorithms and machine learning models that interpret patterns, detect anomalies and generate predictive insights. These analytics tools allow clinicians to track a patient's progress over time, identify deviations from expected recovery trajectories and adjust therapy plans dynamically based on objective metrics.

In addition to motion tracking, many systems integrate mobile applications and cloud platforms that enable remote monitoring and tele-rehabilitation. Patients can perform prescribed exercises at home while their data is automatically uploaded and analyzed. Clinicians can then review this information and make data-driven decisions about exercise intensity, frequency and duration.

Emerging technologies also include augmented reality/virtual reality (AR/VR) systems for immersive rehabilitation and AI-powered chatbots or digital coaches to support patient motivation and adherence. Collectively, these technologies contribute to a more interactive, personalized and efficient physiotherapy experience.

3.3.4. *Benefits*

3.3.4.1. *Enhanced personalization of therapy*

This could refer to tailoring therapeutic interventions to the individual needs of each patient, taking into account their specific conditions, preferences and responses to treatment (Denford et al. 2014).

Personalization ensures that each patient receives the most effective form of care for their unique situation, increasing the likelihood of successful outcomes.

3.3.4.2. *Efficient allocation of clinical resources*

By using data-driven insights, clinicians can prioritize and allocate resources more effectively.

This means that clinical time, equipment and staff can be used where they are needed most, leading to better outcomes and more cost-effective care.

3.3.4.3. *Data-driven adjustments to interventions*

With real-time data collection and analysis, therapies and interventions can be dynamically adjusted to optimize their effectiveness (Heesen et al. 2024).

This flexibility ensures that if a particular approach is not working, adjustments can be made quickly to improve results without waiting for scheduled reviews.

3.4. Predictive analytics in oncology

3.4.1. *Application areas*

– *Prognosis estimation*: survival analysis based on tumor characteristics and patient history.

– *Treatment optimization*: predicting response to chemotherapy, radiotherapy and immunotherapy.

– *Early detection*: identifying high-risk individuals for certain cancers.

3.4.2. *Case studies*

– *Breast cancer*: genomic data used to predict chemotherapy response.

– *Lung cancer*: imaging and EHR data combined to estimate progression risk.

3.4.3. *Technologies used*

Integration of imaging data (radiomics), genomic sequencing and EHRs with machine learning tools such as support vector machines and deep learning (Tong et al. 2023).

3.4.4. *Benefits*

– Tailored treatment plans.

– Reduced side effects through personalized dosing.

– Improved survival rates through early intervention.

3.5. Methodological framework

3.5.1. *Data sources*

– EHRs.

– Wearable devices.

– Genomic and proteomic data.

– Imaging data.

3.5.2. *Data preprocessing*

It includes data cleaning, normalization, feature selection and handling missing data.

3.5.3. *Model development*

Common models include logistic regression, random forests, neural networks and ensemble models. Training involves splitting data into training and testing sets, followed by validation using metrics such as accuracy, precision, recall and area under the receiver operating characteristic curve (AUC-ROC) (Varoquaux and Colliot 2023).

3.5.4. *Implementation in clinical settings*

– Integration with clinical decision support systems (CDSSs).

– Real-time analytics for adaptive treatment adjustments.

3.6. Challenges and limitations

3.6.1. *Data quality and standardization*

One of the primary challenges in deploying AI in clinical settings is ensuring high-quality, standardized data. Inconsistent data entry practices, use of heterogeneous data formats and variations in clinical terminologies across institutions can significantly impact model performance and generalizability (Sargiotis 2024). For example, differences in how diagnoses or procedures are coded (e.g. International Classification of Diseases, Tenth Revision (ICD-10) vs. Systematized Nomenclature of Medicine – Clinical Terms (SNOMED CT)) may lead to misinterpretation or missed patterns by AI systems.

Moreover, missing or incomplete data, such as unrecorded comorbidities or medication histories, can skew model outcomes and limit the reliability of predictions. Establishing data interoperability and adopting standardized data schemas (e.g. FHIR, Fast Healthcare Interoperability Resources) are crucial steps towards improving model robustness and reproducibility across diverse healthcare settings.

3.6.2. *Model interpretability*

While advanced machine learning models, particularly deep learning architectures, have demonstrated superior predictive capabilities, they often function as "black boxes", making it difficult to understand the rationale behind their outputs (Hassija et al. 2024). This lack of transparency can be a significant barrier to clinical adoption, as healthcare providers must be able to trust and validate the recommendations or diagnoses made by AI systems. Without clear interpretability, clinicians may be reluctant to rely on AI for critical decision-making. Efforts to improve model explainability – through techniques such as SHAP (SHapley Additive exPlanations), LIME (Local Interpretable Model-agnostic Explanations) or attention mechanisms – are ongoing but still face limitations in practical deployment.

3.6.3. *Ethical and legal considerations*

The integration of AI into healthcare raises numerous ethical and legal concerns. Ensuring patient data privacy and security is paramount, especially given the sensitive nature of health information (Ibrahim et al. 2024). There are also concerns about informed consent when patient data is used to train AI models, particularly when data is shared across institutions or used for secondary purposes. Moreover, biases embedded in training data can lead to discriminatory outcomes, disproportionately affecting minority or underserved populations. Legal accountability remains ambiguous: if an AI system provides incorrect guidance resulting in harm, it is unclear whether the liability lies with the developer, the clinician or the institution (Frank 2019). Regulatory bodies are working to address these challenges, but clear frameworks are still evolving.

3.6.4. *Integration with clinical workflow*

Effective integration of AI tools into existing clinical workflows remains a major hurdle. Many healthcare providers are already burdened with heavy workloads and may resist adopting new technologies that require additional time or effort to learn (Singh et al. 2020). The introduction of AI systems often necessitates comprehensive staff training, updates to health IT infrastructure and changes in standard operating procedures; all of which require time, financial investment and institutional buy-in. Moreover, poorly designed interfaces or tools that do not align with clinical needs can reduce efficiency rather than enhance it. For AI to be successfully adopted, it must be seamlessly embedded into the daily routines of healthcare professionals, providing value without creating additional friction.

3.7. Ethical and regulatory considerations

3.7.1. *Patient consent and data privacy*

Ensuring transparent policies for data use and securing sensitive health information (Alhasan 2025).

3.7.2. *Algorithmic fairness*

Models must be regularly audited for biases that may adversely affect certain patient groups.

3.7.3. *Compliance with regulations*

Adherence to frameworks such as GDPR (General Data Protection Regulation), HIPAA (Health Insurance Portability and Accountability Act) and Food and Drug Administration (FDA) guidelines for AI in healthcare.

3.8. Future directions

3.8.1. *Federated learning to enable multi-institutional model training without data sharing*

Federated learning allows institutions to collaboratively train machine learning models without needing to exchange patient data (Cheung et al. 2023). This is especially important in healthcare, where data privacy and security are critical. It enables multiple institutions to create more robust and diverse models while maintaining the confidentiality of patient information.

3.8.2. *Integration with digital twins for real-time simulation of treatment outcomes*

Digital twins are virtual replicas of physical entities. In this case, patients or their treatment processes (Bruynseels et al. 2018). By integrating these models with clinical data, healthcare providers can simulate how a treatment will likely impact a patient in real-time, optimizing decision-making and personalizing interventions. This could lead to more accurate predictions of treatment efficacy and better management of chronic conditions.

3.8.3. *Expansion into preventive physiotherapy and early cancer screening*

Moving from reactive treatment to preventive healthcare is a crucial step. In preventive physiotherapy, predictive analytics could identify at-risk patients and deliver interventions before conditions like musculoskeletal injuries develop (Larsson et al. 2020). Similarly, early cancer screening powered by AI tools could enable the identification of cancer in its earliest stages, improving survival rates and reducing treatment costs.

3.8.4. *Enhanced interpretability through explainable AI tools*

As AI becomes more integrated into clinical decision-making, understanding how AI systems arrive at their conclusions becomes crucial (Yang et al. 2019) Explainable AI (XAI) focuses on making AI's decisions transparent and interpretable to healthcare professionals. This will help clinicians trust and effectively use AI-driven insights in their practice, ensuring that decisions are both accurate and understandable.

3.9. Conclusion

Predictive analytics has the potential to greatly improve physiotherapy and oncology by making healthcare more personalized, efficient and proactive. By analyzing large amounts of patient data, predictive tools help healthcare professionals not just respond to problems but anticipate and prevent them. In physiotherapy, this means spotting risks for injury or slow recovery early, so that treatment plans can be adjusted to better suit each patient. In oncology, it allows for earlier detection of cancer and more accurate predictions of how patients will respond to treatment, leading to better outcomes and higher survival rates.

As technology advances, using predictive analytics in everyday clinical care is becoming more practical. This shift will help tailor treatments to individual needs, improve patient outcomes and reduce costs by avoiding complications and unnecessary hospital visits. However, adopting these tools must be done carefully. It is important to protect patient data, avoid bias in algorithms and follow privacy laws like GDPR and HIPAA. To fully benefit from predictive analytics, collaboration is essential: healthcare providers, data scientists, regulators and tech developers must work together! By doing so, we can build a healthcare system that is smarter, more responsive and more accessible, while still protecting patients' rights. The future of physiotherapy and oncology will depend on combining innovation with responsibility.

3.10. References

Alhasan, T.K. (2025). Managing legal risks in health information exchanges: A comprehensive approach to privacy, consent, and liability. *Journal of Healthcare Risk Management*, 44(4), 12–24.

Athar, M. (2024). Potentials of artificial intelligence in familial hypercholesterolemia: Advances in screening, diagnosis, and risk stratification for early intervention and treatment. *International Journal of Cardiology*, 412, 132315.

Bruynseels, K., Santoni de Sio, F., Van den Hoven, J. (2018). Digital twins in health care: Ethical implications of an emerging engineering paradigm. *Frontiers in Genetics*, 9, 31.

Burgess, E., Hassmén, P., Pumpa, K.L. (2017). Determinants of adherence to lifestyle intervention in adults with obesity: A systematic review. *Clinical Obesity*, 7(3), 123–135.

Cheung, A.T., Nasir-Moin, M., Kwon, Y.J., Guan, J., Liu, C., Jiang, L., Raimondo, C., Chotai, S., Chambless, L., Ahmad, H.S. et al. (2023). Methods and impact for using federated learning to collaborate on clinical research. *Neurosurgery*, 92(2), 431–438.

Dalzell, M.A., Smirnow, N., Sateren, W., Sintharaphone, A., Ibrahim, M., Mastroianni, L., Zambrano, L.V., O'Brien, S. (2017). Rehabilitation and exercise oncology program: Translating research into a model of care. *Current Oncology*, 24(3), e191.

Denford, S., Frost, J., Dieppe, P., Cooper, C., Britten, N. (2014). Individualisation of drug treatments for patients with long-term conditions: A review of concepts. *BMJ Open*, 4(3), e004172.

Díaz, S., Stephenson, J.B., Labrador, M.A. (2019). Use of wearable sensor technology in gait, balance, and range of motion analysis. *Applied Sciences*, 10(1), 234.

Ditton, E., Johnson, S., Hodyl, N., Flynn, T., Pollack, M., Ribbons, K., Walker, F.R., Nilsson, M. (2020). Improving patient outcomes following total knee arthroplasty: Identifying rehabilitation pathways based on modifiable psychological risk and resilience factors. *Frontiers in Psychology*, 11, 1061.

Dixit, P., Kommula, A.V., Rathore, P.S. (2024). Learning from scarcity: Unlocking healthcare insights with few-shot machine learning. In *Applying Machine Learning Techniques to Bioinformatics: Few-Shot and Zero-Shot Methods*, Lilhore, U.K., Kumar, A., Simaiya, S., Vyas, N., Dutt, V. (eds). IGI Global, Hershey, PA.

Edoh, N.L., Chigboh, V.M., Zouo, S.J., Olamijuwon, J. (2024). Improving healthcare decision-making with predictive analytics: A conceptual approach to patient risk assessment and care optimization. *International Journal of Scholarly Research in Medicine and Dentistry*, 3(2), 1–10.

Fatima, G., Siddiqui, Z., Parvez, S. (2024). AI and precision medicine: Paving the way for future treatment. *Preprints*. doi: 10.20944/preprints202412.0036.v1.

Frank, X. (2019). Is Watson for oncology per se unreasonably dangerous? Making a case for how to prove products liability based on a flawed artificial intelligence design. *American Journal of Law & Medicine*, 45(2–3), 273–294.

George, A.H., Shahul, A., George, A.S. (2023). Wearable sensors: A new way to track health and wellness. *Partners Universal International Innovation Journal*, 1(4), 15–34.

Haider, A.H., Herrera-Escobar, J.P., Al Rafai, S.S., Harlow, A.F., Apoj, M., Nehra, D., Kasotakis, G., Brasel, K., Kaafarani, H.M., Velmahos, G. et al. (2020). Factors associated with long-term outcomes after injury: Results of the functional outcomes and recovery after trauma emergencies (FORTE) multicenter cohort study. *Annals of Surgery*, 271(6), 1165–1173.

Hassija, V., Chamola, V., Mahapatra, A., Singal, A., Goel, D., Huang, K., Scardapane, S., Spinelli, I., Mahmud, M., Hussain, A. (2024). Interpreting black-box models: A review on explainable artificial intelligence. *Cognitive Computation*, 16(1), 45–74.

Heesen, P., Schelling, G., Birbaumer, M., Jäger, R., Bode, B., Studer, G., Fuchs, B. (2024). Real-world-time data and RCT synergy: Advancing personalized medicine and sarcoma care through digital innovation. *Cancers*, 16(14), 2516.

Ibrahim, A.M., Abdel-Aziz, H.R., Mohamed, H.A., Zaghamir, D.E., Wahba, N.M., Hassan, G.A., Shaban, M., El-Nablaway, M., Aldughmi, O.N., Aboelola, T.H. (2024). Balancing confidentiality and care coordination: Challenges in patient privacy. *BMC Nursing*, 23(1), 564.

Keim-Malpass, J. and Kausch, S.L. (2023). Data science and precision oncology nursing: Creating an analytic ecosystem to support personalized supportive care across the trajectory of illness. *Seminars in Oncology Nursing*, 39(3), 151432.

Kumar, A., Rathore, P.S., Ahuja, S., Lilhore, U.K. (eds) (2025). *Integrating Neurocomputing with Artificial Intelligence*. John Wiley & Sons, New York.

Larsson, M.E., Nordeman, L., Holmgren, K., Grimby-Ekman, A., Hensing, G., Björkelund, C., Bergman, S., Ekhammar, A., Dottori, M., Bernhardsson, S. (2020). Prevention of sickness absence through early identification and rehabilitation of at-risk patients with musculoskeletal pain (PREVSAM): A randomised controlled trial protocol. *BMC Musculoskeletal Disorders*, 21(1), 1–3. doi: 10.1186/s12891-020-03790-5.

Marques, L., Costa, B., Pereira, M., Silva, A., Santos, J., Saldanha, L., Silva, I., Magalhães, P., Schmidt, S., Vale, N. (2024). Advancing precision medicine: A review of innovative in silico approaches for drug development, clinical pharmacology and personalized healthcare. *Pharmaceutics*, 16(3), 332.

Meyer, M.J., Pereira, S., McClure, A., Teasell, R., Thind, A., Koval, J., Richardson, M., Speechley, M. (2015). A systematic review of studies reporting multivariable models to predict functional outcomes after post-stroke inpatient rehabilitation. *Disability and Rehabilitation*, 37(15), 1316–1323.

Muller, B., Yabroudi, M.A., Lynch, A., Popchak, A.J., Lai, C.L., van Dijk, C.N., Fu, F.H., Irrgang, J.J. (2022). Return to preinjury sports after anterior cruciate ligament reconstruction is predicted by five independent factors. *Knee Surgery, Sports Traumatology Arthroscopy*, 30(1), 84–92.

Negm, A.M., Salopek, A., Zaide, M., Meng, V.J., Prada, C., Chang, Y., Zanwar, P., Santos, F.H., Philippou, E., Rosario, E.R. et al. (2022). Rehabilitation at the time of pandemic: Patient journey recommendations. *Frontiers in Aging Neuroscience*, 14, 781226.

Oluwole, O., Emmanuel, O., Ogbuagu, O.O., Alemede, V., Adefolaju, I. (2024). Pharmaceutical supply chain optimization through predictive analytics and value-based healthcare economics frameworks. *International Journal of Engineering Technology Research & Management*, 8(2), 88–104.

Patel, A.U., Gu, Q., Esper, R., Maeser, D., Maeser, N. (2024). The crucial role of interdisciplinary conferences in advancing explainable AI in healthcare. *BioMedInformatics*, 4(2), 1363–1383.

Rahim, M.J., Afroz, A., Akinola, O. (2025). Predictive analytics in healthcare: Big data, better decisions. *International Journal of Scientific Research and Modern Technology*, 4(1), 1–21.

Sargiotis, D. (ed.) (2024). Data quality management: Ensuring accuracy and reliability. In *Data Governance: A Guide*. Springer, Cham.

Singh, R.P., Hom, G.L., Abramoff, M.D., Campbell, J.P., Chiang, M.F. (2020). Current challenges and barriers to real-world artificial intelligence adoption for the healthcare system, provider, and the patient. *Translational Vision Science & Technology*, 9(2), 45.

Srivastava, S., Anand, A., Kumar, A., Saini, B., Rathore, P.S. (eds) (2024). *Bio-Inspired Optimization for Medical Data Mining*. John Wiley & Sons, New York.

Sunny, M.N., Saki, M.B., Al Nahian, A., Ahmed, S.W., Shorif, M.N., Atayeva, J., Rizvi, S.W. (2024). Optimizing healthcare outcomes through data-driven predictive modeling. *Journal of Intelligent Learning Systems and Applications*, 16(4), 384–402.

Tong, L., Shi, W., Isgut, M., Zhong, Y., Lais, P., Gloster, L., Sun, J., Swain, A., Giuste, F., Wang, M.D. (2023). Integrating multi-omics data with EHR for precision medicine using advanced artificial intelligence. *IEEE Reviews in Biomedical Engineering*, 17, 80–97.

Varoquaux, G. and Colliot, O. (2023). Evaluating machine learning models and their diagnostic value. In *Machine Learning for Brain Disorders*, Colliot, O. (ed.). Springer, New York.

Westby, M.D., Klemm, A., Li, L.C., Jones, C.A. (2016). Emerging role of quality indicators in physical therapist practice and health service delivery. *Physical Therapy*, 96(1), 90–100.

Yang, Q., Steinfeld, A., Zimmerman, J. (2019). Unremarkable AI: Fitting intelligent decision support into critical, clinical decision-making processes. In *Proceedings of the 2019 CHI Conference on Human Factors in Computing Systems*. ACM, New York.

4

Integrating AI with Physiotherapy to Improve Quality of Life for Cancer Patients

The integration of artificial intelligence (AI) into healthcare has revolutionized various medical fields, including oncology and physiotherapy. This chapter explores the transformative potential of AI-enhanced physiotherapy in improving the quality of life (QoL) for cancer patients throughout their treatment and survivorship journey. It examines how AI technologies contribute to early diagnosis, personalized treatment planning, progress tracking, real-time feedback and outcome prediction. In addition, AI plays a significant role in addressing the psychological and physical rehabilitation needs of cancer survivors, especially in managing treatment-induced complications such as fatigue, neuropathy and reduced mobility. By leveraging advancements in machine learning, deep learning, wearable technology, computer vision and robotics, the study presents a multidisciplinary framework to optimize physiotherapeutic interventions. AI-driven platforms enable continuous monitoring and adaptive therapy adjustments, thereby improving adherence, engagement and clinical outcomes.

This chapter highlights real-world applications, including remote rehabilitation systems, robotic exoskeletons and AI-powered mobile apps, while evaluating their benefits in enhancing accessibility and resource efficiency. This comprehensive investigation also delves into the ethical, technical and clinical challenges associated with integrating AI in cancer care, including issues of data privacy, algorithmic bias, clinical validation and system interoperability. Ultimately, the research advocates for a collaborative, patient-centric and data-driven approach to rehabilitation, suggesting that AI-augmented physiotherapy could significantly elevate the standard of care for oncology patients.

Chapter written by Namrata KADAM, Rashmi GUDUR and Vivek DESHPANDE.

4.1. Introduction

4.1.1. *Background*

Cancer remains one of the leading causes of morbidity and mortality worldwide, with millions of new cases diagnosed annually (Bray et al. 2018). Grateful to significant advances in early detection, surgical interventions, radiation therapy, chemotherapy and targeted drug treatments, survival rates have steadily improved across various types of cancers. As more individuals live beyond their initial diagnoses, there is a growing emphasis on enhancing their post-treatment quality of life (QoL).

Physiotherapy has emerged as a vital component of cancer rehabilitation, addressing a wide range of physical, functional and psychological challenges that survivors face (Silver et al. 2015). It aids in improving mobility, managing chronic pain, alleviating treatment-induced fatigue, preventing lymphedema and enhancing overall functional capacity. In addition, physiotherapy plays a crucial role in supporting mental well-being through structured physical activity and patient empowerment.

However, conventional physiotherapy approaches face several limitations. They are typically resource-intensive, requiring one-on-one supervision and repeated clinical visits, which can strain both healthcare systems and patients (Yan et al. 2022). Furthermore, these interventions often follow standardized protocols, lacking the personalization necessary to account for individual patient variability in health status, cancer type, treatment history and recovery trajectory. As such, innovative solutions are needed to enhance the reach, efficiency and effectiveness of physiotherapeutic care for cancer patients.

4.1.2. *Rationale for AI integration*

AI has demonstrated tremendous potential in transforming healthcare delivery through data-driven insights, predictive analytics and automation. In the context of physiotherapy, AI offers an unprecedented ability to personalize rehabilitation plans by analyzing individual patient profiles, including medical history, current health status and specific therapeutic needs (Pham et al. 2024). It enables the creation of dynamic, adaptive treatment protocols that evolve based on patient progress, ensuring more effective and timely interventions.

Furthermore, AI systems equipped with wearable sensors and mobile applications can monitor patient progress in real-time, capturing metrics such as range of motion, heart rate and physical activity levels. This allows healthcare providers to remotely track adherence to prescribed exercises, detect deviations from expected recovery patterns and intervene promptly when necessary. Such real-time feedback mechanisms not only enhance clinical outcomes but also empower patients by providing continuous support and motivation (Tsiouris et al. 2020).

AI's predictive capabilities are equally valuable, as machine learning (ML) models can forecast potential complications or setbacks based on early indicators, enabling preemptive adjustments to treatment plans. Moreover, the integration of AI reduces the burden on physiotherapists by automating routine assessments and documentation, thereby allowing them to focus on delivering high-quality, hands-on care. Overall, incorporating AI into physiotherapy presents a strategic advancement that aligns with the broader goals of precision medicine and holistic cancer rehabilitation (Alagappan et al. 2024).

4.1.3. *Objectives*

This research aims to:

– analyze the current role of physiotherapy in cancer care and its impact on QoL for patients undergoing treatment or in remission;

– investigate and catalogue the various AI technologies currently used in healthcare and assess their applicability in physiotherapeutic contexts;

– examine the potential benefits and limitations of integrating AI into physiotherapy specifically for cancer patients, including aspects of accessibility, personalization and engagement;

– develop a comprehensive framework for AI-augmented physiotherapy interventions tailored to different stages of cancer treatment and recovery;

– encourage interdisciplinary collaboration between oncologists, physiotherapists, AI researchers and healthcare policymakers to ensure safe, effective and ethical implementation of AI technologies in rehabilitation (Shawli et al. 2024).

4.2. Role of physiotherapy in cancer care

4.2.1. *Importance of physiotherapy*

Physiotherapy helps address a range of issues associated with cancer and its treatments, including pain, lymphedema, fatigue, neuropathy and decreased mobility. It also contributes to mental health by reducing anxiety and depression through physical activity.

4.2.2. *Types of physiotherapeutic interventions*

The common interventions include:

– *Exercise therapy*: includes aerobic conditioning, strength training and flexibility exercises. It helps improve cardiovascular health, muscle strength and joint mobility (Micheo et al. 2012). Tailored exercise programs can also reduce cancer-related fatigue and enhance functional independence.

– *Manual therapy*: involves hands-on techniques such as massage, mobilization and manipulation to alleviate pain, reduce stiffness and improve range of motion. It also promotes lymphatic drainage and helps manage lymphedema.

– *Electrotherapy*: uses electrical modalities such as Transcutaneous Electrical Nerve Stimulation (TENS), ultrasound and interferential therapy to manage pain, stimulate muscle activity and facilitate tissue healing (Ariel et al. 2022).

– *Hydrotherapy*: incorporates water-based exercises in a pool or aquatic environment. The buoyancy and resistance properties of water make it ideal for low-impact rehabilitation, especially for patients with joint pain or limited mobility.

– *Breathing exercises*: focus on improving lung function, especially in patients who have undergone thoracic surgery or radiation (Langer 2021). Techniques such as diaphragmatic breathing and incentive spirometry help in reducing dyspnea and promoting relaxation.

– *Balance and coordination training*: designed to prevent falls and improve stability, particularly beneficial for patients experiencing chemotherapy-induced peripheral neuropathy or muscle weakness.

– *Postural training and gait retraining*: helps restore normal posture and walking patterns, often disrupted due to surgery, pain or neurological complications (Singh et al. 2024).

– *Education and self-management support*: equips patients with knowledge about managing symptoms, practicing safe exercises at home and integrating physical activity into daily life.

These interventions are often used in combination and tailored to each patient's condition, type of cancer, stage of treatment and overall functional goals (Silver et al. 2013). A multidisciplinary approach ensures optimal rehabilitation outcomes.

4.2.3. *Challenges in conventional physiotherapy*

Traditional methods face limitations such as:

– *Lack of personalization*: standardized treatment protocols may not address the unique needs of individual cancer patients, leading to suboptimal outcomes and decreased patient satisfaction.

– *Inefficient monitoring*: without continuous oversight, it is challenging to accurately track patient progress, exercise adherence or deviations in recovery, especially between clinical visits (Kajzar 2024).

– *Resource constraints*: limited availability of trained physiotherapists, especially in rural or under-resourced areas, can result in delayed or inadequate care.

– *Inconsistent adherence*: many patients struggle to maintain consistent participation in prescribed rehabilitation programs due to fatigue, lack of motivation or logistical challenges such as travel and time constraints.

– *Subjective assessments*: traditional evaluation methods often rely on clinician observation, which can be subjective and lack precision, potentially affecting treatment decisions (Blonde et al. 2018).

– *Accessibility issues*: cancer patients with physical limitations, transportation difficulties or immunocompromised conditions may find it difficult to attend in-person sessions regularly.

– *Limited scalability*: conventional physiotherapy models are not easily scalable, making it difficult to meet the growing demand for cancer rehabilitation services in a cost-effective and sustainable manner.

4.3. AI technologies in healthcare

4.3.1. *ML and deep learning*

ML algorithms analyze large datasets to identify patterns, classify data and make data-driven predictions (Strielkowski et al. 2023). These algorithms are highly effective in applications such as risk stratification, treatment response prediction and rehabilitation outcome forecasting. In cancer rehabilitation, ML models can be trained on patient demographics, treatment regimens and physiotherapy records to predict which interventions are likely to be most beneficial.

Deep learning, a subset of ML, uses multilayered neural networks capable of processing unstructured and high-dimensional data such as radiological scans, videos of patient movement and wearable sensor data. In physiotherapy, deep learning can detect subtle changes in posture, gait and range of motion from video recordings or sensor streams. This enables the generation of precise, automated feedback that can guide therapy adjustments. For example, convolutional neural networks (CNNs) are frequently used to analyze visual data, while recurrent neural networks (RNNs) or long short-term memory (LSTM) models are used to interpret time-series data from sensors (Tsironi et al. 2017). As these models learn from more data, they continuously improve their accuracy and adaptability, allowing for real-time, personalized care delivery in cancer rehabilitation contexts.

4.3.2. *Natural language processing*

Natural language processing (NLP) enables machines to understand, interpret and generate human language, making it particularly useful in the healthcare domain. In the context of cancer physiotherapy, NLP can extract meaningful insights from unstructured data sources such as electronic health records (EHRs), clinical notes, discharge summaries and patient-reported outcomes (Sim et al. 2024). This facilitates the creation of comprehensive patient profiles, helping healthcare providers design more informed and personalized rehabilitation plans.

Furthermore, NLP-powered chatbots and virtual assistants can engage with patients to gather feedback on symptoms, monitor emotional well-being and provide motivational prompts or exercise reminders. Sentiment analysis tools can also detect early signs of anxiety or depression by analyzing patient communication, allowing timely psychological interventions. NLP algorithms can summarize lengthy clinical documents, highlight key findings and ensure critical information is not overlooked during care planning.

By enhancing communication between patients and healthcare teams, streamlining documentation and supporting decision-making, NLP significantly contributes to improving the efficiency, accuracy and responsiveness of physiotherapy services for cancer patients.

4.3.3. *Computer vision*

Computer vision allows machines to interpret visual information, enabling real-time movement analysis and posture correction. In the context of cancer rehabilitation, computer vision is particularly useful in facilitating contactless, objective assessment of functional movement (Hellsten et al. 2021). This technology enables AI systems to track joint angles, detect gait abnormalities and evaluate exercise performance through video input or live camera feeds, thereby reducing the reliance on manual observation by therapists.

Advanced algorithms can recognize and quantify physical impairments, offering detailed biomechanical feedback that aids in refining rehabilitation programs. For example, pose estimation models can map the human skeletal structure and monitor changes in posture, balance or symmetry, which is especially valuable for patients with surgery-related musculoskeletal issues or chemotherapy-induced motor deficits.

In addition, computer vision can be integrated into virtual reality (VR) or augmented reality (AR) environments to provide immersive, gamified physiotherapy sessions. These applications not only make therapy more engaging but also provide clinicians with real-time performance data, enabling remote supervision and timely adjustments to treatment plans. By eliminating geographic and logistical barriers, computer vision broadens access to high-quality physiotherapy and supports continuous recovery for cancer patients.

4.3.4. *Robotics and wearable devices*

Robotics and wearable technologies are revolutionizing physiotherapy by augmenting human capabilities and enabling more precise, continuous rehabilitation for cancer patients (Mehmood et al. 2025). Robotics enhances physical rehabilitation through advanced exoskeletons and assistive devices that support limb movement, assist with gait retraining and facilitate task-specific repetitive exercises. These robotic systems can be programmed to deliver high-intensity, controlled therapy sessions, particularly beneficial for patients with severe motor impairments due to surgery, chemotherapy or cancer-related neuropathy.

Exoskeletons provide mechanical support for walking and limb movement, enabling patients to regain mobility and independence faster. Robotic therapy devices, such as robotic arms or treadmills with body weight support, offer dynamic feedback and adjust resistance or support levels in real time based on patient performance.

Wearable devices, on the contrary, continuously monitor physiological and biomechanical parameters such as heart rate, steps, joint angles, posture and muscle activity (De Fazio et al. 2023). Devices such as smartwatches, inertial measurement units (IMUs) and Electromyography (EMG) sensors can provide valuable data to clinicians and AI systems, which can then analyze trends, assess recovery progress and optimize therapy protocols. Wearables also promote patient engagement by offering feedback, reminders and gamified exercise tasks that can be completed at home.

Together, robotics and wearables bridge the gap between in-clinic and at-home therapy, enabling hybrid rehabilitation models. This integration supports a continuum of care, reduces the need for frequent hospital visits and empowers cancer patients to actively participate in their recovery journey.

4.4. Integrating AI into physiotherapy for cancer patients

4.4.1. *Personalized treatment planning*

AI can analyze patient history, tumor characteristics, comorbidities and prior treatment responses to design highly customized physiotherapy plans tailored to each cancer patient's unique condition (Khansari et al. 2024). By leveraging data from EHRs, wearable sensors and imaging modalities, AI algorithms can identify patterns that inform optimal exercise types, intensity levels and progression schedules. These systems can also adapt plans dynamically in response to changes in a patient's physical capacity, symptom burden or treatment phase (e.g. prehabilitation, post-surgery or long-term survivorship). This continuous personalization helps reduce the risk of overexertion or injury, enhances therapeutic outcomes and aligns rehabilitation goals with the patient's evolving needs and preferences.

4.4.2. *Real-time monitoring and feedback*

Wearable sensors and mobile applications powered by AI provide real-time feedback on exercise form, intensity and adherence, ensuring that cancer patients perform physiotherapy exercises correctly and consistently (Birla et al. 2025).

These systems use data from accelerometers, gyroscopes, heart rate monitors and EMG sensors to track movement patterns and physiological responses during physical activity. AI algorithms then analyze this data to detect deviations from prescribed exercise techniques, offering immediate corrective feedback through visual, auditory or haptic cues.

Real-time monitoring enables dynamic adaptation of exercise routines based on patient fatigue, pain levels or biomechanical performance, thereby preventing injury and promoting safety. Mobile applications often include dashboards that present performance metrics and progress trends, helping both patients and healthcare providers stay informed.

In addition, AI-powered platforms can send personalized reminders, motivational messages and adaptive goals to encourage adherence and long-term engagement with therapy programs.

This continuous monitoring and feedback loop enhances the effectiveness of physiotherapy by making it more responsive, engaging and aligned with each patient's capabilities and recovery pace.

4.4.3. *Predictive analytics*

AI models predict treatment outcomes and potential complications, allowing proactive adjustments in therapy. By analyzing historical patient data, including treatment regimens, physiological responses and recovery trajectories, AI systems can identify patterns that signal potential adverse events or delayed recovery (Finkelstein et al. 2024). This predictive capability enables healthcare providers to intervene early, modify therapy protocols and allocate resources more efficiently. For example, predictive analytics can forecast the likelihood of chemotherapy-induced neuropathy or post-surgical mobility limitations, guiding preemptive physiotherapeutic strategies.

In addition, AI can help stratify patients based on risk levels, enabling a more targeted and efficient delivery of care. These predictive insights not only enhance clinical decision-making but also improve patient safety and satisfaction by anticipating and addressing complications before they escalate.

4.4.4. *Remote rehabilitation*

AI-driven platforms facilitate telerehabilitation, extending care to patients in remote areas and improving accessibility (Calabrò and Mojdehdehbaher 2025). These systems utilize video conferencing, motion tracking and AI-powered feedback tools to deliver supervised physiotherapy sessions outside traditional clinical settings. Through mobile apps and web interfaces, patients can access personalized exercise routines, track their progress and communicate with therapists in real time, all from the comfort of their homes.

AI enhances remote rehabilitation by analyzing patient performance through computer vision or wearable data and offering corrective guidance on posture, range of motion and exercise quality. This remote monitoring ensures that patients maintain proper technique, thereby reducing the risk of injury and enhancing the effectiveness of each session.

Moreover, these platforms allow clinicians to review analytics dashboards summarizing session compliance, fatigue indicators and mobility improvements, which can inform timely therapy modifications.

Telerehabilitation also plays a vital role in continuity of care for immunocompromised cancer patients who may be unable or unwilling to attend in-person sessions due to infection risk (Danandeh and Rezaei 2024). By minimizing geographical and logistical barriers, AI-powered remote physiotherapy increases equity in access, supports patient autonomy and fosters long-term engagement in rehabilitation programs.

4.5. Case studies and applications

4.5.1. *AI-powered rehabilitation*

Platforms Systems such as Kaia Health and Reflexion Health use AI to deliver guided exercises, monitor performance and adapt programs based on progress. Kaia Health uses ML algorithms and smartphone sensors to offer real-time corrective feedback, enabling users to perform physical therapy exercises accurately without the need for constant supervision (Biebl et al. 2021). Its platform adapts exercise routines based on user input, performance trends and pain levels, which is especially beneficial for cancer patients managing fatigue or discomfort.

Reflexion Health combines motion-tracking software and AI-driven analytics to monitor patients' rehabilitation exercises in real time. Using a Kinect-enabled device, it assesses movement precision, adherence and recovery progress and transmits data to healthcare providers for review. These insights enable clinicians to adjust therapy plans proactively and communicate with patients remotely.

Other platforms, such as SWORD Health and Physitrack, also demonstrate the expanding landscape of AI-powered rehabilitation. SWORD Health uses wearable sensors and AI to guide patients through exercises and identify improper form, offering voice-guided corrections and progress metrics. Physitrack integrates AI with telehealth tools to allow therapists to create customized rehab plans supported by video demonstrations, outcome tracking and patient engagement features (Archer and Ellis 2024).

These platforms underscore the transformative potential of AI in delivering scalable, personalized and accessible physiotherapy solutions for cancer patients, particularly in overcoming geographic, logistical and staffing constraints.

4.5.2. *Robotics in cancer rehabilitation*

Robotic exoskeletons help patients regain mobility post-surgery or chemotherapy-induced muscle weakness.

4.5.3. *Wearable technology*

Devices such as Fitbit, Apple Watch and other wearable technologies monitor physical activity, physiological signals and sleep patterns, providing continuous streams of data for AI algorithms to assess recovery progress and detect early signs of complications. These devices track metrics such as step count, heart rate variability, oxygen saturation and caloric expenditure, offering a comprehensive picture of a cancer patient's physical status throughout rehabilitation (Nascimento et al. 2020).

AI systems use this data to identify trends and deviations from expected recovery trajectories. For instance, a sudden drop in activity levels or heart rate fluctuations may signal fatigue, depression or treatment side effects, prompting clinicians to intervene. Wearables can also aid in establishing baseline functional levels and tracking improvements over time, thus allowing for more accurate, responsive therapy planning.

In addition to passive monitoring, many wearables are integrated with mobile apps that deliver real-time feedback, motivational prompts and personalized exercise suggestions. Some platforms use gamification elements to increase engagement and encourage adherence to rehabilitation routines.

Moreover, data from these devices can be seamlessly transmitted to healthcare providers, enhancing remote patient management and reducing the need for frequent in-person assessments.

By enabling continuous, noninvasive monitoring and supporting data-driven decision-making, wearable technologies play a critical role in enhancing the accessibility, personalization and effectiveness of physiotherapy for cancer patients (Mehmood et al. 2025).

4.6. Benefits of AI-enhanced physiotherapy

4.6.1. *Improved personalization*

AI enables the development of patient-specific rehabilitation plans, enhancing effectiveness by considering a wide array of personal health variables such as cancer type, stage, treatment history, comorbidities and current physical function. By leveraging data from EHRs, wearable sensors and patient-reported outcomes, AI algorithms create nuanced, dynamic rehabilitation protocols tailored to each individual (Reddy 2025). This deep level of personalization ensures that exercises are not only appropriate and safe but also optimized for efficacy.

Furthermore, AI systems can continuously update these personalized plans in real time as new data becomes available: adjusting intensity levels, recommending new therapeutic activities or pausing certain exercises based on changes in fatigue, pain or functional ability. This dynamic responsiveness prevents overexertion and reduces the likelihood of setbacks, especially in patients undergoing intensive treatments such as chemotherapy or surgery.

Personalization through AI also includes psychological and motivational factors, such as integrating preferred exercise modalities or gamified challenges that align with a patient's interests. This holistic, patient-centric approach leads to better engagement, more consistent participation, and ultimately improved rehabilitation outcomes and QoL.

4.6.2. *Enhanced engagement and adherence*

Gamification and interactive feedback significantly increase patient motivation and adherence to physiotherapy regimens. By transforming repetitive exercises into engaging activities with goals, rewards and progression levels, gamification introduces an element of fun and competition, which encourages sustained participation (Christopoulos and Mystakidis 2023). This is particularly beneficial for cancer patients, who often face fatigue and psychological stress that can hinder consistent therapy engagement.

Interactive feedback systems, often powered by AI and computer vision, provide real-time visual or auditory cues to correct movement and reinforce proper technique. These systems can simulate a personal trainer or therapist by offering immediate encouragement and performance analysis, fostering a sense of accountability and accomplishment. Some applications also allow patients to track their achievements and progress over time, which enhances motivation through visible, tangible outcomes.

In addition, social features embedded in some digital platforms – such as leaderboards, community support or collaborative challenges – can boost morale and reduce feelings of isolation, which are common in cancer rehabilitation. Through these methods, AI-driven engagement strategies help patients adhere to their treatment plans more effectively, leading to improved rehabilitation outcomes and a higher QoL (Bachina and Kanagala 2023).

4.6.3. *Better outcomes*

Continuous monitoring and adaptive interventions lead to improved functional outcomes and QoL.

4.6.4. *Resource optimization*

Automation reduces the burden on healthcare providers and allows for scalable solutions.

4.7. Ethical and technical challenges

4.7.1. *Data privacy and security*

Handling sensitive health data requires robust encryption and compliance with regulations such as the Health Insurance Portability and Accountability Act (HIPAA) and the General Data Protection Regulation (GDPR) (Herath et al. 2024).

4.7.2. *Algorithm bias and transparency*

AI models must be trained on diverse datasets to avoid biased outcomes. Transparency in decision-making is essential.

4.7.3. *Clinical validation*

AI tools must undergo rigorous clinical trials to ensure safety and efficacy.

4.7.4. *Integration with existing systems*

Seamless integration with EHRs and clinical workflows is crucial for adoption (Kumar et al. 2025).

4.8. Proposed framework for AI-integrated physiotherapy

4.8.1. *Data collection*

Use wearable devices and mobile apps to collect physiological and performance data.

4.8.2. *Data analysis*

Use ML algorithms to interpret data and identify trends.

4.8.3. *Treatment recommendation*

Generate personalized physiotherapy plans based on AI analysis.

4.8.4. *Continuous monitoring*

Provide real-time feedback and adjust therapy as needed.

4.8.5. *Outcome evaluation*

Measure progress using clinical metrics and patient-reported outcomes.

4.9. Future directions

4.9.1. *Advancements in AI*

Continued innovation in AI models will improve accuracy and personalization.

4.9.2. *Multidisciplinary collaboration*

Collaboration among oncologists, physiotherapists, data scientists and engineers is vital for success (Ottenbacher et al. 2019).

4.9.3. *Patient education and involvement*

Empowering patients to understand and participate in their care enhances outcomes (Fumagalli et al. 2015).

4.9.4. *Policy and regulation*

Clear guidelines and standards are needed to ensure ethical use and integration of AI in healthcare.

4.10. Conclusion

Integrating AI with physiotherapy offers a transformative approach to enhancing the QoL of cancer patients. By enabling personalized care, real-time monitoring and predictive analytics, AI addresses numerous challenges inherent in conventional rehabilitation, such as variability in treatment adherence, personalized adaptation of therapeutic plans and the need for continuous performance tracking. AI-driven systems offer customized interventions, ensuring patients receive the right support at the right time, optimizing recovery and preventing setbacks. Moreover, AI technologies empower both patients and healthcare providers, improving the efficiency and accessibility of rehabilitation interventions. With the ability to continuously monitor patient progress and predict potential complications, AI facilitates more proactive, data-driven decisions that can lead to better outcomes in both physical and psychological recovery. However, despite the promising benefits, several ethical and technical hurdles remain. Concerns around data privacy, algorithmic bias and the integration of AI systems with existing healthcare infrastructure must be addressed. Collaboration among oncologists, physiotherapists, data scientists and policymakers is crucial for overcoming these challenges. As AI technology continues to evolve, it holds immense potential to redefine cancer rehabilitation practices, ensuring a more personalized, patient-centric approach to care. Ultimately, AI-enhanced physiotherapy can significantly elevate the standard of care for cancer patients, ensuring more effective, equitable and compassionate treatment throughout their journey.

4.11. References

Alagappan, T.R., Bhansali, N., Bhansali, T. (2024). Navigating the integration of technology in physiotherapy: A balanced approach. *IP Journal of Surgery and Allied Sciences*, 6(1), 1–4.

Archer, K.R. and Ellis, T.D. (2024). Advances in rehabilitation technology to transform health. *Physical Therapy*, 104(2), pzae008.

Ariel, E., Levkovitz, Y., Goor-Aryeh, I., Motti, R. (2022). The effects of TENS, interferential stimulation, and combined interferential stimulation and pulsed ultrasound on patients with disc herniation-induced radicular pain. *Journal of Back and Musculoskeletal Rehabilitation*, 35(2), 363–371.

Bachina, L. and Kanagala, A. (2023). Health revolution: AI-powered patient engagement. *International Journal of Chemical and Biochemical Sciences*, 24(5), 2023.

Biebl, J.T., Rykala, M., Strobel, M., Kaur Bollinger, P., Ulm, B., Kraft, E., Huber, S., Lorenz, A. (2021). App-based feedback for rehabilitation exercise correction in patients with knee or hip osteoarthritis: Prospective cohort study. *Journal of Medical Internet Research*, 23(7), e26658.

Birla, M., Rajan, R., Roy, P.G., Gupta, I., Malik, P.S. (2025). Integrating artificial intelligence-driven wearable technology in oncology decision-making: A narrative review. *Oncology*, 103(1), 69–81.

Blonde, L., Khunti, K., Harris, S.B., Meizinger, C., Skolnik, N.S. (2018). Interpretation and impact of real-world clinical data for the practicing clinician. *Advances in Therapy*, 35(11), 1763–1774.

Bray, F., Ferlay, J., Soerjomataram, I., Siegel, R.L., Torre, L.A., Jemal, A. (2018). Global cancer statistics 2018: GLOBOCAN estimates of incidence and mortality worldwide for 36 cancers in 185 countries. *CA: A Cancer Journal for Clinicians*, 68(6), 394–424.

Calabrò, R.S. and Mojdehdehbaher, S. (2025). AI-driven telerehabilitation: Benefits and challenges of a transformative healthcare approach. *AI*, 6(3), 62.

Christopoulos, A. and Mystakidis, S. (2023). Gamification in education. *Encyclopedia*, 3(4), 1223–1243.

Danandeh, K. and Rezaei, N. (2024). Rehabilitation program can save lives. In *Handbook of Cancer and Immunology*, Rezaei, N. (ed.). Springer, Cham.

De Fazio, R., Mastronardi, V.M., De Vittorio, M., Visconti, P. (2023). Wearable sensors and smart devices to monitor rehabilitation parameters and sports performance: An overview. *Sensors*, 23(4), 1856.

Finkelstein, J., Smiley, A., Echeverria, C., Mooney K. (2024). AI-driven prediction of symptom trajectories in cancer care: A deep learning approach for chemotherapy management. *Bioengineering*, 11(11), 1172.

Fumagalli, L.P., Radaelli, G., Lettieri, E., Masella, C. (2015). Patient empowerment and its neighbours: Clarifying the boundaries and their mutual relationships. *Health Policy*, 119(3), 384–394.

Hellsten, T., Karlsson, J., Shamsuzzaman, M., Pulkkis, G. (2021). The potential of computer vision-based marker-less human motion analysis for rehabilitation. *Rehabilitation Process and Outcome*, 10. doi: 11795727211022330.

Herath, H.M., Herath, H.M., Madhusanka, B.G., Guruge, L.G. (2024). Data protection challenges in the processing of sensitive data. In *Data Protection: The Wake of AI and Machine Learning*, Hewage, C., Yasakethu, L., Jayakody, D.N.K. (eds). Springer, Cham.

Kajzar, M. (2024). Wearable devices for training and patient monitoring: A comprehensive review. *Quality in Sport*, 29, 55667.

Khansari, N. (2024). AI machine learning improves personalized cancer therapies. *Australasian Medical Journal (Online)*, 17(2), 1166–1173.

Kumar, A., Rathore, P.S., Ahuja, S., Lilhore, U.K. (eds) (2025). *Integrating Neurocomputing with Artificial Intelligence*. John Wiley & Sons, New York.

Langer, D. (2021). Addressing the changing rehabilitation needs of patients undergoing thoracic surgery. *Chronic Respiratory Disease*, 18. doi: 1479973121994783.

Mehmood, F., Mumtaz, N., Mehmood, A. (2025). Next-generation tools for patient care and rehabilitation: A review of modern innovations. *Actuators*, 14(3), 133. doi: 10.3390/act14030133.

Micheo, W., Baerga, L., Miranda, G. (2012). Basic principles regarding strength, flexibility, and stability exercises. *PM&R*, 4(11), 805–811.

Nascimento, L.M., Bonfati, L.V., Freitas, M.L., Mendes Junior, J.J., Siqueira, H.V., Stevan Jr, S.L. (2020). Sensors and systems for physical rehabilitation and health monitoring – A review. *Sensors*, 20(15), 4063.

Ottenbacher, K.J., Graham, J.E., Fisher, S.R. (2019). Data science in physical medicine and rehabilitation: Opportunities and challenges. *Physical Medicine and Rehabilitation Clinics of North America*, 30(2), 459–471.

Pham, T.D., Teh, M.T., Chatzopoulou, D., Holmes, S., Coulthard, P. (2024). Artificial intelligence in head and neck cancer: Innovations, applications, and future directions. *Current Oncology*, 31(9), 5255–5290.

Reddy, K.J. (ed.) (2025). Patient perspectives and outcomes. In *Innovations in Neurocognitive Rehabilitation: Harnessing Technology for Effective Therapy*. Springer, Cham.

Shawli, L., Alsobhi, M., Chevidikunnan, M.F., Rosewilliam, S., Basuodan, R., Khan F. (2024). Physical therapists' perceptions and attitudes towards artificial intelligence in healthcare and rehabilitation: A qualitative study. *Musculoskeletal Science and Practice*, 73, 103152. doi: 10.1016/j.msksp.2024.103152.

Silver, J.K., Baima, J., Mayer, R.S. (2013). Impairment-driven cancer rehabilitation: An essential component of quality care and survivorship. *CA: A Cancer Journal for Clinicians*, 63(5), 295–317.

Silver, J.K., Raj, V.S., Fu, J.B., Wisotzky, E.M., Smith, S.R., Kirch, R.A. (2015). Cancer rehabilitation and palliative care: Critical components in the delivery of high-quality oncology services. *Supportive Care in Cancer*, 23, 3633–3643.

Sim, J.A., Huang, X., Horan, M.R., Baker, J.N., Huang, I.C. (2024). Using natural language processing to analyze unstructured patient-reported outcomes data derived from electronic health records for cancer populations: A systematic review. *Expert Review of Pharmacoeconomics & Outcomes Research*, 24(4), 467–475.

Singh, S., Masood, H., Singh, D., Ansari, A.H., Pratap R. (2024). Neurological injury: Consequences and restorative and rehabilitation strategies. In *Evidence-Based Neurological Disorders*, Upadhyay, S.K., Dan, S., Ali, S.A. (eds). Jenny Stanford Publishing, New Delhi.

Strielkowski, W., Vlasov, A., Selivanov, K., Muraviev, K., Shakhnov, V. (2023). Prospects and challenges of the machine learning and data-driven methods for the predictive analysis of power systems: A review. *Energies*, 16(10), 4025.

Tsiouris, K.M., Tsakanikas, V.D., Gatsios, D., Fotiadis, D.I. (2020). A review of virtual coaching systems in healthcare: Closing the loop with real-time feedback. *Frontiers in Digital Health*, 2, 567502.

Tsironi, E., Barros, P., Weber, C., Wermter, S. (2017). An analysis of convolutional long short-term memory recurrent neural networks for gesture recognition. *Neurocomputing*, 268, 76–86.

Yan, A.F., Chen, Z., Wang, Y., Campbell, J.A., Xue, Q.L., Williams, M.Y., Weinhardt, L.S., Egede, L.E. (2022). Effectiveness of social needs screening and interventions in clinical settings on utilization, cost, and clinical outcomes: A systematic review. *Health Equity*, 6(1), 454–475.

5

AI-driven Imaging in Oncology and Physiotherapy: A Game-changer in Diagnostics

The integration of artificial intelligence (AI) into medical imaging represents a paradigm shift in diagnostic strategies within both oncology and physiotherapy. This transformation is driven by the convergence of advanced computational methods – particularly deep learning, convolutional neural networks (CNNs) and computer vision technologies – with increasingly sophisticated imaging modalities such as MRI (Magnetic Resonance Imaging), CT (Computed Tomography), PET (Positron Emission Tomography), ultrasound and digital radiography. These AI tools are capable of rapidly analyzing complex datasets with an accuracy and consistency that can exceed human capabilities, thereby addressing longstanding limitations in traditional imaging interpretation such as inter-observer variability, diagnostic delay and human error. In oncology, AI-driven imaging plays a critical role in the early detection and classification of tumors by identifying subtle radiographic patterns that may be imperceptible to the human eye. Through radiomics and predictive analytics, AI algorithms extract high-dimensional features from images, enabling precise tumor characterization and subtyping. This has significant implications for personalized medicine, as AI can predict treatment responses and potential outcomes by integrating imaging biomarkers with clinical and molecular data.

Furthermore, AI supports radiation therapy planning by automating contouring and dose distribution, ultimately improving therapeutic precision while minimizing harm to surrounding healthy tissues. Longitudinal imaging, enhanced by AI, enables dynamic monitoring of disease progression or remission, allowing clinicians to adjust treatment protocols in near real time. In the realm of physiotherapy, AI has opened new frontiers in functional imaging and patient-specific rehabilitation. By leveraging tools such as motion capture, dynamic ultrasound and electromyographic imaging, AI systems can perform real-time biomechanical assessments. These systems not only detect musculoskeletal abnormalities and movement disorders but also facilitate the quantification of rehabilitation progress.

Chapter written by S. ANANDH, Anand GUDUR and Manoj VAIRALKAR.

5.1. Introduction

Medical imaging has long stood as a central pillar of diagnostic medicine, offering clinicians a non-invasive window into the internal anatomy and physiological functions of the human body (Jeyaseelan 2024). In fields such as oncology and physiotherapy, where understanding tissue integrity, function and pathology is crucial, imaging plays an indispensable role. Technologies such as X-rays, magnetic resonance imaging (MRI), computed tomography (CT), positron emission tomography (PET) and ultrasound have revolutionized the detection, characterization and monitoring of diseases, ranging from malignant tumors to soft tissue injuries. However, despite these technological advancements, conventional image interpretation still relies heavily on human expertise, which is inherently limited by cognitive bias, inter-observer variability, diagnostic fatigue and the overwhelming volume of imaging data produced in modern clinical practice (Chen and Friis 2024). In response to these limitations, the integration of artificial intelligence (AI) into medical imaging has emerged as one of the most disruptive and promising developments in contemporary healthcare. AI, particularly in the form of machine learning (ML) and deep learning (DL) algorithms, has the capacity to analyze large-scale visual datasets with unprecedented precision, speed and consistency (Rane et al. 2024). These intelligent systems are capable of learning complex patterns from annotated imaging data, performing image classification, segmentation and enhancement tasks with minimal human input and continuously improving through progressive learning. This convergence of AI and imaging has ushered in a new era of diagnostics, one that is not only more accurate and efficient but also more personalized and predictive.

In oncology, AI-driven imaging is being used to detect tumors at their earliest stages, often before symptoms manifest or traditional tools would detect them (Dubey and Sikarwar 2025). Through the application of convolutional neural networks (CNNs), AI systems can highlight suspicious regions on radiographs, identify tumor subtypes and distinguish malignant from benign lesions with high sensitivity and specificity. Moreover, AI enables automated tumor segmentation and volumetric analysis, facilitating precise treatment planning, such as in radiation therapy, where dosage and targeting are critical. AI also plays an increasing role in radiogenomics and radiomics, where quantitative imaging features are linked with molecular and genetic data to forecast prognosis and guide therapeutic choices. These capabilities position AI as a central force in the movement toward precision oncology – an approach that tailors cancer care based on individual biological profiles and treatment responses (Tippur 2023).

In the domain of physiotherapy, AI's application is equally transformative, particularly in functional diagnostics and personalized rehabilitation. Traditional assessments of musculoskeletal conditions and physical function often rely on visual observation, patient feedback and manual measurements. AI-enhanced imaging tools now offer objective and dynamic assessments through real-time motion capture, electromyography and musculoskeletal ultrasound. These systems are capable of analyzing gait, posture, joint movement and muscle activity with exceptional accuracy, enabling early diagnosis of biomechanical dysfunctions and real-time feedback during rehabilitation (Das et al. 2022).

Furthermore, AI allows for the longitudinal tracking of patient recovery, quantifying improvements in range of motion, muscle strength and coordination. This data-driven approach enhances treatment personalization, optimizes exercise regimens and improves patient adherence and engagement.

Despite its promise, the integration of AI into diagnostic imaging is not without its challenges. Issues such as algorithmic bias, model transparency (often referred to as the "black box" problem), data privacy and the lack of standardized regulatory frameworks must be addressed to ensure the safe, ethical and equitable use of AI in clinical settings. In addition, technical barriers such as the need for large, high-quality annotated datasets and concerns regarding interoperability with existing hospital systems continue to pose hurdles to widespread adoption (Shen et al. 2025). Clinician acceptance and trust in AI systems are also crucial, as these tools are intended to augment – not replace – human expertise.

The present study aims to comprehensively examine the evolving landscape of AI-driven imaging in oncology and physiotherapy, with a focus on its diagnostic applications, clinical utility and transformative potential. It explores how AI technologies are integrated into existing imaging modalities, highlights real-world case studies and applications and evaluates the advantages and limitations associated with AI adoption (Lastrucci et al. 2024).

Furthermore, this chapter outlines the ethical, technical and regulatory considerations that accompany the use of AI in medical imaging and discusses emerging innovations that may shape the future of healthcare diagnostics.

By investigating the synergy between AI and medical imaging across two distinct yet interrelated fields – oncology and physiotherapy – this chapter seeks to underscore how these technological innovations are revolutionizing diagnostic pathways, improving clinical decision-making and advancing the goals of precision medicine.

As AI continues to evolve, its responsible and thoughtful integration into diagnostic imaging promises not only to enhance the accuracy and efficiency of disease detection but also to fundamentally transform the way clinicians approach patient care across the continuum of diagnosis, treatment and rehabilitation.

5.2. The role of medical imaging in oncology and physiotherapy

5.2.1. *Imaging in oncology*

In oncology, imaging plays a vital role across the entire patient care continuum – from screening and diagnosis to staging, treatment monitoring and follow-up (Young et al. 2020).

Common imaging modalities include:

– *X-ray and mammography*: often used for initial cancer screening, particularly in breast and lung cancer.

– *CT*: provides detailed cross-sectional images of the body to detect tumors, assess metastasis and guide biopsies (Lebastchi et al. 2020).

– *MRI*: offers high-resolution images of soft tissues, useful in brain, spinal cord and prostate cancers.

– *PET*: helps assess tumor metabolism and monitor response to therapy.

– *Ultrasound*: non-invasive imaging used in liver, ovarian and thyroid cancer diagnostics.

5.2.2. *Imaging in physiotherapy*

In physiotherapy, imaging is crucial for diagnosing musculoskeletal injuries, evaluating postural abnormalities and monitoring rehabilitation (Sadineni 2024).

Key modalities include:

– *Ultrasound imaging*: used to assess muscle and tendon function.

– *Motion capture systems*: employ cameras and sensors to analyze gait and movement patterns.

– *MRI and CT scans*: provide insights into joint integrity and soft tissue injuries.

– *Thermal imaging*: evaluates inflammation and circulation changes.

Traditionally, the interpretation of these images relied on human expertise. However, integrating AI into these processes has opened new dimensions in diagnostic precision and clinical decision-making.

5.3. AI technologies in medical imaging

AI-driven imaging primarily leverages ML and DL, particularly CNNs, to interpret and analyze visual data (Nazir et al. 2024). These algorithms are trained on large datasets of labeled images to detect patterns, classify abnormalities and make predictions.

5.3.1. *CNNs*

CNNs are the foundational architecture for image-based DL. Inspired by the structure of the human visual cortex, CNNs are designed to automatically and adaptively learn spatial hierarchies of features from input images. In healthcare imaging, CNNs have demonstrated superior performance in various diagnostic tasks.

– *Image classification*: CNNs are extensively used to classify medical images, such as determining whether a CT or MRI scan contains a tumor, lesion or musculoskeletal injury (Kijowski et al. 2020). For instance, in oncology, CNNs can differentiate between benign and malignant tumors with high accuracy by learning subtle patterns and textures not easily discernible by radiologists.

– *Segmentation*: semantic and instance segmentation with CNNs allows for precise delineation of anatomical structures. In radiotherapy planning, CNN-based segmentation of tumor boundaries and adjacent organs-at-risk ensures accurate radiation dose delivery while minimizing damage to healthy tissue. In physiotherapy, segmentation assists in analyzing joint and muscle architecture for movement assessment.

– *Detection*: object detection frameworks such as Faster R-CNN and YOLO (You Only Look Once) enable CNNs to localize anomalies such as microcalcifications in mammograms or disc herniation in spinal MRIs (Albuquerque et al. 2025). These models support early diagnosis and triaging of critical cases in clinical workflows.

CNNs are particularly advantageous due to their scalability, robustness and adaptability across various imaging modalities such as X-rays, MRIs, PET scans and ultrasound.

5.3.2. *Natural language processing*

Natural language processing (NLP) plays a complementary role to visual AI by interpreting the vast amounts of unstructured textual data generated in clinical settings. In radiology and physiotherapy, clinical narratives, imaging reports and physician notes often contain crucial context that is not directly visible in the image itself.

– *Report summarization and information extraction*: NLP algorithms extract structured information from radiology reports, identifying key entities such as anatomical locations, disease severity and diagnostic impressions (Steinkamp et al. 2019). This facilitates integration with imaging data, enabling AI systems to form more holistic interpretations.

– *Image-text correlation*: advanced models align textual findings with specific regions in images, enhancing diagnostic transparency and validation. For example, NLP can highlight image sections corresponding to terms such as "spiculated mass" or "anterior cruciate ligament tear", aiding both radiologists and referring physicians.

– *Decision support and documentation automation*: NLP tools can assist in auto-generating structured reports, reducing administrative burden while improving consistency and compliance with diagnostic protocols (Al Shuraiqi et al. 2024).

Through NLP, AI bridges the gap between visual and narrative data, enabling richer, multimodal diagnostic models.

5.3.3. *Generative adversarial networks*

Generative adversarial networks (GANs) consist of two neural networks – a generator and a discriminator – engaged in a competitive learning process. In the context of medical imaging, GANs are employed for:

– *Synthetic data generation*: GANs can create realistic synthetic medical images that resemble true patient scans. This is particularly valuable in training AI models for rare cancers or uncommon musculoskeletal disorders where acquiring large labeled datasets is challenging.

– *Data augmentation*: GANs enhance existing datasets by generating variations of real images, improving model generalizability and reducing overfitting (Bowles et al. 2018).

– *Image enhancement and reconstruction*: in low-dose CT scans or noisy ultrasound images, GANs can reconstruct high-quality images, thereby maintaining diagnostic fidelity while minimizing patient exposure to radiation or improving imaging in resource-limited settings.

GANs are crucial in overcoming data scarcity and enabling robust model development in domains where annotated imaging data is expensive and difficult to obtain.

5.3.4. *Reinforcement learning and transfer learning*

Both reinforcement learning (RL) and transfer learning are essential for optimizing AI models under real-world constraints, such as limited labeled data or changing imaging environments.

– *RL*: in RL, an agent learns to make decisions by receiving feedback (rewards or penalties) from its environment. In medical imaging, RL is used for tasks such as automated navigation of 3D volumetric scans to locate tumors or pathologies efficiently (Hu et al. 2023). RL can also guide robotic physiotherapy devices in adapting to a patient's progress or resistance.

– *Transfer learning*: transfer learning allows models pre-trained on large datasets (e.g. ImageNet or large radiology repositories) to be fine-tuned for specific medical imaging tasks with limited data. This technique is especially useful in physiotherapy imaging or in rare oncological cases where data availability is constrained. Transfer learning drastically reduces training time while maintaining performance, making AI adoption feasible even in smaller healthcare institutions.

5.4. AI-driven imaging in oncology: applications and benefits

5.4.1. *Early detection and diagnosis*

AI systems can identify subtle anomalies in imaging scans that may be overlooked by human radiologists, enabling earlier diagnosis of cancers such as lung, breast, prostate and colorectal cancers (Aftab et al. 2025). For example, Google's DeepMind demonstrated AI performance comparable to expert radiologists in breast cancer screening using mammography.

5.4.2. *Tumor segmentation and classification*

Precise delineation of tumor boundaries is essential for surgical planning and radiation therapy. AI models can segment tumors from healthy tissues in CT and MRI scans with high accuracy, aiding in effective treatment planning.

5.4.3. *Monitoring treatment response: quantitative imaging for real-time evaluation*

Monitoring treatment response is a critical component of cancer management, influencing decisions about whether to continue, modify or halt a therapeutic regimen. Traditionally, radiologists assess treatment response through qualitative comparisons of baseline and follow-up scans using criteria such as RECIST (Response Evaluation Criteria in Solid Tumors) (Lathrop and Kaklamani 2018). However, these methods are often subjective, time-consuming and limited by inter-observer variability.

AI significantly enhances this process by offering quantitative, consistent and reproducible assessments of treatment efficacy.

DL models can automatically extract imaging biomarkers and compare changes in tumor characteristics such as:

– *Size and volume*: CNN-based segmentation allows for precise volumetric analysis of tumor regression or progression across serial scans, offering more accurate measurements than manual methods (Xu et al. 2024).

– *Texture and density*: AI can detect subtle changes in tumor heterogeneity, vascularization or necrosis, which may indicate response or resistance to treatment before changes in size become evident.

– *Metabolic activity*: in PET imaging, AI algorithms analyze Standard Uptake Values (SUVs) and identify metabolic shifts, providing early indicators of functional response in cancers such as lymphoma or non-small cell lung cancer.

Moreover, multi-modal AI platforms can integrate data from different imaging techniques (e.g. PET-CT and MRI) and correlate them with clinical variables and lab results to deliver a more comprehensive response profile. These models facilitate adaptive therapy by allowing oncologists to dynamically adjust treatment strategies based on real-time evidence, thereby improving outcomes and minimizing unnecessary toxicity (Ekundayo 2024).

5.4.4. *Predictive modeling: forecasting recurrence, progression and survival*

Beyond assessing current response, AI has shown great promise in predictive modeling, helping clinicians forecast future outcomes based on imaging and clinical data. These models aim to identify patterns that correlate with recurrence, metastasis and overall survival (OS) – factors critical for long-term cancer management and personalized care planning.

AI-driven predictive analytics use techniques such as:

– *Radiomics and radiogenomics*: by extracting hundreds to thousands of features (shape, texture, intensity, spatial relationships) from images, radiomics models can stratify patients into high- and low-risk categories (Zhang et al. 2022). When integrated with genomic data (radiogenomics), the predictive power is further enhanced, revealing associations between imaging phenotypes and molecular alterations.

– *Survival analysis*: DL models such as DeepSurv adapt traditional Cox proportional hazards models into neural networks capable of handling high-dimensional, nonlinear data. These tools predict progression-free survival (PFS) and OS with improved accuracy.

– *Recurrence prediction*: AI systems trained on longitudinal datasets can learn patterns indicative of recurrence in cancers such as breast, prostate or colorectal cancer, sometimes months before clinical symptoms or radiological evidence emerge (Moglia et al. 2025).

Importantly, these models can be used to support shared decision-making, offering patients and physicians data-driven insights into potential outcomes. This can lead to more proactive surveillance strategies, tailored therapeutic regimens and informed end-of-life care planning when necessary.

By transforming imaging data into prognostic intelligence, AI is not only changing how we detect and monitor cancer but also how we anticipate its trajectory, therefore paving the way for more strategic, personalized oncology care:

– *Size and volume*: CNN-based segmentation allows for precise volumetric analysis of tumor regression or progression across serial scans, offering more accurate measurements than manual methods.

– *Texture and density*: AI can detect subtle changes in tumor heterogeneity, vascularization or necrosis, which may indicate response or resistance to treatment before changes in size become evident (Xie et al. 2024).

– *Metabolic activity*: in PET imaging, AI algorithms analyze SUVs and identify metabolic shifts, providing early indicators of functional response in cancers such as lymphoma or non-small cell lung cancer.

Moreover, multi-modal AI platforms can integrate data from different imaging techniques (e.g. PET-CT and MRI) and correlate them with clinical variables and lab results to deliver a more comprehensive response profile.

These models facilitate adaptive therapy by allowing oncologists to dynamically adjust treatment strategies based on real-time evidence, thereby improving outcomes and minimizing unnecessary toxicity.

5.5. AI in physiotherapy imaging: a revolution in functional assessment

5.5.1. *Real-time biomechanical analysis*

AI-enabled motion capture systems analyze joint angles, muscle activation and gait patterns in real time. These systems help detect abnormalities, customize therapy and track recovery milestones.

5.5.2. *Musculoskeletal imaging interpretation*

DL models can interpret musculoskeletal ultrasound and MRI scans to detect soft tissue injuries, ligament tears and degenerative changes (Fritz and Fritz 2022).

5.5.3. *Personalized rehabilitation*

Imaging data analyzed by AI can inform the design of tailored rehabilitation programs. For example, a patient recovering from ACL surgery can receive exercise recommendations based on AI analysis of knee joint kinematics.

5.5.4. *Remote assessment and telerehabilitation*

Wearable sensors and AI imaging systems enable remote monitoring of patient movements, supporting physiotherapy delivery in home-based or telehealth settings.

5.5.5. *Fall risk and postural analysis*

AI tools can assess posture and balance from video or motion sensor data, helping prevent falls in elderly patients or those with neuromuscular conditions (González-Castro et al. 2024).

5.6. Limitations, challenges and ethical considerations in implementing AI-driven imaging in clinical practice

5.6.1. *Data scarcity, quality issues and bias in training AI models*

AI performance depends on large, diverse datasets. Bias in data (e.g. underrepresentation of certain ethnicities) can result in inequitable care.

5.6.2. *Lack of transparency in AI decision-making and the "black box" problem*

DL models often lack interpretability, making it difficult for clinicians to understand how decisions are made, raising concerns about reliability (Şahin et al. 2025).

5.6.3. *Ethical and legal implications of AI use in medical imaging*

The implications discussed include issues such as data privacy, informed consent for AI use, algorithmic accountability and potential malpractice concerns in case of AI errors.

5.6.4. *Regulatory challenges and the need for standardization*

There is a lack of consistent regulatory frameworks for clinical AI tools. Agencies such as the FDA (Food and Drug Administration) and EMA (European Medicines Agency) are evolving standards, but many tools remain unvalidated.

5.6.5. *Integration with hospital systems and clinician resistance to AI adoption*

Challenges include compatibility with Electronic Health Records (EHRs), steep learning curves and clinician skepticism toward machine recommendations (Aliferis and Simon 2024).

5.7. Future directions and innovations in AI imaging for oncology and physiotherapy

5.7.1. *Developing multimodal AI models combining imaging, genomic and clinical data*

The next frontier in AI-driven oncology and physiotherapy lies in the integration of diverse datasets – including radiological imaging, genomic profiles and EHRs – into cohesive, multimodal AI models. These advanced systems will enable comprehensive analyses that transcend the limitations of isolated data sources (Asch et al. 2018). For example, combining tumor imaging features with gene expression profiles and patient history can enhance risk stratification, personalize treatment plans and improve prognostic accuracy. Such integrative models will likely use DL architectures capable of processing heterogeneous data types, including CNNs for imaging and transformer-based models for sequential clinical data. This approach promises a more nuanced understanding of disease progression and response to therapy, paving the way for precision oncology and individualized physiotherapy interventions.

5.7.2. *Federated learning and edge computing for decentralized and secure AI training*

One of the most pressing challenges in healthcare AI is the tension between the need for large, diverse datasets and the imperative to protect patient privacy. Federated learning offers a transformative solution by enabling decentralized training of AI models across multiple institutions without sharing raw data (Banabilah et al. 2022). In this paradigm, local models are trained on-site and only model updates – not sensitive patient information – are aggregated centrally. This methodology enhances data security, preserves compliance with data protection regulations such as HIPAA and GDPR and promotes collaboration across healthcare systems.

When coupled with edge computing, AI models can be deployed directly on imaging devices or local servers, ensuring real-time processing while minimizing latency and reliance on central cloud infrastructures. This decentralized approach is particularly advantageous in scenarios where data transmission is costly, unreliable or legally constrained.

5.7.3. *Integration of AI with robotics and smart rehabilitation technologies*

The fusion of AI with robotic systems and smart physiotherapy technologies represents a major leap forward in post-treatment rehabilitation, especially for oncology patients experiencing physical impairments due to surgery, chemotherapy or radiation (Anesi et al. 2014). AI algorithms can analyze real-time imaging and biomechanical data to adaptively control robotic exoskeletons and other assistive devices.

These systems can tailor exercise intensity, range of motion and support levels based on patient-specific performance metrics and recovery trajectories. For instance, AI-driven gait analysis can inform the dynamic adjustment of robotic limbs during ambulation therapy. Moreover, these technologies offer remote monitoring capabilities, facilitating continuous rehabilitation in home settings while maintaining clinician oversight. This integration not only improves functional recovery but also enhances patient motivation and adherence through interactive, feedback-driven exercises.

5.7.4. *AI-enabled portable imaging devices for low-resource settings and remote care*

The democratization of advanced imaging technologies is crucial for expanding healthcare access, particularly in rural, remote or underserved areas (Chandramohan et al. 2024). AI-enabled portable imaging devices – such as handheld ultrasound machines and mobile X-ray units – are revolutionizing diagnostic capabilities in these settings. Equipped with onboard AI algorithms, these devices can automatically interpret images, detect abnormalities and suggest diagnostic insights in real time, reducing reliance on specialist radiologists. Their portability and cost-effectiveness make them ideal for use in community health centers, disaster response zones and developing nations. Furthermore, cloud connectivity can allow remote experts to validate findings and assist in decision-making, creating a hybrid model of care that bridges geographical and expertise gaps. This innovation stands to significantly improve early detection, timely treatment and overall patient outcomes in resource-limited environments.

5.7.5. *Continuous learning systems and AI evolution through real-world imaging feedback*

Traditional AI models are often static, trained on fixed datasets and unable to adapt post-deployment. The future of AI in imaging lies in continuous learning systems – algorithms that evolve by incorporating real-world data and feedback from clinical use (Oyeniyi and Oluwaseyi 2024). These systems can refine their predictions based on actual patient outcomes, clinician corrections and longitudinal imaging data, thereby improving over time. Implementing mechanisms such as RL, active learning and online model updates ensures that AI remains relevant and accurate as medical practices, patient populations and imaging technologies evolve. Importantly, maintaining the integrity and safety of continuously learning systems requires robust governance frameworks, including rigorous validation pipelines and clinician-in-the-loop feedback mechanisms. Such adaptability will be vital in ensuring long-term utility, trustworthiness and clinical efficacy of AI tools in oncology and physiotherapy.

5.8. Conclusion

AI-driven imaging represents a transformative milestone in the fields of oncology and physiotherapy, fundamentally reshaping how clinicians detect, monitor and manage complex health conditions. Through advanced algorithms that interpret radiological and biomechanical data with remarkable precision and speed, AI has significantly elevated the diagnostic capabilities of conventional imaging modalities. In oncology, AI aids in the early detection of malignancies, accurate tumor characterization and treatment planning by extracting and analyzing high-dimensional imaging features. Similarly, in physiotherapy, AI technologies enable dynamic movement analysis, real-time feedback during rehabilitation and personalized treatment adjustments – marking a departure from generalized care to data-driven, patient-specific interventions. By enhancing diagnostic accuracy, reducing inter-observer variability and alleviating clinical workloads, AI holds the promise of more efficient and equitable healthcare delivery. It empowers clinicians to make faster and better-informed decisions while also offering patients timely and tailored care pathways. However, the path to widespread clinical adoption is not without obstacles. Critical challenges remain, including concerns over algorithmic bias, data privacy, explainability, regulatory compliance and the need for high-quality, diverse training datasets. These challenges underscore the necessity for robust interdisciplinary collaboration among healthcare professionals, AI researchers, ethicists and policymakers. Importantly, the role of AI in medical imaging should be viewed not as a replacement for human expertise but as a complementary tool that enhances clinical judgment. When thoughtfully developed and ethically implemented, AI systems can extend the reach of skilled practitioners, improve diagnostic equity across populations and streamline healthcare workflows.

5.9. References

Aftab, M., Mehmood, F., Zhang, C., Nadeem, A., Dong, Z., Jiang, Y., Liu, K. (2025). AI in oncology: Transforming cancer detection through machine learning and deep learning applications. *arXiv:2501.15489*. doi: 10.48550/arXiv.2501.15489.

Al Shuraiqi, S., Aal Abdulsalam, A., Masters, K., Zidoum, H., AlZaabi, A. (2024). Automatic generation of medical case-based multiple-choice questions (MCQs): A review of methodologies, applications, evaluation, and future directions. *Big Data and Cognitive Computing*, 8(10), 139.

Albuquerque, C., Henriques, R., Castelli, M. (2025). Deep learning-based object detection algorithms in medical imaging: Systematic review. *Heliyon*, 11(1), e41137.

Aliferis, C. and Simon, G. (eds) (2024). Lessons learned from historical failures, limitations and successes of AI/ML in healthcare and the health sciences. Enduring problems, and the role of best practices. In *Artificial Intelligence and Machine Learning in Health Care and Medical Sciences: Best Practices and Pitfalls*. Springer, Cham, 543–606.

Anesi, A., Palumbo, C., Ferretti, M., Salvatori, R., Cavani, F., Chiarini, L. (2014). Preliminary findings of a potenziated piezosurgergical device at the rabbit skull. In *22° Congress of European Association for Cranio-Maxillofacial Surgery (EACMFS)*, Prague, 23–26 September.

Asch, M., Moore, T., Badia, R., Beck, M., Beckman, P., Bidot, T., Bodin, F., Cappello, F., Choudhary, A., De Supinski, B. et al. (2018). Big data and extreme-scale computing: Pathways to convergence toward a shaping strategy for a future software and data ecosystem for scientific inquiry. *The International Journal of High Performance Computing Applications*, 32(4), 435–479.

Banabilah, S., Aloqaily, M., Alsayed, E., Malik, N., Jararweh, Y. (2022). Federated learning review: Fundamentals, enabling technologies, and future applications. *Information Processing & Management*, 59(6), 103061.

Bowles, C., Chen, L., Guerrero, R., Bentley, P., Gunn, R., Hammers, A., Dickie, D.A., Hernández, M.V., Wardlaw, J., Rueckert, D. (2018). Gan augmentation: Augmenting training data using generative adversarial networks. *arXiv:1810.10863*. doi: 10.48550/arXiv.1810.10863.

Chandramohan, A., Krothapalli, V., Augustin, A., Kandagaddala, M., Thomas, H.M., Sudarsanam, T.D., Jagirdar, A., Govil, S., Kalyanpur, A. (2024). Teleradiology and technology innovations in radiology: Status in India and its role in increasing access to primary health care. *The Lancet Regional Health-Southeast Asia*, 23, 100195.

Chen, R. and Friis, J.K. (2024). Vision, body and interpretation in medical imaging diagnostics. *Medicine, Health Care and Philosophy*, 27(2), 253–266.

Das, R., Paul, S., Mourya, G.K., Kumar, N., Hussain, M. (2022). Recent trends and practices toward assessment and rehabilitation of neurodegenerative disorders: Insights from human gait. *Frontiers in Neuroscience*, 16, 859298.

Dubey, S. and Sikarwar, S.S. (2025). Applications of AI in cancer detection – A review of the specific ways in which AI is being used to detect and diagnose various types of cancer. In *AI in Disease Detection: Advancements and Applications*, Singh, R., Gehlot, A., Rathour, N., Akram, S.V. (eds). John Wiley & Sons, Hoboken.

Ekundayo, F. (2024). Reinforcement learning in treatment pathway optimization: A case study in oncology. *International Journal of Science and Research Archive*, 13(2), 2187–2205.

Fritz, B. and Fritz, J. (2022). Artificial intelligence for MRI diagnosis of joints: A scoping review of the current state-of-the-art of deep learning-based approaches. *Skeletal Radiology*, 51(2), 315–329.

González-Castro, A., Leirós-Rodríguez, R., Prada-García, C., Benítez-Andrades, J.A. (2024). The applications of artificial intelligence for assessing fall risk: Systematic review. *Journal of Medical Internet Research*, 26, e54934.

Hu, M., Zhang, J., Matkovic, L., Liu, T., Yang, X. (2023). Reinforcement learning in medical image analysis: Concepts, applications, challenges, and future directions. *Journal of Applied Clinical Medical Physics*, 24(2), e13898.

Jeyaseelan, C., Haque, R., Verma, R. (2024). Introduction to medical imaging and artificial intelligence. In *Revolutionising Medical Imaging with Computer Vision and Artificial Intelligence*, Bhatnagar, S., Narad, P., Das, R., Paul, D. (eds). Cambridge Scholars Publishing, Newcastle upon Tyne.

Kijowski, R., Liu, F., Caliva, F., Pedoia, V. (2020). Deep learning for lesion detection, progression, and prediction of musculoskeletal disease. *Journal of Magnetic Resonance Imaging*, 52(6), 1607–1619.

Lastrucci, A., Wandael, Y., Barra, A., Ricci, R., Maccioni, G., Pirrera, A., Giansanti, D. (2024). Exploring augmented reality integration in diagnostic imaging: Myth or reality? *Diagnostics*, 14(13), 1333.

Lathrop, K. and Kaklamani, V. (2018). The response evaluation criteria in solid tumors (RECIST). In *Predictive Biomarkers in Oncology: Applications in Precision Medicine*, Badve, S. and Kumar, G.L. (eds). Springer, Cham, 501–511.

Lebastchi, A.H., Gupta, N., DiBianco, J.M., Piert, M., Davenport, M.S., Ahdoot, M.A., Gurram, S., Bloom, J.B., Gomella, P.T., Mehralivand, S. et al. (2020). Comparison of cross-sectional imaging techniques for the detection of prostate cancer lymph node metastasis: A critical review. *Translational Andrology and Urology*, 9(3), 1415.

Moglia, V., Johnson, O., Cook, G., de Kamps, M., Smith, L. (2025). Artificial intelligence methods applied to longitudinal data from electronic health records for prediction of cancer: A scoping review. *BMC Medical Research Methodology*, 25(1), 24.

Nazir, A., Hussain, A., Singh, M., Assad, A. (2024). Deep learning in medicine: Advancing healthcare with intelligent solutions and the future of holography imaging in early diagnosis. *Multimedia Tools and Applications*, 84(17), 17677–17740.

Oyeniyi, J. and Oluwaseyi, P. (2024). Emerging trends in AI-powered medical imaging: Enhancing diagnostic accuracy and treatment decisions. *International Journal of Enhanced Research in Science Technology & Engineering*, 13, 2319–7463.

Rane, N., Choudhary, S., Rane, J. (2024). Machine learning and deep learning: A comprehensive review on methods, techniques, applications, challenges, and future directions. *Techniques, Applications, Challenges, and Future Directions*. doi: 10.2139/ssrn.4850451.

Sadineni, H.S. (2024). Orthopedic applications: Advancing physiotherapy in musculoskeletal health. In *Physical Therapy – Towards Evidence-Based Practice*, Nakano, H. (ed.). IntechOpen, London.

Şahin, E., Arslan, N.N., Özdemir, D. (2025). Unlocking the black box: An in-depth review on interpretability, explainability, and reliability in deep learning. *Neural Computing and Applications*, 37(2), 859–965.

Shen, Y., Yu, J., Zhou, J., Hu, G. (2025). Twenty-five years of evolution and hurdles in electronic health records and interoperability in medical research: Comprehensive review. *Journal of Medical Internet Research*, 27, e59024.

Steinkamp, J.M., Chambers, C., Lalevic, D., Zafar, H.M., Cook, T.S. (2019). Toward complete structured information extraction from radiology reports using machine learning. *Journal of Digital Imaging*, 32, 554–564.

Tippur, A. (2023). AI-powered precision oncology: Computational insights redefining therapeutic landscapes. *DHR Proceedings*, 3(S1). doi: 10.47488/dhrp.v3iS1.95.

Xie, T., Huang, A., Yan, H., Ju, X., Xiang, L., Yuan, J. (2024). Artificial intelligence: Illuminating the depths of the tumor microenvironment. *Journal of Translational Medicine*, 22(1), 799.

Xu, Y., Quan, R., Xu, W., Huang, Y., Chen, X., Liu, F. (2024). Advances in medical image segmentation: A comprehensive review of traditional, deep learning and hybrid approaches. *Bioengineering*, 11(10), 1034.

Young, A.M., Charalambous, A., Owen, R.I., Njodzeka, B., Oldenmenger, W.H., Alqudimat, M.R., So, W.K. (2020). Essential oncology nursing care along the cancer continuum. *The Lancet Oncology*, 21(12), e555–e563.

Zhang, X., Zhang, Y., Zhang, G., Qiu, X., Tan, W., Yin, X., Liao, L. (2022). Deep learning with radiomics for disease diagnosis and treatment: Challenges and potential. *Frontiers in Oncology*, 12, 773840.

6

AI-driven Innovations in Biomaterial-based Physiotherapy and Oncology: Personalizing Patient Care Through Intelligent Monitoring and Predictive Rehabilitation

Biomaterials have revolutionized modern medicine and rehabilitation, especially within physiotherapy, by supporting tissue regeneration, replacing damaged structures and enhancing therapeutic outcomes. This chapter explores the intricate interactions between biomaterials and biological tissues, beginning with protein adsorption and the wound healing cascade, following implantation. It emphasizes the importance of biocompatibility and the tissue-specific integration of implants, highlighting the critical influence of surface properties and material composition. This chapter also discusses the significant role of physiotherapists in adapting rehabilitation protocols to account for implant-tissue dynamics, with special attention to movement, functionality and potential complications. Particular focus is placed on metallic implants, especially their susceptibility to corrosion and how degradation can affect physiotherapeutic strategies. Corrosion can alter mechanical stability, release toxic by-products and provoke inflammatory responses – necessitating cautious load-bearing exercises and close monitoring. This chapter further delves into biodegradable and resorbable materials, noting their advantages and challenges in dynamic healing environments. The integration of artificial intelligence (AI) into rehabilitation is presented as a transformative tool. AI algorithms aid in detecting implant degradation, analyzing sensor data, medical imaging and biomarkers to enable early interventions and personalized therapy planning. These advancements offer real-time monitoring and predictive analytics, ensuring safer, more adaptive rehabilitation. Ultimately, this chapter bridges the gap between biomedical engineering and clinical physiotherapy practice, underscoring the need for collaborative, personalized and technology-integrated approaches to optimize patient recovery and implant performance.

Chapter written by G. HIMASHREE, G. VARADHARAJULU and Radhika CHINTAMANI.
For a color version of all of the figures in this chapter, see www.iste.co.uk/kumar/ai-driven1.zip.

6.1. Interactions with blood and proteins

The implantation of a biomaterial often creates a wound, and bleeding generally ensues. Blood thus typically makes a first contact with the implanted biomaterial. Proteins are building blocks of the supporting extracellular matrix (ECM) of many tissues.

Changes in the levels of proteins or the structure of proteins can lead to altered function and are responsible for many diseases. This paved the way for blood screening for certain proteins indicating a diseased or cancer state.

Thus, protein plays a major role in determining the final nature of the tissue–implant interface.

Biomaterials can promote cell/tissue attachment and activity by allowing selective protein absorption or can inhibit tissue interactions by repelling protein.

pH and ionic strength of the biomaterial can alter confirmation of nearby proteins and its functions.

In fact, proteins also experience the structural alterations during interaction with the solid surface of biomaterials and lose some of their activity as the functions get altered (Enderle and Bronzino 2012).

A special protein known as fibrinogen has greater affinity for the biomaterial surface.

6.2. Wound healing response after biomaterial implantation

Disruption of the anatomic continuity of the tissue occurs after implantation, which creates a wound. There are four overlapping phases of wound healing:

1) Hemostasis.
2) Inflammation.
3) Proliferation.
4) Remodeling (Palani et al. 2024).

Figure 6.1 represents the stages (Palani et al. 2024).

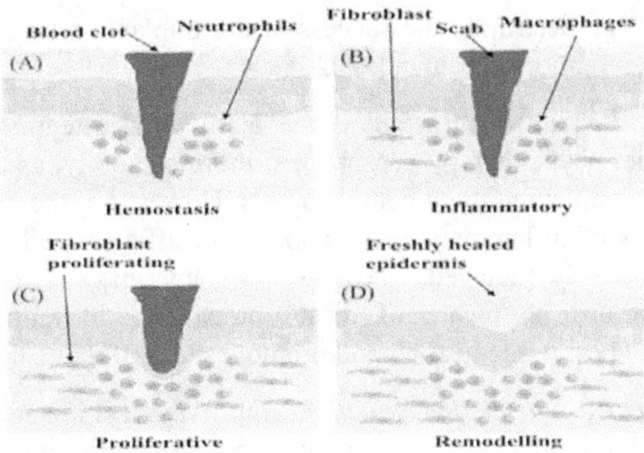

Figure 6.1. *Wound healing response after biomaterial implantation*

Recent advances in wound dressing and the material and skin interaction are vital for in-depth knowledge about skin–material interaction. Metal nanoparticles and biomaterials have recently started playing a pivotal role in encouraging effective wound healing and averting secondary infections or any concern regarding immunogenicity. The tiny size of the healing materials, wound dressing or even the materials inserted topographically onto skin or in vitro have a high surface area, which facilitates enhanced biological interaction. Thus, the materials need to be specifically designed with respect to topography, size, material and biocompatibility. The material used for the therapy/assessment should have an optimal cell-to-cell interaction, fostering an effective purpose (Palani et al. 2024).

Chapter 5 emphasized the type of biomaterial used in the current medical scenario. The targeted approach of this chapter is to emphasize the problems of interaction of the biomaterials commonly encountered in the medical field, especially in physiotherapy subjects. The biomaterials are specially designed to mimic the ECM and the tissue topographically and hence are biocompatible. The type of materials studied in recent evidence are nano-metallic particles and biodegradable materials. The metallic substances are generally known to have a major disadvantage; this is metallic corrosion.

6.3. Effect of tissue biomaterial interaction on physiotherapy

Tissue biomaterial interactions can significantly affect physiotherapy, influencing both the rehabilitation process and the outcomes of treatment. The way a biomaterial interacts with the surrounding tissue – whether it is bone, cartilage, muscle or skin – can determine the success of the implant and the effectiveness of physiotherapy interventions (Pramanik 2024).

1) *Healing and tissue integration*: if the biomaterial integrates well with the surrounding tissue, physiotherapy can proceed more smoothly, as the tissue is able to heal around the implant without complications. However, if the interaction is poor – such as in cases of inflammation, infection or insufficient cellular growth – the healing process can be delayed, and physiotherapy may need to be adjusted. Physiotherapists must be aware of these interactions to tailor exercises and techniques that support tissue healing and promote successful integration.

2) *Impact on movement and function*: the way a biomaterial affects the biomechanics of the body can influence the types of exercises and movements that are safe and beneficial during rehabilitation. For instance, an implant that does not fit properly or causes discomfort may require physiotherapists to modify rehabilitation plans, taking into account the potential for mechanical failure or incorrect loading patterns. Conversely, biomaterials designed to mimic the natural function of tissues – such as joint replacements – may facilitate more traditional therapy techniques aimed at improving strength and mobility.

3) *Risk of complications*: poor biomaterial interactions, such as stress shielding, improper alignment or rejection, can lead to complications such as pain, instability or re-injury, which would affect physiotherapy. For example, in cases where a bone implant fails to integrate properly with the bone, excessive strain could occur during rehabilitation, which could compromise both the healing process and the longevity of the implant. Physiotherapists must adjust their approaches to account for such risks, ensuring exercises are not too strenuous on areas with biomaterial implants.

4) *Adaptation of therapy plans*: the type of biomaterial used (e.g. metal, polymer and ceramic) will influence how physiotherapists approach rehabilitation. For example, a metal implant in a joint may require different mobilization techniques and loading exercises compared to a soft tissue scaffold used in cartilage repair. The physiotherapy plan needs to be dynamic and responsive to how the biomaterial interacts with the surrounding tissue during healing, requiring personalized modifications to ensure the patient's recovery progresses effectively.

5) *Chronic effects and long-term management*: over time, the interaction between biomaterials and tissue may evolve, especially in weight-bearing joints or areas subjected to high levels of stress. Physiotherapists must be aware of the potential for long-term complications, such as wear and tear on the implant or surrounding tissue degeneration, and adjust therapy plans to maintain function and prevent future issues.

In summary, tissue biomaterial interactions play a significant role in determining how effective physiotherapy will be during rehabilitation. Physiotherapists must continuously assess and adapt their approaches based on the type of biomaterial used, the patient's healing progress and the potential challenges presented by tissue–implant interactions.

6.4. Metallic corrosions

Several mechanisms can lead to metallic corrosion; thus, its property, which is corrosive resistance, is one of the important properties of metals used for implants. There are many types of corrosion (Manam et al. 2017):

– *Galvanic (mixed metal) corrosion*: this results when two dissimilar metals in electrical contact are immersed in an electrolyte. There are four important components that must exist for galvanic current to occur: an anode, a cathode, an electrolyte and an external electrical conductor. For example, a patient with two Total Hip Replacements (THRs) made of different alloys is not subject to mixed metal corrosion due to the absence of electrical connection. However, a THR made of two alloys or a fracture plate of one metal fixed with screws of another metal may be susceptible to a mixed metal corrosion (Kumar et al. 2025a).

– *Crevice corrosion*: this can occur in a confined space that is exposed to a chloride solution. The space can be a kind of gasket-type connection between a metal and a nonmetal or between two pieces of metal bolted or clamped together. This is observed in some types of implants where metals come into contact with another metal such as THR devices, screws and plates used in fracture fixation and some orthodontic appliances (Kumar et al. 2025a).

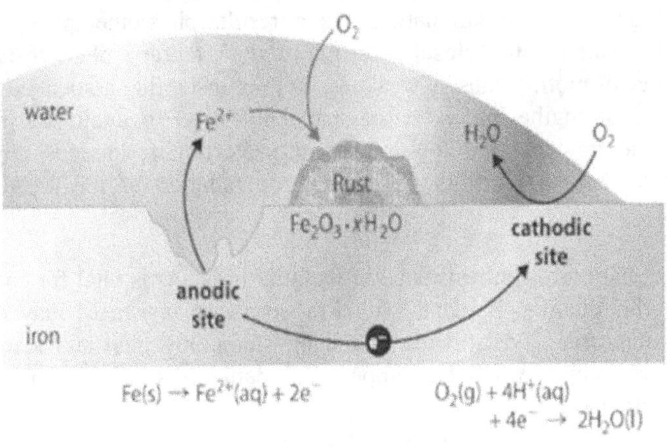

Figure 6.2. *Metallic corrosions*

6.5. Effect of corrosion on physiotherapy regimen

Corrosion resistance is a critical property of metals used for implants, as it directly impacts the longevity and functionality of the implant, which in turn affects the physiotherapy regimen. When metallic implants corrode, they can release harmful ions into the body, leading to tissue damage, inflammation, infection and sometimes implant failure.

Several mechanisms contribute to metallic corrosion, such as electrochemical reactions, mechanical stress and the presence of body fluids, including sweat, blood and lymph. These reactions can weaken the mechanical properties of the implant over time, affecting its stability and performance (Reddy and Srinivasan 2018).

The corrosion resistance of an implant is essential in ensuring the durability of the metal and maintaining its biocompatibility. If an implant is prone to corrosion, it may degrade over time, leading to complications such as implant loosening, infection and systemic toxicity.

In these cases, the physiotherapy regimen must be adapted to take into account the potential for pain, swelling and compromised healing (Naeimirad et al. 2023). For example, if an implant shows signs of corrosion and becomes unstable, the physiotherapist may need to adjust the intensity and type of physical activity to prevent further damage or overloading of the compromised area.

Corrosion can also affect the tissue–implant interface, potentially leading to implant migration or inflammation. As a result, physiotherapists may need to monitor the patient more closely and adjust their therapy plans to accommodate limited range of motion, muscle weakness or joint instability associated with implant degradation. Strengthening exercises may be altered to avoid putting excessive stress on the implant, and low-impact or protective techniques might be used to minimize the risk of injury or further degradation of the metal (Naeimirad et al. 2023).

Overall, the corrosion resistance of metallic implants is vital for ensuring stable and safe rehabilitation. Implants with low corrosion resistance may require more frequent monitoring, adjustments to the physiotherapy plan and sometimes even earlier replacement, which can impact the patient's rehabilitation timeline and recovery outcomes.

Hence, physiotherapists must be aware of the material properties of the implant and adapt their approach accordingly to ensure that the recovery process is not hindered by potential complications from corrosion.

6.6. Biomaterial degradation and resorption

Biomaterials can be permanent within the body or can be degradable. The degradation process may be chemically driven or accomplished by cells. The bioresorbable materials are known to degrade themselves in body over time and be replaced with natural tissue. The goal is to meet strength and cell support while the tissue or cell regenerates (Dewey and Harley 2022).

Biomaterial degradation may lead to chronic non-healing wounds. This may happen if biomaterial degrades too quickly and releases particulate matter, which extends the inflammation stage. If the healing passes through to the fibrous capsule formation stage, there may still be complications (Dewey and Harley 2022).

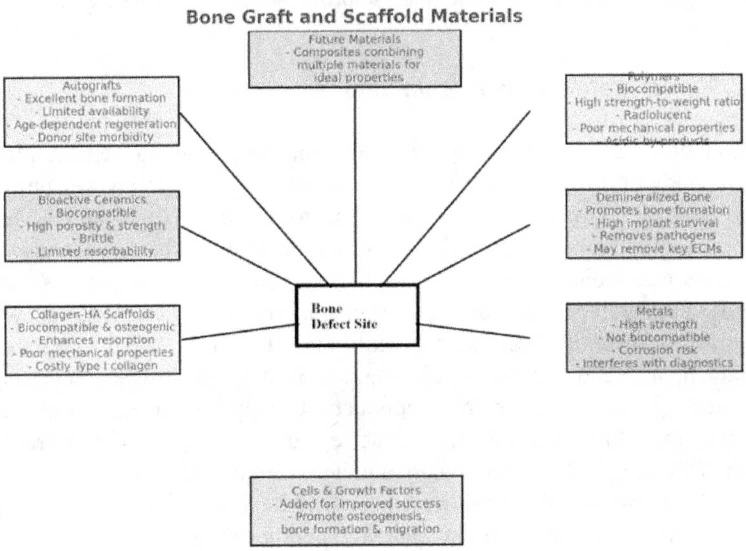

Figure 6.3. *Biomaterial degradation and resorption*

Farraro et al. (2014) conducted an in vitro (goats as subjects of the experiment) study on the potential magnesium-based materials to understand its biodegradable ability and bio-absorbable feature. The authors used magnesium-based orthopedic devices such as interference screw to accelerate ACL graft healing and a ring to aid in the healing of an injured ACL. The authors studied their effect by using a robotic testing system and concluded that both devices could optimally restore the joint functions of the goat. Furthermore, the magnesium-based interference screw reported to non-toxic, even at the 12-week assessment (Farraro et al. 2014).

6.7. Artificial intelligence in detecting biodegradation and absorption

By combining data collecting, pattern recognition and predictive modeling, artificial intelligence (AI) can significantly contribute to the detection of biodegradation and absorption of biomaterials. In order to detect early indications of biomaterial deterioration or absorption, AI algorithms can evaluate data from a variety of sources, including sensor data, medical imaging and patient health monitoring systems. AI can identify these processes in the following ways.

6.7.1. *Sensor data and wearable devices*

AI algorithms can seamlessly integrate data from wearable sensors placed on a patient's body or around the implant site, continuously monitoring key physiological signals such as temperature, pH levels, electrical conductivity and inflammatory markers. These sensors provide real-time insights, where changes in these parameters can indicate early signs of biomaterial degradation or absorption. For example, temperature variations may signal inflammation or infection, which can arise due to material degradation. Shifts in pH levels could suggest material breakdown, as certain biomaterials degrade through hydrolysis, releasing acidic by-products. In addition, electrical conductivity may be altered as the biomaterial degrades, impacting its ability to conduct electrical currents. AI-powered machine learning (ML) algorithms can process and analyze this sensor data over time, detect abnormal patterns and alert healthcare providers to potential complications, enabling timely interventions and improving patient outcomes (Miao et al. 2021).

6.7.2. *Medical imaging analysis*

AI-based imaging techniques, such as X-rays, magnetic reasoning imaging (MRI) or computed tomography (CT) scans, can be used to track physical changes in the implant or surrounding tissue. For instance, AI-driven image processing algorithms can detect subtle changes in implant shape, size or position, which might suggest degradation or absorption. AI algorithms such as convolutional neural networks (CNNs) (Suresh and Kumari 2020) can analyze high-resolution images to identify differences in tissue density or implant erosion, providing a clear visual representation of material degradation. In addition, AI can track bone density changes or assess the integrity of tissue around the implant to determine how well the material is interacting with the body as it degrades.

6.7.3. *Biomarker detection*

Biodegradation of biomaterials often leads to the release of specific biomarkers into the bloodstream or surrounding tissues, signaling the breakdown of the material and its interaction with the body. AI plays a crucial role in detecting these biomarkers by integrating data from various sources such as blood tests, urine analysis or biopsy results, enabling real-time monitoring of the degradation process. This integration allows for a comprehensive view of the biomaterial's behavior within the body. For instance, AI algorithms can process and analyze the complex datasets obtained from these tests to identify subtle changes in biomarker levels that might indicate the early stages of material breakdown (Ravi and Raghavendra 2021). ML algorithms, particularly random forests or support vector machines (SVMs), can be used to classify and predict the stages of degradation by correlating the biomarkers' presence with the specific type of biomaterial used.

Furthermore, deep learning techniques, such as recurrent neural networks (RNNs) or CNNs, can track temporal changes in biomarkers, learning from previous data to predict future degradation patterns and the associated health risks (Reddy and Meena 2020). By using these advanced AI algorithms, healthcare providers can gain real-time insights into the degradation process, potentially preventing complications and enabling more personalized and proactive patient care.

6.7.4. *Predictive modeling and ML*

AI can build predictive models based on patient-specific data, including implant type, patient health history and biomaterial characteristics. By continuously analyzing ongoing data, AI can predict when biodegradation or absorption is likely to occur, based on known degradation rates for different biomaterials and patient responses. For example, ML algorithms such as random forests or SVMs can predict the likelihood of implant failure due to material degradation by identifying patterns in past patients or data collected over time (Natarajan and Varadarajan 2019). In addition, AI can track biomechanical data, such as joint movement or loading patterns, and predict when the implant may begin to weaken or when resorption may occur, enabling more accurate forecasts of potential issues. These predictions, powered by AI algorithms, help healthcare providers to take proactive measures and customize patient care.

6.7.5. *Continuous monitoring and real-time alerts*

AI enables real-time monitoring of a patient's healing process by integrating data from sensors, clinical assessments and imaging into a centralized system. ML algorithms such as random forests or gradient boosting machines (GBMs) analyze sensor data to detect signs of biodegradation or absorption, such as changes in temperature, pH or inflammatory markers. CNNs process imaging data (e.g. X-rays and MRIs) to identify structural changes in implants or tissues (Vishwakarma and Saini 2020). If any issues are detected, AI-powered alert systems notify healthcare providers for timely intervention. In addition, RNNs or long short-term memory (LSTM) networks track recovery over time, predicting when degradation might occur (Sharma et al. 2021a, 2021b). This continuous feedback loop allows for early detection, enabling prompt adjustments to treatment or rehabilitation plans, ensuring better patient outcomes.

6.7.6. *Data integration and multi-modal analysis*

AI systems can integrate multi-modal data, such as mechanical tests, biochemical markers and clinical observations, to create a comprehensive view of the biomaterial degradation process. For example, ML algorithms such as random forests or SVMs can analyze motion data from physiotherapy sessions alongside biodegradation markers to detect if a patient is overloading an implant, which may accelerate its degradation (Kumar et al. 2025c). CNNs can process motion data from wearable sensors to track patient movements, while RNNs or LSTM networks can analyze time-series data to predict the risk of material failure (Li et al. 2025). This integration allows for real-time monitoring and proactive interventions, ensuring optimal rehabilitation and preventing further damage to the implant.

6.7.7. *Patient-specific customization*

By analyzing large datasets of patient-specific information, AI can tailor predictions and treatment plans based on individual healing rates, implant types and biomaterial characteristics. ML algorithms such as random forests or SVMs can identify patterns in a patient's data to predict how their body will respond to specific implants, factoring in variables such as age, health status and biomaterial properties. In addition, deep learning techniques, including CNNs, can process complex imaging data to assess implant position and tissue integration, while RNNs or long LSTM networks can analyze time-series data to track patient recovery over time (Kumar et al. 2025b). This personalized approach optimizes the detection and management of biodegradation or implant absorption by accounting for the unique factors influencing each patient's recovery, allowing for more accurate and effective interventions.

6.8. Effect of biomaterial degradation and resorption on physiotherapy

Biomaterial degradation and resorption significantly impact the effectiveness of physiotherapy regimens, especially when the biomaterial is designed to be resorbed or gradually degrade within the body, such as in cases of biodegradable implants, scaffolds or bone grafts. These processes involve the breakdown of the biomaterial over time, which can alter the mechanical properties of the implant or tissue scaffold and affect the surrounding tissues' ability to function normally (Müller and Sievers 2020). As degradation and resorption progress, the physiotherapy approach must be carefully adjusted to ensure that healing remains optimal and complications are avoided.

6.8.1. *Changes in mechanical stability*

As the biomaterial degrades, it may lose its structural integrity, potentially leading to changes in joint or bone stability. For example, biodegradable bone scaffolds may provide initial support but may become weaker as they degrade, reducing their ability to bear loads and support the surrounding tissue. Physiotherapists need to adjust the rehabilitation regimen to account for this decreasing support, potentially incorporating protective or low-impact exercises that minimize stress on the affected area (Amini et al. 2011).

6.8.2. *Tissue response to degradation*

The resorption of biomaterials can trigger biological responses in the body, such as inflammation or cellular reactions that promote tissue regeneration. While this is often part of the natural healing process, excessive or improper resorption can cause complications, including infection, irritation or delayed healing. Physiotherapists need to monitor the patient for signs of these responses and modify their regimen accordingly. If inflammation is present, treatments such as cold therapy or gentle exercises may be incorporated to reduce irritation and prevent further damage (Ciapetti et al. 2012).

6.8.3. *Loss of load-bearing capacity*

In the case of degradable implants, such as biodegradable screws or rods used in fractures, the implant may initially provide support but gradually lose its strength as it resorbs. This can affect the patient's ability to bear weight and perform functional movements (Spertus and Clark 2017). Physiotherapists must adjust their exercise plan, carefully transitioning from immobilization to functional loading based on the implant's degradation timeline and the body's ability to compensate with natural healing.

6.8.4. *Impact on range of motion and muscle strength*

Degradation or resorption of the biomaterial might alter the biomechanical properties of the affected area, impacting muscle function and joint motion. For example, as bone grafts or tissue scaffolds resorb, there might be an initial loss of stability or alignment, requiring physiotherapists to focus on muscle strengthening and joint mobilization exercises to restore function. If the area becomes unstable, physiotherapy exercises may need to prioritize stabilization techniques to prevent further complications (Cheers et al. 2024).

6.8.5. *Individualized therapy plans*

As biomaterial degradation and resorption occur at different rates for each patient, physiotherapists must continuously assess the patient's healing process and adjust the regimen accordingly. Some biomaterials might degrade faster than others, and this can necessitate early-stage rehabilitation adjustments to ensure the patient does not prematurely stress the resorbing implant or scaffold (Zhao et al. 2020).

6.8.6. *Preventing re-injury*

As the implant resorbs, the surrounding tissues may not have fully adapted to taking on the load, leading to potential re-injury or weakness. Physiotherapists must carefully guide patients in gradual rehabilitation, starting with low-load activities and progressing as the tissue and bone become stronger (Govindaraju et al. 2023).

In summary, the degradation and resorption of biomaterials impact the physiotherapy regimen by altering the mechanical stability of the body part, triggering tissue responses and influencing the patient's ability to safely perform exercises.

Physiotherapists must closely monitor the progress of these processes and adapt their therapy plans to ensure that recovery remains on track while preventing further injury or complications. The goal is to balance the body's natural healing with the changing properties of the biomaterial, ensuring a smooth and effective rehabilitation process.

6.9. AI aiding physiotherapist to diagnose the degradation during intervention

Incorporating AI algorithms into physiotherapy can greatly improve the detection of biomaterial degradation and resorption, allowing physiotherapists to adjust treatment plans early and effectively. AI analyzes real-time data from sources such as wearable sensors, motion tracking and patient-reported outcomes to monitor responses during rehabilitation. For example, ML models can assess changes in joint movement or biomechanical load, detecting abnormalities that suggest biomaterial degradation.

AI can also process biological markers such as swelling or temperature changes to identify early signs of failure. In addition, computer vision algorithms can evaluate gait and movement patterns, detecting subtle issues such as limping that may indicate implant problems. By integrating imaging data and sensor feedback, AI helps physiotherapists track implant integrity and tissue health, guiding more informed decisions about rehabilitation intensity. AI's ability to process real-time data also supports personalized care, adapting to the patient's specific needs and minimizing the risk of complications.

6.9.1. *Data collection (Soni et al. 2020)*

Collect real-time data from multiple sources:

– wearable sensors (e.g. accelerometers, force sensors);

– motion tracking systems;

– patient-reported outcomes;

– biological markers (e.g. swelling, temperature, inflammation).

6.9.2. *Data integration (Rathore and Garg 2021)*

Integrate data into an AI-driven platform, combining:

– biomechanical data (joint movement and load distribution);

– biological data (e.g. swelling and temperature changes);

– imaging data (X-rays, MRIs and ultrasound).

6.9.3. *Data analysis (biomechanical monitoring) (Sharma and Kumar 2020)*

Use AI algorithms such as ML and deep learning models to analyze sensor data:

– track joint movement and biomechanical load distribution;

– detect abnormal patterns, such as uneven joint motion or excessive load-bearing, which may indicate biomaterial degradation.

6.9.4. *Early detection of degradation (predictive modeling) (Reddy and Verma 2021)*

Apply predictive models to forecast how biomaterial may degrade over time:

– correlate changes in biomechanical data with known patterns of degradation (e.g. structural failure and resorption);

– alert a physiotherapist when predefined thresholds (e.g. abnormal load-bearing or joint movement) are exceeded.

6.9.5. *Biological data analysis (Patel and Sharma 2020)*

Use AI to analyze biological markers (e.g. inflammation and temperature changes) that are tracked through sensors:

– apply algorithms such as SVMs or random forests to correlate biomarker changes with biomaterial degradation;

– notify the physiotherapist of early signs of inflammation or other degradation indicators.

6.9.6. *Motion analysis (computer vision) (Rathi and Gupta 2021)*

Use AI-powered computer vision algorithms to analyze video footage or motion capture data:

– assess the patient's gait, posture and movement patterns during therapy sessions;

– detect subtle changes in joint function or posture, such as limping, compensatory motions or instability, indicating potential biomaterial failure.

6.9.7. *Continuous monitoring and reporting (Elfouly and Alouani 2025)*

– Continuously monitor patient responses during rehabilitation exercises;

– detect real-time changes and generate reports on potential issues related to biomaterial degradation or resorption;

– provide the physiotherapist with actionable insights for intervention.

6.9.8. *Adaptive intervention (personalized treatment plan) (Sharma et al. 2021a)*

AI algorithms analyze the patient's specific implant type, biomechanical responses and tissue health:

– adjust the rehabilitation plan based on evolving data to ensure a personalized treatment approach;

– optimize the level of activity and stress the biomaterial can safely handle during rehabilitation.

6.9.9. *Decision support for physiotherapist (Raghav Awasthi et al. 2022)*

Provide the physiotherapist with data-driven insights and decision support tools:

– suggest adjustments to rehabilitation intensity or therapy techniques based on real-time data;

– offer recommendations for preventing further complications or premature biomaterial failure.

6.9.10. *Feedback loop and continuous learning (Patel and Shah 2021)*

Continuously improve the AI model through patient data feedback, refining the algorithm over time:

– adapt the system to better predict and diagnose biomaterial degradation for future patients, improving diagnostic accuracy and treatment outcomes.

A collective AI method for diagnosing biomaterial degradation during physiotherapy could integrate multiple AI algorithms to provide a comprehensive, real-time analysis of patient data. By combining ML models to assess biomechanical data from wearable sensors (e.g. joint movement and load distribution) with deep learning algorithms for motion analysis through video or motion capture, the system can detect subtle changes in gait and posture that may indicate biomaterial failure. In addition, predictive models can forecast potential degradation patterns based on biological markers (e.g. inflammation and swelling) using SVMs or random forest algorithms while integrating imaging data (X-rays and MRIs) to track implant integrity. This multi-algorithm approach would allow physiotherapists to receive continuous, data-driven insights, enabling early detection of issues and the adaptation of personalized rehabilitation plans for improved outcomes.

6.10. Conclusion

The success of modern physiotherapy, especially in post-implant rehabilitation, increasingly depends on a comprehensive understanding of biomaterial behavior and its dynamic interaction with biological systems. The tissue–biomaterial interface plays a crucial role in determining the rate and quality of healing, influencing rehabilitation protocols. Poor integration or adverse interactions can lead to inflammation, instability or implant failure, which in turn necessitates tailored and cautious physiotherapeutic intervention. Corrosion of metallic implants remains a critical concern due to its ability to release ions and compromise mechanical integrity. Physiotherapists must be vigilant in detecting early signs of implant deterioration, modifying exercises to reduce stress on affected areas, and adopting protective strategies to minimize the risk of re-injury. Similarly, biodegradable and resorbable biomaterials, while offering promising regenerative outcomes, introduce variability in mechanical support over time. This necessitates an adaptive therapy model aligned with the biomaterial's degradation timeline and tissue regeneration rate. AI has emerged as a game-changing asset in this domain, enabling real-time data integration from sensors, imaging and patient-reported outcomes. Through predictive modeling and motion analysis, AI facilitates early detection of biomaterial degradation and supports physiotherapists in customizing interventions to the patient's unique healing trajectory. This ensures both safety and efficiency in the rehabilitation process. Thus, an interdisciplinary approach – merging biomedical engineering, physiotherapy and AI – is essential to advancing patient care. Embracing technology, material science and personalized medicine will shape the future of physiotherapy and implant-based rehabilitation, ensuring better outcomes, fewer complications and improved quality of life for patients.

6.11. References

Amini, A.R., Wallace, J.S., Nukavarapu, S.P. (2011). Short-term and long-term effects of orthopedic biodegradable implants. *Journal of Long-term Effects of Medical Implants*, 21(2).

Cheers, G.M., Weimer, L.P., Neuerburg, C., Arnholdt, J., Gilbert, F., Thorwächter, C., Holzapfel, B.M., Mayer-Wagner, S., Laubach, M. (2024). Advances in implants and bone graft types for lumbar spinal fusion surgery. *Biomaterials Science*, 12(19), 4875-4902.

Ciapetti, G., Granchi, D., Baldini, N. (2012). The combined use of mesenchymal stromal cells and scaffolds for bone repair. *Current Pharmaceutical Design*, 18(13), 1796–1820.

Dewey, M.J. and Harley, B.A.C. (2022). Biomaterial design principles to accelerate bone tissue engineering. In *Bone Tissue Engineering*, Guastaldi, F.P. and Mahadik, B. (eds). Springer, Cham.

Elfouly, T. and Alouani, A. (2025). A comprehensive survey on wearable computing for mental and physical health monitoring. *Electronics*, 14(17), 3443.

Enderle, J. and Bronzino, J. (eds) (2012). *Introduction to Biomedical Engineering*, 2nd edition. Elsevier Academic Press, New York.

Farraro, K.F., Kim, K.E., Woo, S.L., Flowers, J.R., McCullough, M.B. (2014). Revolutionizing orthopaedic biomaterials: The potential of biodegradable and bioresorbable magnesium-based materials for functional tissue engineering. *Journal of Biomechanics*, 47(9), 1979–1986. doi: 10.1016/j.jbiomech.2013.12.003.

Govindaraju, D.T., Chen, C.H., Shalumon, K.T., Kao, H.H., Chen, J.P. (2023). Bioactive nanostructured scaffold-based approach for tendon and ligament tissue engineering. *Nanomaterials*, 13(12), 1847.

Kumar, A., Rathore, P.S., Ahuja, S. and Lilhore, U.K. (eds) (2025a). *Integrating Neurocomputing with Artificial Intelligence*. John Wiley & Sons, New York.

Kumar, A., Das, P., Rathore, P.S., Ahuja, S., Sharma, C. (eds) (2025b). *Targeted Chemotherapy with Personalized Immunotherapy: An AI Approach*. John Wiley & Sons, New York.

Kumar, A., Batta, P., Rathore, P.S., Ahuja, S. (2025c). Secure healthcare data sharing and attack detection framework using radial basis neural network. *Scientific Reports*, 15(1), 15432.

Li, X., Yang, Q., Zhao, Z., Li, M., Zhao, Y., Fan., Mao, Q. (2025). Bibliometric review on flexible pressure sensor design strategies. *Nanoscale*, 17(25), 14999–15028.

Manam, N.S., Harun, W.S., Shri, D.A., Ghani, S.A., Kurniawan, T., Ismail, M.H., Ibrahim, M.H. (2017). Study of corrosion in biocompatible metals for implants: A review. *Journal of Alloys and Compounds*, 701, 698–715.

Miao, L., Li, J., Zhang, J. (2021). AI-driven monitoring of biomaterial degradation and absorption using wearable sensors in orthopedic implants. *Journal of Biomedical Materials Research Part B: Applied Biomaterials*, 109(3), 473–481.

Müller, B. and Sievers, B. (2020). Impact of biodegradable implants on rehabilitation and physiotherapy. *Journal of Tissue Engineering and Regenerative Medicine*, 14(5), 641–652.

Naeimirad, M., Krins, B., Gruter, G.J. M. (2023). A review on melt-spun biodegradable fibers. *Sustainability*, 15, 14474.

Natarajan, K. and Varadarajan, R. (2019). Predictive modeling and machine learning algorithms for healthcare applications in biodegradation of biomaterials. *Journal of Biomedical Science and Engineering*, 12(9), 455–463.

Palani, N., Vijayakumar, P., Monisha, P., Ayyadurai, S., Rajadesingu, S. (2024). Electrospun nanofibers synthesized from polymers incorporated with bioactive compounds for wound healing. *Journal of Nanobiotechnology*, 22(1), 211.

Patel, S. and Shah, P. (2021). Enhancing rehabilitation outcomes with continuous learning in AI-driven physiotherapy models. *Indian Journal of Healthcare Informatics*, 12(4), 110–115.

Patel, A.R. and Sharma, S.K. (2020). AI-based biological marker analysis for monitoring biomaterial degradation in physiotherapy rehabilitation. *International Journal of Biomedical Engineering and Technology*, 30(5), 426–438.

Pramanik, K. (2024). *Stem Cell and Tissue Engineering: Bone, Cartilage, and Associated Joint Tissue Defects*. CRC Press, Boca Raton.

Raghav Awasthi, M.S.C., Cywinski, J.B., Khanna, A.K., Tandon, M.D., Pozdeyev, N., Thomas, J., Habboub, G. (2022). Artificial intelligence in healthcare: 2022 year in review. Report, BrainX.

Rathi, S. and Gupta, P. (2021). AI-based motion analysis for physiotherapy: Assessing gait and posture through computer vision. *Journal of Healthcare Engineering*, 1–10.

Rathore, A.S. and Garg, R. (2021). Artificial intelligence in rehabilitation: Integrating biomechanical, biological, and imaging data for personalized physiotherapy. *Journal of Medical Systems*, 45(8), 121.

Ravi, S. and Raghavendra, P. (2021). Artificial intelligence in biomarker detection for biodegradable implant monitoring. *Indian Journal of Medical Informatics*, 7(3), 130–137.

Reddy, K.B. and Meena, R. (2020). Application of deep learning techniques in biomedical signal processing and medical imaging: A review. *Indian Journal of Science and Technology*, 13(16), 1627–1637.

Reddy, P.P. and Srinivasan, S. (2018). Corrosion resistance in metallic implants and its effects on rehabilitation: An overview. *Journal of Biomedical Materials Research*, 35(6), 723–730.

Reddy, P.K. and Verma, M. (2021). Predictive modeling for early detection of biodegradation in biomaterials: Application in physiotherapy. *Journal of Biomaterials Science*, 32(10), 1335–1347.

Schmalzried, T.P. and Dorey, F.J. (2019). Corrosion and mechanical failure of orthopedic implants: Impact on rehabilitation. *Journal of Orthopaedic Research*, 37(3), 487–495.

Sharma, N. and Kumar, R. (2020). Biomechanical monitoring in physiotherapy using machine learning and deep learning techniques. *Journal of Healthcare Engineering*, 7893415.

Sharma, P., Gupta, A., Verma, R. (2021a). AI-driven adaptive intervention for personalized rehabilitation in orthopedic patients. *Indian Journal of Physiotherapy and Occupational Therapy*, 15(3), 101–108.

Sharma, R., Gupta, S., Mishra, R. (2021b). Artificial intelligence for continuous monitoring and early intervention in implantable medical devices. *Artificial Intelligence in Medicine*, 122, 103395.

Soni, P., Yadav, S., Sharma, S. (2020). Wearable sensors and data analytics for physiotherapy rehabilitation. *International Journal of Healthcare and Medical Sciences*, 10(2), 50–57.

Spertus, J.A. and Clark, C.R. (2017). The role of biodegradable implants in fracture healing and load-bearing recovery. *Orthopedic Clinics of North America*, 48(3), 365–372.

Suresh, P. and Kumari, R. (2020). AI-based imaging techniques for monitoring orthopedic implants: A review. *Journal of Orthopaedic Surgery and Research*, 15(1), 184.

Vishwakarma, P. and Saini, A. (2020). AI and machine learning approaches for real-time monitoring of biomaterial degradation in medical implants. *Journal of Medical Systems*, 44(8), 137.

Zhao, X., Li, J., Zhang, Z. (2020). Individualized rehabilitation plans for biodegradable implants: The role of physiotherapy in adaptive healing. *Journal of Orthopedic and Trauma Rehabilitation*, 42(3), 165–172.

7

Telemedicine and AI in Physiotherapy and Oncology: Bridging Gaps in Care

The rapid advancement of digital health technologies has fundamentally reshaped how healthcare services are delivered, particularly in specialized fields such as oncology and physiotherapy. Telemedicine and artificial intelligence (AI) have emerged as powerful tools in addressing long-standing challenges related to accessibility, continuity of care and personalization of treatment. This chapter explores how the integration of telemedicine and AI is revolutionizing care in oncology and physiotherapy, especially in resource-limited and underserved regions. Telemedicine enables the remote delivery of consultations, diagnostics, treatment supervision and rehabilitation, thereby eliminating geographical and logistical barriers. Simultaneously, AI brings computational intelligence to the forefront of care, assisting in early diagnosis through imaging, predicting treatment outcomes, customizing rehabilitation plans and enabling real-time monitoring of patients. Together, these technologies form a synergistic model that enhances the efficiency, quality and reach of healthcare services. In oncology, AI-powered imaging and predictive analytics support early detection and tailored treatment strategies, while teleoncology facilitates remote consultations, symptom monitoring and palliative care. In physiotherapy, AI algorithms analyze patient movement and guide exercise regimens, whereas telerehabilitation ensures continuity and engagement from home. This chapter also discusses implementation models from various global contexts, highlighting successful applications and measurable outcomes. Despite the promising benefits, barriers such as technological infrastructure, digital literacy, data security and regulatory challenges persist. However, with continued innovation, policy evolution and stakeholder collaboration, these hurdles can be effectively addressed.

7.1. Introduction

In the 21st century, healthcare systems around the world are undergoing a profound transformation driven by digital technologies. At the forefront of this

Chapter written by Ankita DURGAWALE, Anand GUDUR and Vivek DESHPANDE.

transformation are telemedicine and artificial intelligence (AI), two pillars of innovation that are redefining how care is delivered, especially in specialized and high-demand fields such as oncology and physiotherapy (Kuo 2023). These technologies are not only improving operational efficiencies and diagnostic accuracy but are also actively bridging long-standing gaps in healthcare access, quality and continuity – challenges that have been particularly pronounced in cancer care and physical rehabilitation.

Oncology, the branch of medicine that deals with the prevention, diagnosis and treatment of cancer, faces significant hurdles in delivering equitable care (Ahmad 2025). The growing global burden of cancer, coupled with an uneven distribution of oncology specialists and healthcare infrastructure, often results in delayed diagnoses, fragmented care and poor health outcomes – particularly for patients in rural, low-income or underserved communities. Physiotherapy, a key component of comprehensive cancer care, plays a crucial role in managing the side effects of treatment, improving functional mobility and enhancing patients' quality of life (Silver et al. 2015). However, access to qualified physiotherapists and personalized rehabilitation programs is similarly limited by geographic and systemic constraints. Telemedicine offers a practical solution to many of these challenges by enabling healthcare professionals to evaluate, diagnose and treat patients remotely using telecommunications technology. It allows continuous, real-time engagement with patients regardless of location, significantly reducing the need for in-person visits and travel. In parallel, AI has introduced a new dimension of intelligence and precision into healthcare. From detecting tumors in imaging scans to creating adaptive physiotherapy plans, AI enables data-driven decision-making that enhances both clinical outcomes and operational efficiency (Oyeniyi and Oluwaseyi 2024).

This chapter explores the convergence of telemedicine and AI in physiotherapy and oncology, arguing that their integration represents not just a technological advancement but a paradigm shift toward more accessible, personalized and value-based care. By analyzing real-world applications, technological innovations and emerging research, this study highlights how these tools are redefining patient experience and outcomes across the cancer care continuum – from diagnosis and treatment to recovery and survivorship.

Furthermore, this chapter addresses the systemic, ethical and infrastructural challenges that must be overcome to unlock the full potential of these technologies (Hoodbhoy 2023). Through this exploration, we seek to provide a comprehensive understanding of how telemedicine and AI can serve as synergistic tools for bridging critical gaps in care.

7.2. The role of telemedicine in oncology and physiotherapy

7.2.1. *Definition and scope*

Telemedicine refers to the use of digital communication technologies – such as video conferencing, mobile apps and wearable sensors – to deliver healthcare services remotely. It encompasses a wide range of activities, including virtual consultations, remote diagnostics, patient monitoring and follow-up care (Behar et al. 2020).

In oncology, telemedicine supports remote consultations with oncologists, remote symptom monitoring, virtual chemotherapy supervision and palliative care services. In physiotherapy, it enables remote assessments, guided exercise sessions and progress monitoring.

7.2.2. *Benefits of telemedicine*

– *Increased access to care*: patients in remote areas gain access to specialized oncology and rehabilitation services (Parikh et al. 2021).

– *Continuity of care*: enables regular follow-ups without the need for physical travel.

– *Cost reduction*: decreases travel and hospital visit costs for patients and operational costs for providers.

– *Patient empowerment*: encourages active patient participation in their care through digital tools.

7.2.3. *Applications in oncology*

Teleoncology has gained momentum in recent years. Key applications include:

– *Virtual tumor boards*: multidisciplinary teams discuss cases remotely.

– *Telechemotherapy supervision*: real-time monitoring of side effects during home-based chemotherapy (Cornall et al. 2024).

– *Symptom management*: remote tracking of pain, fatigue and nausea through digital platforms.

– *Survivorship care*: follow-up care for cancer survivors, including lifestyle counseling and mental health support.

7.2.4. *Applications in physiotherapy*

Telerehabilitation is the remote delivery of physiotherapy using telecommunication tools. Examples include:

– *Video-guided exercises*: therapists supervise patient movements in real-time or via recorded sessions.

– *Progress tracking apps*: patients log their symptoms, pain levels and adherence to exercises.

– *Wearable integration*: sensors and wearables measure range of motion and physical activity levels.

7.3. AI in oncology and physiotherapy: enhancing care delivery

7.3.1. *Overview of AI in healthcare*

AI in healthcare involves machine learning (ML), natural language processing (NLP), computer vision and data analytics to process complex medical data (Rubinger et al. 2023). AI supports clinicians in making better-informed decisions, automates repetitive tasks and enables predictive modeling.

7.3.2. *AI applications in oncology*

AI has revolutionized oncology through innovations such as:

– *Medical imaging*: AI models detect tumors in radiological scans (e.g. magnetic resonance imaging (MRI), computed tomography (CT) and positron emission tomography (PET)) with high sensitivity and specificity.

– *Genomic analysis*: AI processes large-scale genomic data to identify mutations and suggest targeted therapies.

– *Predictive analytics*: algorithms predict treatment response, recurrence and survival rates.

– *Clinical decision support systems (CDSS)*: AI-driven tools assist oncologists in selecting treatment protocols based on patient data.

7.3.3. *AI applications in physiotherapy*

In physiotherapy, AI is applied to the following:

– *Motion analysis*: AI-powered systems evaluate patient movements to detect abnormalities and monitor progress (Bhamidipaty et al. 2025).

– *Personalized exercise plans*: ML algorithms adjust exercise regimens based on patient feedback and performance.

– *Virtual assistants*: chatbots provide exercise instructions and answer common patient questions.

– *Outcome prediction*: predicts rehabilitation outcomes and guides goal-setting.

7.3.4. *Integration with telemedicine*

When combined with telemedicine, AI becomes a powerful enhancer:

– *Real-time feedback*: AI systems analyze patient movement during telerehab sessions and provide immediate correctional feedback.

– *Triage systems*: AI chatbots assess symptoms and direct patients to appropriate care levels.

– *Risk Stratification*: AI identifies high-risk patients needing urgent intervention.

– *NLP*: converts unstructured data from clinical notes and patient messages into actionable insights (Vashishtha and Kapoor 2023).

7.4. Case studies and implementation models

7.4.1. *Remote oncology monitoring: case study from Canada*

Ontario Health's Virtual Care Program integrated teleoncology and AI-powered symptom tracking for breast cancer patients. Patients used mobile apps to report symptoms, which were analyzed by an AI algorithm that flagged urgent cases for oncologist review. Outcomes included reduced ER visits and improved symptom control.

7.4.2. *AI-powered telerehabilitation in stroke survivors*

A UK-based startup developed an AI platform that uses computer vision to monitor post-stroke patients during telephysiotherapy. The system analyzed joint angles and balance, providing feedback and progress reports to physiotherapists. The approach led to higher adherence and improved recovery timelines.

7.4.3. *Hybrid model: India's remote cancer rehab initiative*

In rural India, where access to oncology care is limited, a non-profit organization developed a hybrid model combining teleconsultations with community-based rehabilitation aided by AI-guided mobile apps. Physiotherapists remotely supervised cancer survivors recovering from surgery or chemotherapy using locally stationed digital health workers.

7.5. Benefits of integrating AI and telemedicine in oncology and physiotherapy

The integration of AI and telemedicine represents a revolutionary approach to delivering healthcare services, especially in fields such as oncology and physiotherapy, which demand precision, continuity and personalized care (Adeghe et al. 2024). This combination not only enhances clinical outcomes but also addresses systemic issues related to accessibility, affordability and scalability. Below are key benefits with expanded discussion.

7.5.1. *Improved access and equity*

One of the most immediate and profound benefits of telemedicine, especially when combined with AI, is the ability to expand healthcare access to marginalized populations. Patients in remote, rural or underserved areas often lack proximity to specialized oncology centers or rehabilitation facilities. Telemedicine bridges this gap by enabling consultations, follow-ups and guided therapy sessions from the comfort of a patient's home (George and George 2023).

When AI is layered into this model, it provides intelligent triage systems that can assess patient symptoms and prioritize cases that require urgent intervention. For instance, an AI-driven symptom checker can identify red-flag signs of disease progression in a cancer patient and prompt immediate virtual consultation.

Similarly, AI can interpret wearable sensor data to detect early signs of functional decline in a post-operative physiotherapy patient. These capabilities reduce care delays and ensure that limited healthcare resources are effectively allocated.

Moreover, AI-powered language translation and accessibility features – such as voice-guided navigation for visually impaired users – further promote health equity among linguistically and physically diverse populations (Desai et al. 2023).

7.5.2. *Enhanced diagnostic accuracy*

AI's ability to analyze vast amounts of complex data with high precision has significantly enhanced diagnostic accuracy in both oncology and physiotherapy. In oncology, AI models trained on large datasets of medical images can detect tumors, lesions and abnormalities that may be missed by human observers. Deep learning algorithms applied to mammograms, CT scans or MRI have demonstrated diagnostic accuracy comparable to, or in some cases surpassing, experienced radiologists.

In physiotherapy, AI-enabled motion capture and computer vision tools can assess a patient's posture, range of motion, gait and muscle coordination (Wankhede et al. 2024). These systems can identify subtle impairments or asymmetries that may not be evident in traditional in-person assessments. This level of diagnostic precision supports early intervention, better treatment targeting and prevention of long-term complications.

Furthermore, AI tools can continuously learn from new clinical data, improving diagnostic capabilities over time and minimizing human error. This not only builds clinician confidence but also contributes to safer and more consistent care delivery.

7.5.3. *Personalized care*

Personalization is essential in oncology and physiotherapy, as patient responses to treatment can vary widely depending on genetic, physiological, behavioral and environmental factors (Brajković and Babić 2024). AI supports truly individualized care by analyzing diverse data sources – including electronic health records (EHRs), genomics, imaging, wearable data and patient-reported outcomes – to tailor interventions.

In oncology, AI-driven CDSS can recommend chemotherapy regimens, immunotherapy options or surgical plans that are optimized for a patient's genetic profile and cancer subtype. For example, IBM Watson for Oncology and similar platforms provide evidence-based treatment suggestions derived from global data.

In physiotherapy, ML algorithms can adapt exercise programs in real time based on patient performance, pain levels, fatigue and progress metrics. A patient recovering from breast cancer surgery, for instance, may receive a modified rehabilitation plan if AI analysis shows limited shoulder mobility or lymphedema risk. This dynamic adaptability not only enhances outcomes but also improves patient engagement and satisfaction by making care feel relevant and responsive (Saleh et al. 2024).

7.5.4. *Continuous monitoring and feedback*

Another major benefit is the ability to continuously monitor patients, far beyond the confines of the hospital or clinic. Wearable technologies – such as fitness trackers, smartwatches and biosensors – collect data on physical activity, heart rate, sleep patterns, gait stability and more. AI systems analyze this data in real time to detect deviations from normal patterns, allowing for timely alerts and interventions.

In oncology, continuous monitoring can detect side effects of chemotherapy such as dehydration, cardiovascular issues or abnormal vital signs, enabling early intervention and reducing hospital readmissions (Gri et al. 2023). AI can also flag when a patient fails to adhere to medication regimens or when symptoms worsen.

In physiotherapy, real-time biofeedback systems allow patients to perform exercises at home while receiving corrective feedback through mobile apps or smart mirrors. AI interprets motion data to ensure correct form and prevent injury. This feedback loop empowers patients to self-manage their recovery with confidence, while therapists can remotely review progress and adjust care plans.

In addition, longitudinal tracking facilitated by AI helps clinicians visualize trends over weeks or months, making it easier to evaluate treatment efficacy and adjust goals accordingly.

7.5.5. *Cost efficiency*

The economic advantages of integrating telemedicine and AI are significant for both patients and healthcare systems. Traditional in-person care models involve high operational costs – including physical infrastructure, administrative staffing and patient transport. For patients, the costs associated with traveling to specialists, time off work and extended hospital stays can be substantial.

Telemedicine eliminates many of these expenses by delivering care virtually, while AI automates time-consuming tasks such as documentation, screening and triage (Reis 2025). For example, AI-powered virtual assistants can handle routine inquiries, appointment scheduling and symptom assessments, freeing up clinical staff for more complex cases.

In oncology, early detection facilitated by AI can prevent late-stage interventions, which are often more expensive and less effective. Similarly, AI-guided physiotherapy reduces the need for prolonged rehabilitation periods or expensive surgical interventions by ensuring timely and efficient recovery.

Health systems benefit through resource optimization, reduced readmissions and better population health outcomes. Furthermore, value-based care models that incentivize outcomes over volume are more achievable when treatment is data-driven and proactively managed through tele-AI integration.

7.6. Challenges and barriers

While the integration of telemedicine and AI into oncology and physiotherapy offers transformative potential, several challenges continue to limit widespread adoption. These barriers range from technological and regulatory issues to clinical and patient-level concerns. Understanding and addressing these challenges are crucial for the successful and sustainable implementation of AI-driven telemedicine in specialized care.

7.6.1. *Technological limitations*

Infrastructure gaps

One of the most significant hurdles is the lack of reliable digital infrastructure, especially in rural or low-income areas. High-speed Internet, stable electricity and robust data networks are prerequisites for effective telehealth delivery and AI deployment (high-speed Internet). Unfortunately, many communities still face connectivity issues, which prevent patients from accessing virtual consultations or remote monitoring systems.

In regions where 4G or 5G networks are absent, even basic video calls can be unreliable, disrupting continuity of care. Moreover, healthcare facilities themselves may lack the information technology (IT) infrastructure or bandwidth needed to support AI applications such as imaging analytics or cloud-based data sharing.

Device compatibility

Access to compatible devices is another barrier. Not all patients own smartphones, tablets or wearable health trackers, which are often essential for telemedicine sessions and AI-based monitoring.

Furthermore, older devices may not support necessary applications or updates, creating inconsistencies in care delivery. Interoperability between devices and healthcare platforms is also limited, complicating the integration of AI systems with existing EHRs and remote health tools (Bala Dhandayuthapani 2024).

7.6.2. *Data privacy and security*

Handling sensitive medical data – especially in oncology, where genetic profiles, imaging scans and long-term treatment histories are involved – requires stringent safeguards. The use of AI and telemedicine platforms introduces new cybersecurity risks such as data breaches, unauthorized access and identity theft.

Although regulations such as the Health Insurance Portability and Accountability Act (HIPAA) in the United States and the General Data Protection Regulation (GDPR) in Europe provide frameworks for data protection, they are not uniformly enforced or adapted for AI technologies.

Telemedicine platforms must employ end-to-end encryption, secure cloud storage, multi-factor authentication and regular audits to protect patient information (Kothai et al. 2025). However, many smaller or underfunded providers may lack the resources to implement these protections adequately, increasing vulnerability.

In addition, ethical concerns around AI decision-making – such as algorithmic bias, lack of transparency and unclear accountability – further complicate trust in these systems.

7.6.3. *Regulatory and legal issues*

The regulatory landscape governing telemedicine and AI remains fragmented and underdeveloped. Key legal challenges include:

– *Licensure across jurisdictions*: healthcare providers often face restrictions when delivering care across state or national borders. For example, a physiotherapist licensed in one region may not be permitted to offer telehealth services to a patient in another jurisdiction, limiting the scalability of virtual care.

– *Reimbursement policies*: payment models for telehealth and AI-assisted services are inconsistent across public and private payers (Ingole et al. 2024). While some insurance providers have begun to reimburse virtual consultations, AI services (such as automated diagnostics or remote monitoring) are still often excluded from coverage.

– *Liability and accountability*: determining liability in cases involving AI recommendations remains a grey area. If a misdiagnosis occurs due to an AI error, it is unclear whether the developer, clinician or institution bears responsibility. This uncertainty discourages healthcare providers from fully embracing AI tools in clinical practice.

Efforts are underway in many countries to update telehealth and AI regulations, but until there is clear, standardized governance, widespread adoption will remain cautious.

7.6.4. *Clinical acceptance*

Despite the promise of AI and telemedicine, a substantial proportion of healthcare professionals remain hesitant to adopt these technologies.

Concerns include:

– *Loss of human touch*: clinicians worry that virtual care may erode the therapeutic relationship with patients, reducing empathy, trust and emotional connection – especially critical in oncology care (Wu et al. 2025).

– *Reliability of AI systems*: many clinicians are skeptical of AI-driven diagnostics or treatment suggestions, particularly if the algorithms are "black boxes" that offer no clear rationale for their recommendations.

– *Workflow disruption*: integrating new digital tools into existing clinical workflows can be complex and time-consuming. Without adequate training and support, providers may perceive these technologies as burdensome rather than beneficial.

– *Lack of digital literacy*: some clinicians, especially older professionals or those working in under-resourced settings, may struggle with unfamiliar software, devices or AI interfaces, resulting in reduced efficiency and increased resistance.

Overcoming these barriers requires targeted education, training and the involvement of clinicians in the design and implementation of tele-AI systems to ensure alignment with real-world clinical needs.

7.6.5. *Patient engagement*

Patient engagement is essential for the success of any healthcare model, but digital health approaches face particular challenges:

– *Digital literacy*: many patients, particularly elderly individuals or those from disadvantaged backgrounds, may lack the skills needed to use telemedicine platforms or AI-driven health apps (Perez et al. 2025). Tasks such as downloading applications, entering health data or following video exercise instructions can be overwhelming without proper guidance.

– *Technology anxiety*: patients may fear using new technologies due to concerns about privacy, data misuse or making mistakes. This anxiety can reduce compliance and satisfaction with virtual care.

– *Limited human interaction*: some patients value face-to-face interactions and feel disconnected or less reassured during remote consultations. This is especially true in oncology, where emotional support is a vital part of care.

– *Language and accessibility barriers*: AI tools and telemedicine platforms may not be fully accessible to patients with disabilities or those who speak minority languages, further widening health disparities.

7.7. Future directions and innovations

As digital health technologies continue to evolve, the integration of telemedicine and AI in oncology and physiotherapy is set to reach new levels of sophistication (Borkar et al. 2024). Beyond the current applications, several emerging innovations are poised to further redefine care delivery, patient engagement and clinical decision-making. These advancements not only enhance capabilities but also aim to overcome some of the persistent challenges in access, precision and personalization.

7.7.1. *AI-driven digital twins*

One of the most promising innovations in personalized medicine is the development of AI-driven digital twins – virtual replicas of individual patients that simulate physiological processes, disease progression and treatment responses in real time. These digital models are built using a combination of clinical data, genomics, lifestyle information and real-time sensor inputs (Marabita et al. 2022). In oncology, digital twins can help oncologists test different treatment strategies virtually before applying them in real life, potentially reducing adverse effects and improving therapeutic efficacy. In physiotherapy, digital twins can model musculoskeletal function and recovery trajectories, enabling therapists to tailor exercise regimens that adapt dynamically to each patient's progress. This technology holds great promise for predictive medicine and proactive care planning.

7.7.2. *Augmented reality and virtual reality*

Augmented reality (AR) and virtual reality (VR) technologies are being integrated into oncology and physiotherapy to create immersive, interactive environments that improve patient experience and engagement. In oncology, VR is already being used as a non-pharmacological tool for pain distraction during chemotherapy or procedures, reducing anxiety and improving patient comfort. In physiotherapy, AR-enabled mirrors or VR-based exercise games allow patients to perform therapeutic movements in a virtual setting that feels engaging and motivating (Tiwari et al. 2024). These environments can provide real-time feedback on performance, posture and technique, thereby increasing adherence to rehabilitation plans and enhancing outcomes.

7.7.3. *Federated learning in AI models*

One of the major hurdles in AI development is access to large, diverse and high-quality datasets without compromising patient privacy. Federated learning addresses this by allowing AI algorithms to be trained across multiple decentralized data sources – such as hospitals or clinics – without transferring raw data to a central server (Nguyen et al. 2022). Instead, only model updates are shared. This approach significantly enhances data security while also improving the generalizability and robustness of AI models. In oncology, federated learning can enable collaborative AI development across cancer centers globally. In physiotherapy, it allows the pooling of data from diverse patient populations to better train AI on movement patterns and rehabilitation outcomes.

7.7.4. *Remote robotic-assisted physiotherapy*

Robotics is a growing area of interest in physical rehabilitation, and the future may see remote-controlled robotic systems being used for assisted physiotherapy sessions. These systems, guided by therapists over telehealth platforms, could deliver precise, hands-on manipulation and support to patients in their homes. Technologies such as haptic feedback and telerobotics would allow therapists to “feel” patient resistance or tension remotely, simulating in-person treatment quality (Hilty et al. 2020). Such solutions could revolutionize access to intensive rehabilitation, especially for patients recovering from major surgeries or cancer-related physical decline.

7.7.5. *Policy and education reforms*

The success of technological innovation in healthcare depends not only on tools but also on supportive systems. Future advancement in telemedicine and AI will require comprehensive policy and education reforms. Governments and regulatory bodies must establish clear frameworks for interoperability, data sharing standards and cross-border licensure, ensuring seamless collaboration across institutions and regions (Supangkat et al. 2025). Simultaneously, continuous education and training programs are needed to equip healthcare professionals with digital literacy and familiarity with emerging tools. Medical curricula should include modules on AI ethics, telehealth communication and data interpretation to prepare future clinicians for hybrid digital care models (Wibberly 2024).

In parallel, patient education must not be neglected. Empowering patients to understand and use digital tools confidently will be essential for improving engagement and adherence in both oncology treatment and physiotherapy.

7.8. Conclusion

Telemedicine and AI are no longer futuristic concepts limited to research labs or pilot projects; they are actively reshaping the landscape of modern healthcare. In critical fields such as oncology and physiotherapy, where timely interventions, personalized treatment and continuous monitoring are vital, these technologies have become indispensable. The integration of remote care platforms with intelligent, data-driven systems is fostering a healthcare model that is more inclusive, responsive and cost-effective. Telemedicine has already proven its value in extending care to populations previously marginalized by distance, mobility limitations or specialist shortages. When combined with AI, its potential multiplies – allowing for more accurate diagnoses, customized treatment pathways and real-time monitoring that empowers both clinicians and patients.

AI's capabilities in analyzing complex datasets, detecting anomalies in imaging and adapting rehabilitation plans ensure that care is not just accessible but also precise and personalized. However, the path forward is not without challenges. Technological infrastructure gaps, especially in low-resource settings, remain a significant hurdle. Legal and regulatory ambiguity, concerns over data privacy and variability in digital literacy across patient and clinician populations further complicate implementation. However, these barriers are not insurmountable. With coordinated efforts in policy reform, investment in infrastructure and ongoing training for both providers and patients, these gaps can be bridged. As healthcare systems worldwide continue to adapt to evolving demands – particularly in the wake of the Covid-19 pandemic – the convergence of telemedicine and AI represents a strategic shift toward sustainable, value-based care.

In oncology and physiotherapy, this synergy is poised to redefine not only how care is delivered but also how outcomes are measured and improved. Embracing this digital transformation is essential for building a future where high-quality, patient-centered care is accessible to all, regardless of geography or socioeconomic status.

7.9. References

Adeghe, E.P., Okolo, C.A., Ojeyinka, O.T. (2024). A review of emerging trends in telemedicine: Healthcare delivery transformations. *International Journal of Life Science Research Archive*, 6(1), 137–147.

Ahmad, I.A. (2025). A scoping review of barriers within the cancer care continuum: Addressing screening, diagnosis, and treatment disparities for an equitable future. *Advances in Medicine, Psychology, and Public Health*, 2(2), 96–106.

Bala Dhandayuthapani, V. (2024). Advancing healthcare: AI integration, interoperability and sustainability challenges. *JAIR – Journal of Applied Interdisciplinary Research Special Issue*, 64–74. doi: 10.25929/j3xxxy63.

Behar, J.A., Liu, C., Kotzen, K., Tsutsui, K., Corino, V.D., Singh, J., Pimentel, M.A., Warrick, P., Zaunseder, S., Andreotti, F. et al. (2020). Remote health diagnosis and monitoring in the time of COVID-19. *Physiological Measurement*, 41(10), 10TR01.

Bhamidipaty, V., Bhamidipaty, D.L., Guntoory, I., Bhamidipaty, K.D., Iyengar, K.P., Botchu, B., Botchu, R. (2025). Revolutionizing healthcare: The impact of AI-powered sensors. In *Generative Artificial Intelligence for Biomedical and Smart Health Informatics*, Khamparia, A. and Gupta, D. (eds). Wiley-IEEE Press, Hoboken, NJ, 355–373.

Borkar, S., Chakole, S., Prasad, R., Bansod, S. (2024). Revolutionizing oncology: A comprehensive review of digital health applications. *Cureus*, 16(4), e59203.

Brajković, M. and Babić, D. (2024). Personalized healthcare as an indicator of patient resilience. *Health Bulletin/Zdravstveni Glasnik*, 10(1), 136–152.

Cornall, G., Zhao, E., Luckett, T., Erciyas, E., Monck, D., Glare, P., Wang, A., Lee, Y.C. (2024). Management of pain in cancer patients – lessons from practices during the COVID-19: A qualitative study of cancer care providers' perspectives. *BMC Health Services Research*, 24(1), 232.

Desai, S., Lundy, M., Chin, J. (2023). "A painless way to learn:" Designing an interactive storytelling voice user interface to engage older adults in informal health information learning. In *Proceedings of the 5th International Conference on Conversational User Interfaces*, Eindhoven, 1–16.

Easton, T., Budhiraja, V., He, Y., Zhang, Q., Arora, A., Koutsos, V., Chatzisymeon, E. (2025). Antibiotic adsorption by microplastics: Effect of weathering, polymer type, size, and shape. *Environments*, 12(4), 120.

George, A.S. and George, A.H. (2023). Telemedicine: A new way to provide healthcare. *Partners Universal International Innovation Journal*, 1(3), 98–129.

Gri, N., Longhitano, Y., Zanza, C., Monticone, V., Fuschi, D., Piccioni, A., Bellou, A., Esposito, C., Ceresa, I.F., Savioli, G. (2023). Acute oncologic complications: Clinical–therapeutic management in critical care and emergency departments. *Current Oncology*, 30(8), 7315–7334.

Hilty, D.M., Randhawa, K., Maheu, M.M., McKean, A.J., Pantera, R., Mishkind, M.C., Rizzo, A.S. (2020). A review of telepresence, virtual reality, and augmented reality applied to clinical care. *Journal of Technology in Behavioral Science*, 5, 178–205.

Hoodbhoy, P. (2023). Sustainable infrastructures: Bridging the gap between technological advancements and societal needs. *Liberal Journal of Language and Literature Review*, 1(01), 12–21.

Ingole, B.S., Ramineni, V., Krishnappa, M.S., Jayaram, V. (2024). AI-driven innovation in medicaid: Enhancing access, cost efficiency, and population health management. *arXiv:2410.21284*. doi: 10.48550/arXiv.2410.21284.

Kothai, G., Harish, P., Ajinesh, D., Sathyanarayanan, V. (2025). Optimizing data privacy and scalability in health monitoring: Leveraging, encryption, blockchain, and cloud technologies. In *Critical Phishing Defense Strategies and Digital Asset Protection*, Gupta, B.B. (ed.). IGI Global, Hershey, PA, 245–270.

Kuo, C.L. (2023). Revolutionizing healthcare paradigms: The integral role of artificial intelligence in advancing diagnostic and treatment modalities. *International Microsurgery Journal (IMJ)*, 7(1), 4.

Marabita, F., James, T., Karhu, A., Virtanen, H., Kettunen, K., Stenlund, H., Boulund, F., Hellström, C., Neiman, M., Mills, R. et al. (2022). Multiomics and digital monitoring during lifestyle changes reveal independent dimensions of human biology and health. *Cell Systems*, 13(3), 241–255.

Nguyen, D.C., Pham, Q.V., Pathirana, P.N., Ding, M., Seneviratne, A., Lin, Z., Dobre, O., Hwang, W.J. (2022). Federated learning for smart healthcare: A survey. *ACM Computing Surveys (Csur)*, 55(3), 1–37.

Oyeniyi, J. and Oluwaseyi, P. (2024). Emerging trends in AI-powered medical imaging: Enhancing diagnostic accuracy and treatment decisions. *International Journal of Enhanced Research in Science Technology & Engineering*, 13, 2319–7463.

Parikh, D.R., Diaz, A., Bernardes, C., De Ieso, P.B., Thachil, T., Kar, G., Stevens, M., Garvey, G. (2021). The utilization of allied and community health services by cancer patients living in regional and remote geographical areas in Australia. *Supportive Care in Cancer*, 29, 3209–3217.

Perez, K., Wisniewski, D., Ari, A., Lee, K., Lieneck, C., Ramamonjiarivelo, Z. (2025). Investigation into application of AI and telemedicine in rural communities: A systematic literature review. *Healthcare*, 13(3), 324.

Reis, T.C. (2025). Artificial intelligence and natural language processing for improved telemedicine: Before, during and after remote consultation. *Atención Primaria*, 57(8), 103228.

Rubinger, L., Gazendam, A., Ekhtiari, S., Bhandari, M. (2023). Machine learning and artificial intelligence in research and healthcare. *Injury*, 54, S69–S73.

Saleh, H.H., Abbas, Z.A., Latif, N., Khalil, Z.T. (2024). Agile management in healthcare improving patient outcomes through flexibility and responsiveness. *Journal of Ecohumanism*, 3(5), 633–649.

Silver, J.K., Raj, V.S., Fu, J.B., Wisotzky, E.M., Smith, S.R., Kirch, R.A. (2015). Cancer rehabilitation and palliative care: Critical components in the delivery of high-quality oncology services. *Supportive Care in Cancer*, 23, 3633–3643.

Srivastava, S., Anand, A., Kumar, A., Saini, B., Singh Rathore, P. (eds) (2024). *Bio-Inspired Optimization for Medical Data Mining*. John Wiley & Sons, New York.

Supangkat, S.H., Firmansyah, H.S., Rizkia, I., Kinanda, R. (2025). Challenges in implementing cross-border digital identity systems for global public infrastructure: A comprehensive analysis. *IEEE Access*, 13, 42083–42098.

Tiwari, A., Dubey, A., Yadav, A.K., Bhansali, R., Bagaria, V. (2024). A review of smart future of healthcare in the digital age to improve quality of orthopaedic patient care in metaverse called: The Healthverse!! *Journal of Clinical Orthopaedics and Trauma*, 48, 102340.

Vashishtha, E. and Kapoor, H. (2023). Enhancing patient experience by automating and transforming free text into actionable consumer insights: A natural language processing (NLP) approach. *International Journal of Health Sciences and Research*, 13(10), 275–288.

Wankhede, V., Verma, P., Gahane, S. (2024). A review on the developments in the field of AI-based gait analysis. In *2nd DMIHER International Conference on Artificial Intelligence in Healthcare, Education and Industry (IDICAIEI)*. IEEE, Wardha.

Wibberly, K.H. (2024). Preparing mental health providers for the future: The case for moving beyond the elective telehealth course to integrating telehealth training throughout the curriculum. *Frontiers in Psychology*, 14, 1301569.

Wu, K., Dang Nguyen, M., Rouleau, G., Azavedo, R., Srinivasan, D., Desveaux, L. (2025). Understanding how virtual care has shifted primary care interactions and patient experience: A qualitative analysis. *Journal of Telemedicine and Telecare*, 31(1), 73–81.

8

Personalized AI-based Exercise Therapy for Oncology and Physiotherapy Patients

As the prevalence of cancer and chronic physical conditions continues to rise globally, the demand for personalized, effective and scalable rehabilitation strategies becomes increasingly critical. Exercise therapy has long been recognized as a cornerstone of physiotherapy and supportive cancer care, improving functional capacity, reducing treatment-related side effects and enhancing patients' quality of life. However, conventional therapy approaches often rely on standardized protocols that lack adaptability to individual patient needs, progress and physiological differences. The integration of artificial intelligence (AI) into exercise therapy presents a transformative solution. By leveraging machine learning algorithms, computer vision, wearable technologies and real-time data analytics, AI can design and deliver highly individualized exercise regimens. These systems dynamically adapt to patient feedback, physiological signals, clinical parameters and personal goals, thereby optimizing therapeutic outcomes. In oncology, AI-based exercise therapy can address cancer-related fatigue, monitor post-treatment recovery and support immune health. In general physiotherapy, AI tools can facilitate rehabilitation from musculoskeletal injuries, neurological conditions and post-surgical recovery through data-driven guidance and real-time monitoring.

This chapter explores the technological architecture, clinical applications, benefits and challenges of AI-powered personalized exercise therapy across oncology and physiotherapy settings. We discuss current innovations, clinical evidence, ethical considerations and future directions, emphasizing the critical role of multidisciplinary collaboration in the development and deployment of these systems. While the technology offers enormous promise, challenges such as data privacy, model bias and system integration must be addressed to ensure equitable and effective care. Personalized AI-based exercise therapy stands as a paradigm shift in patient-centered rehabilitation, offering the potential for scalable, adaptive and evidence-based care that aligns with the precision medicine model of the 21st century.

Chapter written by Ankita DURGAWALE, Rashmi GUDUR and Vivek DESHPANDE.

8.1. Introduction

Cancer and musculoskeletal (MSK) disorders represent two of the most pressing healthcare challenges worldwide, significantly contributing to disability, diminished quality of life and healthcare expenditure. Cancer survivors often experience a multitude of side effects from treatments such as chemotherapy, radiotherapy and surgery, including fatigue, muscle wasting, joint stiffness, neuropathy and reduced cardiorespiratory capacity (Emery et al. 2022). Likewise, individuals with MSK conditions, including arthritis, fractures, spinal cord injuries and post-operative complications, frequently require extended rehabilitation to restore mobility, strength and functional independence.

Rehabilitation for these populations is typically centered on physiotherapy and structured exercise therapy. These interventions have been shown to reduce symptom burden, improve functional status and enhance psychological well-being. However, traditional approaches often adopt standardized, one-size-fits-all models that may not fully consider patient-specific variables such as cancer stage, treatment history, pain levels, comorbidities or progression rates. Such generalization can limit therapeutic effectiveness and patient engagement, particularly in complex or fluctuating conditions.

The rise of artificial intelligence (AI) in healthcare presents a groundbreaking opportunity to transform these limitations into strengths (Aminizadeh et al. 2024). With the ability to process large volumes of multi-dimensional data – including biometric signals, imaging results, behavioral patterns and clinical parameters – AI can facilitate real-time personalization of exercise programs. Through machine learning (ML), computer vision (CV), wearable sensors and decision-support systems, AI can monitor patient responses, predict outcomes and dynamically adjust therapy protocols to meet evolving needs.

This chapter explores the convergence of AI technologies with exercise therapy in the domains of oncology and physiotherapy. Specifically, it examines how personalized AI-based exercise therapy can be designed, implemented and scaled to support individualized rehabilitation plans (Deshmukh 2023). This chapter also investigates current innovations, real-world applications, benefits, limitations and future opportunities, emphasizing the potential of AI to redefine patient-centered care and enhance long-term outcomes.

8.2. Background

Personalized rehabilitation is a cornerstone of modern therapeutic care, particularly in complex domains such as oncology and physiotherapy. As healthcare transitions from generalized treatment approaches to precision medicine, it becomes imperative to understand the diverse needs of individual patients and how evolving technologies – especially AI – can support that transition (Ghanem et al. 2024). This section outlines the clinical necessity for personalized rehabilitation in oncology and MSK care while identifying the limitations of traditional exercise therapy that have paved the way for AI-driven solutions.

8.2.1. *Oncology and physiotherapy: the need for personalization*

Cancer and physiotherapy patients often present multifaceted and dynamic clinical profiles that demand individualized care strategies. In oncology, patients undergoing chemotherapy, radiotherapy or surgical interventions frequently experience a spectrum of debilitating side effects, including cancer-related fatigue (CRF), muscle wasting (cachexia), neuropathies, lymphedema and cardiopulmonary dysfunction. These effects not only compromise physical function but also reduce psychological well-being and hinder participation in everyday activities. Exercise therapy, when appropriately prescribed, has been shown to mitigate many of these side effects, enhance immune response, reduce the risk of cancer recurrence and improve overall survival and quality of life.

Similarly, individuals suffering from MSK and neurological conditions – such as osteoarthritis, stroke, spinal cord injuries and orthopedic trauma – require long-term rehabilitation to regain motor control, balance and physical independence (Lo et al. 2021). However, functional recovery in such cases is highly variable and influenced by factors such as age, comorbidities, injury severity and baseline fitness. In both patient groups, a one-size-fits-all rehabilitation approach is inadequate.

Personalization in therapy – tailoring exercise intensity, frequency, type and progression to the specific needs and limitations of each patient – is essential for optimizing outcomes. This level of individualized care, while theoretically ideal, is challenging to achieve consistently in conventional clinical settings due to constraints in time, workforce and resources. This gap has stimulated the exploration of AI as a tool to support and automate the delivery of personalized care at scale.

8.2.2. *Limitations of traditional exercise therapy*

Traditional exercise therapy programs, although grounded in clinical expertise and evidence-based practices, often exhibit systemic limitations that hinder their effectiveness (Chaughule 2024). Among the key shortcomings are:

– *Lack of real-time feedback*: in standard physiotherapy settings, patients typically receive instructions during in-person visits and are expected to continue exercises independently at home. This disconnection results in poor exercise fidelity and an inability to correct improper techniques, which can lead to suboptimal outcomes or even injury.

– *Generalized treatment protocols*: many rehabilitation plans are designed using population-level data and do not account for individual variability in recovery speed, tolerance, preferences or comorbid conditions. Such generalization may overlook nuanced patient needs and result in decreased efficacy or engagement.

– *Limited monitoring outside clinical settings*: once discharged or between sessions, patients often perform exercises unsupervised, and clinicians lack visibility into their progress, adherence or challenges. This absence of remote oversight limits timely interventions and adjustments.

– *Difficulty in long-term adherence*: sustained adherence to rehabilitation programs is notoriously low, especially when patients do not perceive immediate benefits or when they encounter pain, fatigue or logistical difficulties (Shaw 2015). Without continuous support or motivation, long-term compliance suffers.

These limitations create a pressing need for a solution that can bridge the gap between clinical supervision and independent rehabilitation. AI, through its capabilities in sensing, interpreting and responding to data in real time, holds significant promise in transforming how personalized exercise therapy is delivered.

AI systems can track patient performance continuously, provide adaptive recommendations, issue real-time corrections and maintain patient motivation through interactive feedback mechanisms.

8.3. Technological framework

The integration of AI into personalized exercise therapy relies on a robust technological infrastructure that brings together advanced software, multidimensional data inputs and smart hardware (Mikolajewska et al. 2025). AI-based therapeutic systems function as intelligent, adaptive platforms that assess patient-specific data, interpret real-time physiological and biomechanical signals and deliver personalized exercise recommendations dynamically.

This section outlines the foundational AI technologies, essential data inputs and hardware ecosystem that collectively enable the successful deployment of AI-driven exercise therapy in oncology and physiotherapy.

8.3.1. *Key AI technologies*

8.3.1.1. *ML*

ML is the backbone of most AI applications in healthcare. It enables systems to analyze large datasets, identify patterns and predict outcomes based on patient-specific variables (Lorenzo et al. 2024). In exercise therapy, ML algorithms are used to:

– predict individual recovery trajectories based on treatment history and response;

– classify patient types and risk profiles;

– continuously optimize exercise prescriptions through ongoing data input.

Supervised learning can be used for initial patient profiling, while unsupervised learning can help cluster similar rehabilitation patterns. Deep learning, a subfield of ML, further enhances the system's ability to recognize complex patterns in movement and biometric data.

8.3.1.2. *CV*

CV plays a pivotal role in monitoring and evaluating patient movement. By processing data from cameras or wearable motion sensors, CV algorithms can:

– detect and assess posture, joint angles and motion trajectories;

– identify deviations from ideal movement patterns;

– provide real-time corrective feedback to prevent injury or compensate for impairment.

This technology is especially beneficial in home-based rehabilitation settings, allowing for precise movement assessment without the need for continuous human supervision (Sassi et al. 2024).

8.3.1.3. *Natural language processing*

Natural language processing (NLP) enables AI systems to understand and respond to spoken or written language, allowing patients to communicate naturally with digital therapeutic assistants. NLP functionalities include:

– voice-controlled exercise instructions;

– conversational feedback and motivational support;

– automatic generation of progress reports based on user input.

This enhances accessibility for elderly patients or those with visual or motor impairments, facilitating a more intuitive and user-friendly interface.

8.3.1.4. *Reinforcement learning*

Reinforcement learning (RL) allows AI systems to adapt exercise recommendations based on trial-and-error learning (Doherty et al. 2024). The algorithm receives feedback based on patient performance and modifies its strategy to maximize therapeutic benefit. In this context, RL can:

– adjust exercise intensity, duration and complexity in response to adherence and outcome data;

– reward consistent patient engagement with tailored motivational prompts;

– learn from diverse user profiles to enhance personalization across patient populations.

Together, these AI technologies form a multidimensional, responsive system capable of delivering dynamic and personalized rehabilitation programs that evolve in real time.

8.3.2. *Data inputs for personalization*

The personalization of exercise therapy hinges on the quality, variety and granularity of data fed into the AI system. Key data inputs include:

– *Patient demographics*: age, gender, weight, height and socioeconomic background influence baseline fitness, risk factors and engagement styles (Simmons et al. 2014).

– *Cancer type/stage or injury type*: AI systems tailor therapy differently for patients with breast cancer, colorectal cancer or MSK injuries based on clinical guidelines and individualized risk profiles.

– *Treatment history*: information about past surgeries, chemotherapy, radiation exposure or previous physiotherapy outcomes helps inform exercise tolerances and contraindications.

– *Functional mobility scores*: data from validated tools such as the Timed Up and Go (TUG), 6-Minute Walk Test (6MWT) or manual muscle testing informs baseline function and progress (Fessele and Syrkin 2024).

– *Pain and fatigue levels*: real-time tracking of pain (via numerical rating scales or self-reports) and fatigue (using PROMs or wearable proxies) helps adjust intensity to avoid overexertion.

– *Wearable sensor data*: biometric data such as heart rate variability, respiratory rate, gait symmetry and step count offer continuous insight into physical performance and recovery dynamics.

The integration of these diverse data streams allows AI to make precise, evidence-based decisions in real time, ensuring therapy remains appropriate and effective at every stage of recovery.

8.3.3. *Hardware ecosystem*

The delivery of AI-based personalized exercise therapy is supported by a diverse range of hardware devices that collect, process and display data (Khalid et al. 2024). Key components include:

8.3.3.1. *Wearables*

Wearable technologies such as smartwatches, fitness bands, chest straps and sensor-embedded clothing capture real-time physiological data.

Key metrics include:

– heart rate and variability;

– blood oxygen saturation (SpO_2);

– step count and cadence;

– sleep quality and activity levels.

These metrics are used to assess physical readiness, recovery status and adherence to prescribed activity levels.

8.3.3.2. *Cameras and motion sensors*

High-resolution cameras, depth sensors (e.g. Microsoft Kinect) and inertial measurement units (IMUs) capture and interpret body movements, providing:

– joint angle measurements;

– motion path tracking;

– balance and coordination assessments.

This hardware enables real-time movement correction, performance evaluation and long-term progress tracking (Saez et al. 2018).

8.3.3.3. *Smartphones and tablets*

These devices serve as the central interface for patient interaction with the AI system. Their roles include:

– displaying exercise routines via videos or augmented reality (AR);

– delivering audio-visual feedback;

– hosting virtual consultations with healthcare providers;

– collecting self-reported symptoms, pain levels and adherence logs.

8.4. Application in oncology

Oncology patients experience a diverse array of symptoms and functional impairments that vary widely based on cancer type, stage, treatment modality and individual response (Reeve et al. 2014). Exercise therapy has proven to be a safe and effective non-pharmacological intervention for managing cancer-related side effects, improving physical function and enhancing psychological well-being. However, prescribing and monitoring exercise in oncology care is complex due to the fluctuating physical states and vulnerabilities of patients, particularly during active treatment phases. AI-based systems, through continuous data interpretation and real-time adaptation, present a promising solution for delivering precision exercise interventions tailored to the unique needs of cancer patients.

8.4.1. *CRF and AI exercise solutions*

CRF is one of the most prevalent and debilitating symptoms experienced by oncology patients, affecting up to 80% of individuals undergoing chemotherapy or radiotherapy (Hussey and Gupta 2022). CRF is characterized by persistent physical, emotional and cognitive exhaustion that is not alleviated by rest and significantly impairs daily functioning.

AI systems offer targeted solutions to manage CRF through individualized exercise regimens that are dynamically adjusted based on real-time data.

Key applications include:

– *Personalized intensity regulation*: ML models analyze physiological parameters such as heart rate variability, sleep patterns and patient-reported energy levels to recommend light-to-moderate-intensity aerobic or resistance exercises. These may include walking, yoga or aquatic therapy, selected based on fatigue thresholds and safety considerations.

– *Fatigue prediction and pattern recognition*: by continuously processing behavioral and physiological data from wearables and patient logs, AI systems can identify early indicators of impending fatigue episodes (Kakhi et al. 2024). Predictive models help in adjusting therapy schedules and integrating adequate rest days to prevent exacerbation.

– *Adaptive programming*: AI platforms use RL to adapt exercise types and durations dynamically. For instance, if a patient demonstrates reduced performance due to fatigue, the system can suggest lower-impact exercises (e.g. stretching or breathing routines) and incorporate guided rest intervals to maintain engagement without overexertion.

Such intelligent responsiveness ensures that exercise remains both safe and beneficial, promoting adherence and enhancing the therapeutic value of physical activity in the management of CRF (Dong 2024).

8.4.2. *Post-surgical rehabilitation*

Surgical interventions for cancer treatment – such as mastectomy, prostatectomy, colectomy and limb amputation – often result in functional impairments that require targeted rehabilitation. These may include reduced range of motion (ROM), lymphedema, muscle weakness and postural asymmetries. Traditional rehabilitation protocols are effective but resource-intensive and often rely on in-clinic visits, which may not be feasible for all patients due to travel limitations, immunosuppression or financial constraints (Durrand 2022).

AI-based exercise therapy platforms support post-surgical recovery through the following mechanisms:

– *Tailored range-of-motion exercises*: using CV and sensor-based motion analysis, AI systems can guide patients through customized stretching and mobility exercises. For example, breast cancer survivors post-mastectomy may receive progressive shoulder and arm exercises, adapted to their pain tolerance and healing stage.

– *Prevention of overexertion*: real-time monitoring through wearable devices allows AI algorithms to detect signs of fatigue, elevated heart rate or poor movement quality, prompting automatic modifications to exercise intensity or scheduling rest periods to avoid injury and support tissue healing.

– *Image-based swelling and asymmetry detection*: through the integration of smartphone cameras and AI-powered image analysis, patients can routinely scan affected areas (Oyeniyi and Oluwaseyi 2024). These systems can detect early signs of lymphedema, inflammation or asymmetrical movement patterns, allowing clinicians to promptly intervene.

Incorporating AI in post-surgical rehabilitation not only reduces dependence on physical visits but also provides patients with a sense of autonomy and safety while ensuring continuity of care during vulnerable recovery periods.

8.4.3. *Immune system monitoring*

Immunosuppression is a common consequence of chemotherapy, targeted therapy and hematopoietic stem cell transplantation. During periods of reduced immune function, patients are at elevated risk of infection, making exercise prescription particularly delicate. Although moderate exercise may enhance immune surveillance and resilience, overexertion can have the opposite effect, especially in immunocompromised individuals.

AI platforms integrated with physiological monitoring systems can play a pivotal role in safeguarding the immune health of cancer patients through:

– *Detection of infection indicators*: wearables can continuously track parameters such as skin temperature, resting heart rate, respiratory rate and sleep disturbances (Prieto-Avalos et al. 2022). AI algorithms analyze these data for subtle deviations that may indicate the onset of fever, infection or inflammation. Such early warnings allow for rapid clinical intervention.

– *Exercise risk assessment*: by correlating biometric data with environmental variables (e.g. temperature and air quality), AI systems can assess whether it is safe for a patient to engage in physical activity on a given day. If biomarkers suggest elevated systemic stress or immune compromise, the system may recommend rest, light stretching or supervised activity under medical oversight.

– *Clinical alerts and integration*: AI-driven platforms can be programmed to send automated alerts to clinicians if signs of physiological distress or infection risk are detected, ensuring timely triage and intervention (Li et al. 2025). These systems can be integrated with electronic health records (EHRs) to support holistic care coordination.

8.5. Application in physiotherapy

Physiotherapy spans a wide range of patient populations with varying rehabilitation goals, including recovery from injury, surgery, neurological insult or chronic pain. The traditional physiotherapeutic model, though effective, faces limitations such as therapist-to-patient ratios, subjectivity in assessment and patient adherence issues. The incorporation of AI into physiotherapy offers innovative solutions that enhance personalization, objectivity and remote care (Mohapatra et al. 2024). This section explores the practical application of AI-based exercise therapy in physiotherapy domains, focusing on MSK disorders, neurological rehabilitation and pain management.

8.5.1. *MSK disorders*

MSK disorders such as osteoarthritis, ligament injuries, tendonitis and post-fracture conditions are among the most common reasons for seeking physiotherapy. Recovery requires carefully planned exercise regimens to restore strength, flexibility and function while avoiding re-injury.

AI systems can transform MSK rehabilitation by leveraging biomechanical data to design and adjust exercise protocols in real time:

– *Exercise prescription based on biomechanics*: using data collected from motion sensors, force plates and CV systems, AI algorithms can analyze joint angles, gait patterns and load distribution (Han et al. 2025). Based on these insights, the system can recommend specific strengthening, stretching or balance exercises tailored to the affected anatomical region (e.g. quadriceps for knee osteoarthritis or scapular muscles for rotator cuff injuries).

– *Monitoring load and ROM*: real-time tracking of movement parameters allows the system to assess whether patients are performing exercises within safe limits. For example, AI can measure shoulder abduction angles during rehabilitation of a rotator cuff injury and ensure that ROM progression aligns with tissue healing timelines.

– *Adaptive protocols*: as healing progresses, AI platforms can autonomously increase resistance, repetition or complexity based on patient performance metrics (Kuo 2023). If a decline in motion quality or adherence is detected, the system can regress the exercise plan or alert the therapist.

By integrating AI into MSK physiotherapy, clinicians can ensure that exercise protocols are precise, progressive and aligned with the patient's individual recovery trajectory.

8.5.2. *Neurological conditions*

Patients with neurological conditions such as stroke, spinal cord injury (SCI), traumatic brain injury (TBI) or multiple sclerosis (MS) present unique challenges in physiotherapy due to impaired motor control, muscle tone abnormalities and cognitive deficits. Neurorehabilitation aims to retrain the nervous system through repetitive, targeted motor activities that promote neuroplasticity (Braun and Wittenberg 2021).

AI offers significant benefits in neurological physiotherapy by enhancing both the intensity and specificity of training:

– *Tracking neurological recovery markers:* AI systems collect data on muscle activation patterns (e.g. via EMG sensors), coordination, reaction times and movement symmetry. These markers provide insight into the progression of neurological recovery and can be used to set realistic and individualized therapy goals.

– *Repetitive motor learning*: repetition is critical for neuroplastic changes. AI-based platforms can automate task-oriented exercises such as grasping, reaching or walking and ensure that patients engage in high-repetition, high-intensity activities (Dell'Eva 2024). Gamified interfaces or virtual environments can be used to maintain motivation and focus.

– *Immediate corrective feedback*: real-time feedback, both visual and auditory, allows patients to adjust movements during execution. For instance, if a post-stroke patient demonstrates abnormal synergy patterns during a reach task, the system can cue corrective motions, promoting motor relearning.

In this context, AI serves as an assistive coach, augmenting the physiotherapist's role and enabling intensive, data-driven rehabilitation that fosters neurological recovery.

8.5.3. *Pain management*

Pain is a complex and multifactorial condition that often coexists with MSK and neurological disorders. Chronic pain, in particular, can lead to fear-avoidance behaviors, deconditioning and psychological distress, complicating rehabilitation efforts (Vincent et al. 2013).

AI-based exercise systems provide a novel approach to pain management by integrating biomechanics, behavior analysis and patient-reported outcomes:

– *Movement pattern analysis*: using CV and wearable sensors, AI platforms can identify compensatory or dysfunctional movement patterns that may contribute to pain. For example, an abnormal gait or lumbar flexion pattern may be linked to lower back pain.

– *Exercise modification*: based on the identification of pain-related movements, the AI system can suggest alternative exercises or modifications to reduce strain.

For instance, replacing high-impact squats with chair-assisted sit-to-stand exercises for a patient with knee pain.

– *Predictive pain alerts*: by analyzing trends in biometric data and movement quality, AI systems can predict potential flare-ups or exacerbations. Early warning alerts can prompt the patient to rest, apply pain relief modalities or consult their therapist, thus preventing the worsening of symptoms.

Moreover, AI-driven platforms can incorporate self-report tools such as visual analogue scales (VAS) and pain diaries, allowing patients to log their discomfort in real time (El-Tallawy et al. 2024). These data enhance the system's learning capability and inform therapy adjustments, resulting in a truly patient-centered approach to pain management.

8.6. Clinical implementation

The translation of AI-based exercise therapy from research to routine clinical use involves overcoming technological, administrative and human factors. For these systems to be truly impactful, they must be seamlessly integrated into existing healthcare infrastructures, support remote and decentralized care models, and demonstrate efficacy through empirical validation. This section outlines the core requirements for clinical implementation, focusing on system integration, remote monitoring capabilities and selected real-world applications and case studies.

8.6.1. *Integration into healthcare systems*

For AI-driven exercise therapy platforms to be clinically viable, they must align with the operational, regulatory and technological standards of contemporary healthcare systems. Key considerations include:

– *Interoperability with EHRs*: AI platforms must be able to both extract and contribute relevant patient data to EHRs (Kaswan et al. 2021). This ensures continuity of care, enables clinicians to make informed decisions and allows patient progress to be viewed in the context of their full medical history. Standards such as HL7 and FHIR (Fast Healthcare Interoperability Resources) are essential for achieving seamless data exchange between systems.

– *Cloud-based infrastructure*: a cloud-based ecosystem facilitates centralized data storage, real-time accessibility and collaborative care. Cloud platforms allow AI systems to receive updates, learn from a global dataset and deploy personalized recommendations across multiple care locations. Security protocols such as HIPAA (Health Insurance Portability and Accountability Act) compliance in the United States or GDPR adherence in the EU are essential to protect patient data.

– *Clinician training and adoption*: the successful deployment of AI tools depends on clinician confidence and competence in interpreting AI-driven insights (Esmaeilzadeh 2024). Structured training programs, user-friendly interfaces and decision-support features must be incorporated to empower physiotherapists, oncologists and primary care providers to integrate AI-generated exercise prescriptions into patient management workflows effectively.

Incorporating AI tools into clinical care is not merely a technological upgrade; it necessitates a shift in culture, workflow redesign and continuous evaluation to ensure clinical relevance and ethical application.

8.6.2. *Remote monitoring and telerehabilitation*

One of the most transformative aspects of AI-based exercise therapy is its compatibility with telemedicine and remote care models. Particularly in the context of oncology or post-surgical recovery, where frequent clinic visits may be impractical or risky, remote monitoring ensures continuous care without compromising safety.

– *Real-time remote supervision*: AI platforms equipped with motion sensors or smartphone cameras can assess patients' movements in real time. CV and ML algorithms evaluate biomechanical accuracy, flag incorrect postures and provide immediate feedback (Roggio et al. 2024). This mimics the role of an in-clinic physiotherapist, ensuring that patients perform exercises safely at home.

– *Dynamic routine adjustments*: by analyzing biometric and performance data on a daily or weekly basis, the system can automatically update exercise intensity, frequency and type. For example, if wearable data show improved gait symmetry, the platform may advance the patient from assisted walking to resistance band exercises.

– *Virtual consultations and reporting*: AI systems can generate automated progress reports summarizing adherence, performance trends and any deviations from expected recovery pathways. These reports facilitate efficient virtual check-ins between patients and clinicians. Platforms can also integrate teleconsultation features, allowing for video sessions when clinical reassessment or manual input is necessary.

Through remote monitoring, AI not only increases accessibility and convenience but also supports proactive, responsive care that adjusts to the patient's changing condition in real time (Bhambri and Khang 2024).

8.6.3. *Case studies and trials*

Several AI-based platforms and research initiatives have demonstrated the clinical feasibility and therapeutic benefits of AI-guided exercise therapy. Below are notable examples:

– *Stanford's Mobilize Center*: this interdisciplinary research center has pioneered ML models that analyze gait patterns, joint kinematics and physical activity levels. Their work has been particularly impactful in patients with neurological or orthopedic impairments. The center's algorithms are capable of real-time activity recognition, enabling precise identification of walking abnormalities and personalized exercise interventions.

– *Kaia Health*: Kaia Health has developed an AI-powered mobile application that delivers personalized exercise therapy for chronic back pain. The platform uses CV to track movement through a smartphone camera, offering feedback on form and performance. In clinical trials, users of Kaia Health reported significant reductions in pain and improvements in function, comparable to those achieved with traditional physiotherapy.

– *Sword Health*: this digital therapy company offers AI-assisted telerehabilitation solutions that combine wearable motion sensors and a virtual therapist (Senadheera et al. 2024). Clinical outcomes have demonstrated improved adherence and recovery metrics in orthopedic post-operative patients.

– *RehabKit and BioX Systems*: these emerging platforms integrate wearable sensors with AI to guide patients through MSK and neurological rehabilitation programs. Data from early pilot studies show promise in terms of usability, safety and efficacy in both inpatient and home-based settings.

These real-world applications underscore the clinical validity and scalability of AI-based exercise interventions. However, broader implementation will require larger randomized controlled trials (RCTs), longitudinal outcome data and regulatory approvals to move from pilot programs to standard care pathways.

8.7. Benefits and outcomes

8.7.1. *Improved personalization*

Patients receive routines tailored to their:

– physical condition;

– daily performance;

– psychological state.

8.7.2. *Enhanced engagement and adherence*

Gamification, reminders and visual feedback from AI interfaces can improve patient adherence to prescribed regimens.

8.7.3. *Real-time feedback and safety*

AI ensures patients perform exercises safely, reducing the risk of injury through posture correction and fatigue monitoring (Zhang 2025).

8.7.4. *Scalability and accessibility*

AI tools democratize access to expert-level therapy for rural and underserved communities by removing geographical and financial barriers.

8.8. Challenges and limitations

8.8.1. *Data privacy and security*

Handling sensitive health data raises issues regarding:

– HIPAA compliance;

– data encryption;

– patient consent for AI model training.

8.8.2. *Bias in AI models*

AI systems trained on homogeneous datasets may:

– misinterpret data from underrepresented populations;

– lead to ineffective or unsafe recommendations.

8.8.3. *Technical limitations*

– Inaccuracies in movement recognition;

– dependence on device compatibility and internet access;

– hardware malfunctions or sensor misreading.

8.8.4. *Ethical considerations*

– Replacing human interaction with machines may reduce empathy and emotional support;

– questions regarding AI decision accountability.

8.9. Future directions

8.9.1. *Integrating genomics and AI*

Genomic data can further personalize exercise prescriptions, especially in oncology, by understanding how genes affect fatigue, recovery and response to therapy.

8.9.2. *Multi-modal AI systems*

Combining video, audio, sensor and textual data to create a more holistic understanding of patient status.

8.9.3. *AI-coach hybrid models*

AI as a support system to human physiotherapists rather than a replacement, ensuring empathy and contextual judgment remain central.

8.9.4. *AI in pediatric and geriatric oncology*

Age-specific AI algorithms can better cater to unique developmental or degenerative challenges in these populations.

8.10. Conclusion

Personalized AI-based exercise therapy represents a transformative advancement in the intersection of oncology, physiotherapy and digital health. By harnessing the capabilities of ML, CV and wearable technologies, AI is enabling the development of dynamic, responsive and patient-centric rehabilitation programs that evolve with individual needs and progress. This represents a significant shift away from traditional "one-size-fits-all" rehabilitation methods, introducing a dynamic model that is adaptive, evidence-based and continually refined in real time. In oncology,

AI-driven exercise interventions address the complex challenges experienced by patients undergoing chemotherapy, radiation or surgery, including CRF, immune suppression and post-operative mobility impairments. Similarly, in physiotherapy for MSK and neurological conditions, AI enhances clinical care by providing personalized feedback, real-time performance monitoring and predictive insights into healing and pain patterns. These innovations collectively contribute to improved adherence, reduced risk of complications and enhanced functional outcomes. Furthermore, the integration of AI with telerehabilitation platforms expands access to care, particularly for patients in remote or underserved regions. Through remote monitoring, automated progress tracking and virtual consultations, AI supports continuity of care outside clinical settings – an essential factor in long-term rehabilitation success. However, the deployment of AI in clinical practice is not without challenges. Issues such as data privacy, algorithmic bias, user trust and regulatory compliance must be carefully addressed. The effectiveness of AI-based exercise therapy also depends on rigorous clinical validation, interdisciplinary collaboration and training of healthcare professionals to interpret and integrate AI insights into patient care. Moreover, ensuring that AI systems are designed with diverse, inclusive datasets is vital to avoid health disparities and ensure equitable outcomes.

8.11. References

Aminizadeh, S., Heidari, A., Dehghan, M., Toumaj, S., Rezaei, M., Navimipour, N.J., Stroppa, F., Unal, M. (2024). Opportunities and challenges of artificial intelligence and distributed systems to improve the quality of healthcare service. *Artificial Intelligence in Medicine*, 149, 102779.

Bhambri, P. and Khang, A. (2024). Managing and monitoring patient's healthcare using AI and IoT technologies. In *Driving Smart Medical Diagnosis Through AI-Powered Technologies and Applications*, Khang, A. (ed.). IGI Global, Hershey, PA, 1–23.

Braun, R.G. and Wittenberg, G.F. (2021). Motor recovery: How rehabilitation techniques and technologies can enhance recovery and neuroplasticity. *Seminars in Neurology*, 41(2), 167–176.

Chaughule, S. (2024). Satisfaction and adherence to exercises in patients for home health vs clinic, PT. *Journal of Physical Medicine Rehabilitation Studies & Reports*, 6(12). doi: 10.47363/JPMRS/2024(6)e102.

Dell'Eva, F.R. (2024). Neuromodulation approaches for the rehabilitation of neurological diseases: Noninvasive electrical stimulation and robotics. PhD Thesis, Politecnico di Milano, Milan.

Deshmukh, P. (2023). The integration of artificial intelligence and wearable technology in personalized physical therapy programs for enhanced recovery outcomes. *International Journal of Artificial Intelligence, Data Science, and Machine Learning*, 4(3), 1–9.

Doherty, C., Lambe, R., O'Grady, B., O'Reilly-Morgan, D., Smyth, B., Lawlor, A., Hurley, N., Tragos, E. (2024). An evaluation of the effect of app-based exercise prescription using reinforcement learning on satisfaction and exercise intensity: Randomized crossover trial. *JMIR mHealth and uHealth*, 12, e49443.

Dong, L.A. (2024). Evaluating the therapeutic impact of exercise on chronic diseases: A comprehensive review and future directions. *International Journal of Education and Humanities*, 4(3), 359–369.

Durrand, J.W. (2022). Development of a remotely supervised digitally facilitated multibehavioural prehabilitation intervention for patients approaching major surgery. Thesis, University of Northumbria, Newcastle.

El-Tallawy, S.N., Pergolizzi, J.V., Vasiliu-Feltes, I., Ahmed, R.S., LeQuang, J.K., El-Tallawy, H.N., Varrassi, G., Nagiub, M.S. (2024). Incorporation of "artificial intelligence" for objective pain assessment: A comprehensive review. *Pain and Therapy*, 13(3), 293–317.

Emery, J., Butow, P., Lai-Kwon, J., Nekhlyudov, L., Rynderman, M., Jefford, M. (2022). Management of common clinical problems experienced by survivors of cancer. *The Lancet*, 399(10334), 1537–1550.

Esmaeilzadeh, P. (2024). Challenges and strategies for wide-scale artificial intelligence (AI) deployment in healthcare practices: A perspective for healthcare organizations. *Artificial Intelligence in Medicine*, 151, 102861.

Fessele, K.L. and Syrkin, G. (2024). Mobility assessment instruments. *Seminars in Oncology Nursing*, 40(4), 151660.

Ghanem, M., Ghaith, A.K., Bydon, M. (2024). Artificial intelligence and personalized medicine: Transforming patient care. In *The New Era of Precision Medicine*, Bydon, M. (ed.). Academic Press, San Diego, 131–142.

Han, X., Guffanti, D., Brunete, A. (2025). A comprehensive review of vision-based sensor systems for human gait analysis. *Sensors*, 25(2), 498.

Hussey, C. and Gupta, A. (2022). Exercise interventions to combat cancer-related fatigue in cancer patients undergoing treatment: A review. *Cancer Investigation*. 40(9), 822–838.

Kakhi, K., Jagatheesaperumal, S.K., Khosravi, A., Alizadehsani, R., Acharya, U.R. (2024). Fatigue monitoring using wearables and AI: Trends, challenges, and future opportunities. *arXiv:2412.16847*. doi: 10.48550/arXiv.2412.16847.

Kaswan, K.S., Gaur, L., Dhatterwal, J.S., Kumar, R. (2021). AI-based natural language processing for the generation of meaningful information electronic health record (EHR) data. In *Advanced AI Techniques and Applications in Bioinformatics*, Gaur, L., Solanki, A., Wamba, S.F., Jhanjhi, N.Z. (eds). CRC Press, Boca Raton, 41–86.

Khalid, U.B., Naeem, M., Stasolla, F., Syed, M.H., Abbas, M., Coronato, A. (2024). Impact of AI-powered solutions in rehabilitation process: Recent improvements and future trends. *International Journal of General Medicine*, 17, 943–969.

Kuo, C.L. (2023). Revolutionizing healthcare paradigms: The integral role of artificial intelligence in advancing diagnostic and treatment modalities. *International Microsurgery Journal (IMJ)*, 7(1). doi: 10.24983/scitemed.imj.2023.00177.

Li, F., Wang, S., Gao, Z., Qing, M., Pan, S., Liu, Y., Hu, C. (2025). Harnessing artificial intelligence in sepsis care: Advances in early detection, personalized treatment, and real-time monitoring. *Frontiers in Medicine*, 11, 1510792.

Lo, J., Chan, L., Flynn, S. (2021). A systematic review of the incidence, prevalence, costs, and activity and work limitations of amputation, osteoarthritis, rheumatoid arthritis, back pain, multiple sclerosis, spinal cord injury, stroke, and traumatic brain injury in the United States: A 2019 update. *Archives of Physical Medicine and Rehabilitation*, 102(1), 115–131.

Lorenzo, G., Ahmed, S.R., Hormuth II, D.A., Vaughn, B., Kalpathy-Cramer, J., Solorio, L., Yankeelov, T.E., Gomez, H. (2024). Patient-specific, mechanistic models of tumor growth incorporating artificial intelligence and big data. *Annual Review of Biomedical Engineering*, 26(1), 529–560.

Mikolajewska, E., Mikoiajewski, D., Mikolajczyk, T., Paczkowski, T. (2025). A breakthrough in producing personalized solutions for rehabilitation and physiotherapy thanks to the introduction of AI to additive manufacturing. *Applied Sciences*, 15(4), 2219.

Mohapatra, A., Mohanty, P., Pattnaik, M., Padhan, S. (2024). Physiotherapy in the digital age: A narrative review of the paradigm shift driven by the integration of artificial intelligence and machine learning. *Physiotherapy – The Journal of Indian Association of Physiotherapists*, 18(2), 63–71.

Oyeniyi, J. and Oluwaseyi, P. (2024). Emerging trends in AI-powered medical imaging: Enhancing diagnostic accuracy and treatment decisions. *International Journal of Enhanced Research In Science Technology & Engineering*, 13, 2319–7463.

Prieto-Avalos, G., Cruz-Ramos, N.A., Alor-Hernandez, G., Sánchez-Cervantes, J.L., Rodriguez-Mazahua, L., Guarneros-Nolasco, L.R. (2022). Wearable devices for physical monitoring of heart: A review. *Biosensors*, 12(5), 292.

Reeve, B.B., Mitchell, S.A., Dueck, A.C., Basch, E., Cella, D., Reilly, C.M., Minasian, L.M., Denicoff, A.M., O'Mara, A.M., Fisch, M.J. et al. (2014). Recommended patient-reported core set of symptoms to measure in adult cancer treatment trials. *JNCI: Journal of the National Cancer Institute*, 106(7), dju129.

Roggio, F., Di Grande, S., Cavalieri, S., Falla, D., Musumeci, G. (2024). Biomechanical posture analysis in healthy adults with machine learning: Applicability and reliability. *Sensors*, 24(9), 2929.

Saez, M., Maturana, F.P., Barton, K., Tilbury, D.M. (2018). Real-time manufacturing machine and system performance monitoring using Internet of Things. *IEEE Transactions on Automation Science and Engineering*, 15(4), 1735–1748.

Sassi, M., Villa Corta, M., Pisani, M.G., Nicodemi, G., Schena, E., Pecchia, L., Longo, U.G. (2024). Advanced home-based shoulder rehabilitation: A systematic review of remote monitoring devices and their therapeutic efficacy. *Sensors*, 24(9), 2936.

Senadheera, I., Hettiarachchi, P., Haslam, B., Nawaratne, R., Sheehan, J., Lockwood, K.J., Alahakoon, D., Carey, L.M. (2024). AI applications in adult stroke recovery and rehabilitation: A scoping review using AI. *Sensors (Basel, Switzerland)*, 24(20), 6585.

Shaw, K.L. (2015). Patient education, motivation, compliance, and adherence to physical activity, exercise, and rehabilitation. In *Pathology and Intervention in Musculoskeletal Rehabilitation*, Magee, D.J., Zachazewski, J.E., Quillen, W.S., Manske, R.C. (eds). Saunders/Elsevier, Maryland Heights, MO, 1–24.

Simmons, L.A., Wolever, R.Q., Bechard, E.M., Snyderman, R. (2014). Patient engagement as a risk factor in personalized health care: A systematic review of the literature on chronic disease. *Genome Medicine*, 6, 1–3.

Vincent, H.K., Adams, M.C., Vincent, K.R., Hurley, R.W. (2013). Musculoskeletal pain, fear avoidance behaviors, and functional decline in obesity: Potential interventions to manage pain and maintain function. *Regional Anesthesia & Pain Medicine*, 38(6), 481–491.

Zhang, W. (2025). Enhancing rehabilitation assessment with artificial intelligence: A comprehensive investigation of posture quality prediction using machine learning. *ITM Web of Conferences*, 70, 02025.

9

Neural Networks in Physiotherapy and Oncology: Enhancing Recovery Pathways

The integration of neural networks into the fields of physiotherapy and oncology marks a pivotal advancement in modern healthcare. These sophisticated artificial intelligence systems, inspired by the structure and function of the human brain, possess the ability to learn from complex medical data, adapt to patient-specific needs and offer real-time, actionable insights. In oncology, neural networks have shown significant potential in improving diagnostic accuracy by enabling more precise image recognition, tumor classification and interpretation of genomic data. They also contribute significantly to personalized treatment planning, enabling oncologists to tailor therapies based on predicted patient responses and expected outcomes. In physiotherapy, neural networks are reshaping recovery strategies by enabling precise motion analysis, optimizing rehabilitation exercises and predicting functional recovery. Coupled with wearable devices and telemedicine platforms, these technologies empower both clinicians and patients to monitor progress remotely, increasing accessibility to care and improving long-term adherence to therapeutic programs.

Moreover, in cross-disciplinary applications such as oncology rehabilitation and cognitive recovery, neural networks bridge the gap between data-driven decision-making and patient-centered care. Despite their transformative potential, challenges remain and these issues surrounding data privacy, model interpretability and clinical integration must be addressed to ensure the responsible and effective deployment of neural networks. Future directions, including explainable AI, federated learning and multimodal data fusion, promise to enhance the transparency, security and versatility of these systems. This chapter provides an in-depth exploration of how neural networks are enhancing recovery pathways in oncology and physiotherapy. By evaluating current applications, clinical impacts and technological advancements, it outlines a roadmap for future integration of AI in patient recovery processes, ultimately contributing to more efficient, personalized and proactive healthcare delivery.

Chapter written by Chandrakant PATIL, Dhiraj Kumar MANE, Anand GUDUR and Kalpana MALPE.

9.1. Introduction

9.1.1. *Background*

The increasing complexity of cancer treatment and the multifaceted nature of physiotherapy underscore the need for more intelligent, adaptable and data-driven healthcare solutions. Traditional clinical approaches, while effective in many respects, often rely on generalized treatment protocols that may not fully accommodate the unique physiological and recovery needs of individual patients (Beutler et al. 2016). These standardized models can fall short in addressing patient-specific variables such as genetic makeup, tumor heterogeneity, comorbidities and variations in treatment response and rehabilitation potential.

The advent of artificial intelligence (AI), particularly neural networks, presents a promising paradigm shift. Neural networks, modeled after the interconnected structure of neurons in the human brain, possess the ability to learn from vast and diverse datasets, identify complex and often imperceptible patterns and make predictive decisions with high accuracy. Unlike rule-based systems, neural networks improve over time as they are exposed to more data, allowing for dynamic adaptation and continuous learning. These features make them especially valuable in healthcare domains where precision and personalization are paramount.

In oncology, neural networks have demonstrated exceptional promise in enhancing diagnostic imaging, predicting treatment outcomes and stratifying patient risk with unprecedented granularity (Khalighi et al. 2024). For instance, convolutional neural networks (CNNs) are increasingly used to analyze radiological images such as CT scans, MRIs and PET scans, facilitating early detection of malignancies and reducing diagnostic errors. Similarly, recurrent neural networks (RNNs) are being used to model disease progression and support long-term treatment planning through temporal data analysis.

In physiotherapy, neural networks enable personalized rehabilitation programs by processing real-time patient data from wearable sensors, motion capture systems and electronic health records. They can detect subtle deviations in movement patterns, assess recovery progress and suggest adjustments to therapy regimens to optimize outcomes (Harbourne and Stergiou 2009). Moreover, neural networks can integrate data from multiple modalities – biomechanical, physiological and behavioral – to create a holistic view of patient health, thereby supporting more informed and responsive clinical decisions.

The convergence of AI and healthcare, particularly through neural networks, signals a transformative era wherein treatment strategies are no longer confined to one-size-fits-all models but are instead tailored to the individual (Garg et al. 2025).

This shift holds significant promise for oncology and physiotherapy, two disciplines that demand high levels of personalization and precision. As research and technology continue to evolve, the integration of neural networks into these fields is poised to not only enhance clinical effectiveness but also improve patient quality of life and long-term health outcomes.

9.1.2. *Significance of neural networks in healthcare*

In recent years, neural networks have increasingly been incorporated into clinical practice due to their superior ability to process complex, high-dimensional and nonlinear medical data. As healthcare systems around the world strive to deliver more precise, efficient and personalized care, neural networks have emerged as a cornerstone of modern medical innovation, particularly in fields requiring intricate data interpretation such as oncology and physiotherapy (Thacharodi et al. 2024).

In oncology, neural networks have revolutionized diagnostic processes by significantly improving the accuracy and speed of image analysis. Convolutional neural networks (CNNs), in particular, excel in analyzing radiological images – including MRI, CT and PET scans – by detecting minute abnormalities that may be overlooked by human observers. These networks are capable of identifying malignancies at earlier stages, classifying tumors based on subtle histopathological features and differentiating between benign and malignant lesions with high sensitivity and specificity. Moreover, they are instrumental in genomic-based oncology, where deep learning models analyze vast amounts of genetic and molecular data to predict tumor behavior, therapeutic response and potential for metastasis (Ortega-Batista et al. 2025). This capability enhances personalized medicine by aligning treatment plans with the unique genetic profile of each patient.

Beyond diagnostics, neural networks also support prognostic modeling and treatment optimization. By integrating clinical data, imaging results and molecular profiles, these systems can forecast disease progression, estimate survival rates and assist in the selection of the most effective therapy regimens. This predictive power enables oncologists to make informed decisions, reduce overtreatment or undertreatment, and improve overall patient outcomes (Walsh et al. 2019).

In physiotherapy, neural networks play a pivotal role in the development of individualized rehabilitation protocols. Using data from motion capture systems, inertial measurement units (IMUs), wearable sensors and real-time patient feedback, these networks assess biomechanical function, identify compensatory movement patterns and monitor recovery trajectories. By recognizing deviations from optimal motion or strength benchmarks, neural networks provide actionable insights that help therapists adjust interventions dynamically and with greater precision.

9.1.3. *Objective*

The primary objective of this chapter is to examine the integration and impact of neural networks in physiotherapy and oncology, with a focus on how these technologies enhance patient recovery pathways (Alagappan et al. 2024). Specifically, this study explores how neural networks contribute to improved diagnostic precision, personalized treatment strategies and predictive modeling of recovery outcomes. By analyzing current applications, technological innovations and future prospects, this chapter seeks to provide a comprehensive understanding of the transformative role of neural networks in modern therapeutic and oncological care.

9.2. Neural networks: concepts and healthcare integration

9.2.1. *Basics of neural networks*

Neural networks are sophisticated computational frameworks inspired by the architecture of the human brain (Parhi and Unnikrishnan 2020). They consist of multiple layers of interconnected processing units, known as neurons, which simulate biological neural activity. Each neuron receives input, applies a weighted transformation, passes the result through an activation function and transmits the output to neurons in the subsequent layer. Through iterative training and backpropagation, neural networks learn to identify complex, nonlinear patterns in large datasets, making them particularly suited for diverse applications in healthcare.

9.2.2. *Types of neural networks in healthcare*

Various neural network architectures are used in healthcare, each tailored to specific types of data and clinical objectives:

– *Convolutional neural networks (CNNs)*: CNNs excel in image-related tasks and are extensively used in medical imaging for disease detection, segmentation and classification of radiographic data such as MRI, CT and X-rays.

– *Recurrent neural networks (RNNs)*: RNNs are well-suited for sequential or time-series data. In healthcare, they are instrumental in modeling patient recovery trajectories, monitoring vital signs over time and predicting disease progression (Li-wei et al. 2014).

– *Deep neural networks (DNNs)*: these are multi-layered networks capable of processing high-dimensional and complex data. In personalized medicine, DNNs facilitate the integration of genomic, clinical and lifestyle data to tailor treatments to individual patients.

– *Generative adversarial networks (GANs)*: GANs are used to generate synthetic medical images for training purposes and to augment limited datasets. They enhance the robustness of diagnostic algorithms by simulating diverse pathological conditions.

9.2.3. *Integration with healthcare systems*

The integration of neural networks into healthcare infrastructure is transforming clinical workflows (Maleki Varnosfaderani and Forouzanfar 2024). These models are embedded within electronic health records (EHRs), diagnostic imaging platforms and wearable health monitoring devices. By enabling real-time data analysis and predictive modeling, neural networks support clinical decision-making, automate routine tasks and personalize patient care. Their deployment across various domains – from diagnostics and therapeutics to patient monitoring and administrative functions – highlights their potential to improve efficiency, accuracy and outcomes in modern healthcare systems.

9.3. Neural networks in oncology

9.3.1. *Early detection and diagnosis*

Early detection significantly improves cancer prognosis. Convolutional neural networks (CNNs) have emerged as powerful tools in identifying malignancies at their nascent stages by analyzing complex radiological data such as CT scans, MRIs and PET images. These networks learn from vast annotated datasets to detect abnormal tissue patterns with remarkable precision, often matching or surpassing the diagnostic accuracy of expert radiologists (Singh et al. 2024).

Case study

A CNN model trained on over 200,000 mammographic images achieved a sensitivity rate of 94.5% in detecting breast cancer. It notably reduced false positives and expedited clinical decision-making, demonstrating the technology's potential to alleviate radiologist workload and enhance screening outcomes.

9.3.2. *Tumor classification and staging*

Accurate classification and staging of tumors are vital for determining prognosis and tailoring therapeutic approaches. Neural networks, especially those incorporating multi-modal data, can integrate imaging results with histopathological and genomic information. This comprehensive data fusion enables more precise differentiation between benign and malignant lesions, tumor subtypes and cancer stages, supporting more nuanced and effective clinical interventions (Steyaert et al. 2023).

For example, hybrid models combining CNNs with genomic data have shown promise in classifying gliomas into subtypes associated with distinct molecular profiles, influencing both prognosis and therapy selection.

9.3.3. *Treatment personalization*

One of the most impactful applications of neural networks in oncology is the personalization of cancer therapy. Deep neural networks (DNNs) and ensemble learning models analyze diverse patient data – including gene expression profiles, tumor microenvironment characteristics and physiological markers – to predict individual responses to chemotherapy, immunotherapy and radiation. By identifying patients who are likely to benefit from specific interventions, these models help minimize toxic side effects and optimize therapeutic efficacy.

Emerging research demonstrates that AI models can recommend treatment plans that align with oncologist decisions in over 90% of cases, offering an assistive second opinion in complex scenarios.

9.3.4. *Prognostic modeling*

Neural networks are increasingly used in modeling cancer progression and estimating survival probabilities (Yan et al. 2024).

Recurrent neural networks (RNNs), due to their proficiency in handling temporal sequences, are well-suited for analyzing longitudinal patient data, such as follow-up imaging and lab results. These models can predict recurrence risk, progression timelines and survival outcomes based on dynamic patient variables.

Prognostic tools powered by DNNs have been shown to outperform traditional statistical models such as Cox regression in predicting five-year survival in cancers such as non-small cell lung carcinoma and colorectal cancer. These forecasts are instrumental in guiding decisions about treatment intensity, surveillance scheduling and palliative care planning.

9.4. Neural networks in physiotherapy

9.4.1. *Automated movement analysis*

Neural networks have become pivotal in transforming physiotherapy assessment methods by enabling automated movement analysis (neural networks). Leveraging data from motion capture systems, inertial measurement units (IMUs) and other wearable sensors, neural networks – particularly CNNs and LSTMs (Long Short-Term Memory networks) – can accurately classify biomechanical movement patterns, detect gait asymmetries and identify functional impairments in patients with neurological or musculoskeletal conditions.

Application

In post-stroke rehabilitation, CNNs have demonstrated high precision in analyzing gait kinematics. These systems can assess spatiotemporal parameters such as step length, symmetry and cadence, offer real-time feedback and recommend exercise modifications tailored to the patient's motor capabilities. This continuous monitoring supports both clinicians and patients in achieving more effective and objective rehabilitation outcomes.

9.4.2. *Personalized rehabilitation programs*

One of the transformative contributions of neural networks in physiotherapy is their ability to develop and continuously adapt personalized rehabilitation programs (Calderone et al. 2024). By incorporating diverse patient data – including age, injury type, baseline mobility, comorbidities and real-time performance metrics – these models can prescribe customized exercise plans that evolve with patient progress.

Adaptive neural network models optimize exercise intensity, frequency and type based on ongoing feedback, increasing patient engagement and therapy adherence. In clinical trials, AI-driven rehabilitation plans have demonstrated faster recovery times and improved functional scores compared to standard protocols.

9.4.3. *Predictive recovery modeling*

Forecasting a patient's recovery trajectory is essential for setting realistic expectations and optimizing care (Muscedere et al. 2024). Neural networks – especially DNNs and RNNs – analyze early therapy data, such as initial assessments, early mobility scores and response to treatment, to predict long-term functional outcomes.

These predictive models enable clinicians to:

– identify patients at a risk of slow or incomplete recovery;

– modify treatment strategies proactively;

– allocate resources more effectively.

For instance, RNNs trained on post-operative recovery data in orthopedic patients have accurately predicted six-week and twelve-week mobility milestones, guiding both clinical interventions and patient counseling.

9.4.4. *Telerehabilitation and virtual coaching*

With the rise of digital health, neural networks play an essential role in expanding access to physiotherapy through telerehabilitation platforms (Mohapatra et al. 2024). Integrated into mobile applications and wearable devices, neural networks process real-time sensor and video data to provide personalized feedback, corrective cues and motivational coaching remotely.

These AI-enhanced systems support:

– remote supervision by therapists;

– real-time posture and form correction;

– exercise progression monitoring;

– data-driven alerts for deviations or noncompliance.

Telerehabilitation powered by neural networks has shown significant promise in rural and underserved areas, where access to in-person therapy is limited (Surya and Someshwar 2025). Moreover, studies indicate that such systems can maintain clinical efficacy comparable to traditional face-to-face sessions while improving convenience and reducing healthcare costs.

9.5. Cross-disciplinary applications

9.5.1. *Oncology rehabilitation*

Cancer survivors frequently contend with long-term physical impairments, such as fatigue, reduced mobility, neuropathy and musculoskeletal dysfunction, often resulting from surgery, chemotherapy or radiation therapy (Nekhlyudov et al. 2021). Neural networks offer a powerful toolset for enhancing rehabilitation outcomes in this population by enabling continuous, adaptive care.

Key applications include:

– *Fatigue and mobility monitoring*: wearable sensors combined with CNNs and RNNs can track movement patterns, postural stability and energy expenditure, offering real-time data on patient function and fatigue levels.

– *Complication prediction*: neural networks trained on clinical, demographic and treatment data can predict risks for musculoskeletal complications, such as lymphedema or joint stiffness, enabling proactive intervention.

– *Adaptive physiotherapy protocols*: AI-driven systems use real-time performance feedback to dynamically adjust exercise intensity and type, ensuring that therapy remains effective and aligned with patient capabilities (Deshmukh 2023).

These capabilities support personalized, responsive rehabilitation strategies that improve quality of life and reduce long-term disability for cancer survivors.

9.5.2. *Pain management*

Chronic pain is a common comorbidity in both oncology and post-operative physiotherapy. Neural networks are increasingly used to model and predict pain experiences by analyzing multimodal inputs, including behavioral signals (e.g. facial expressions, movement hesitancy, etc.), physiological data (e.g. heart rate variability, EMG signals, etc.) and patient-reported outcomes.

AI systems can:

– identify patterns correlating with high-pain episodes;

– distinguish between neuropathic, inflammatory and musculoskeletal pain sources;

– suggest evidence-based interventions ranging from therapeutic exercises and neuromodulation to pharmacologic adjustments.

This approach facilitates a more nuanced and individualized pain management plan, reducing reliance on opioids and enhancing patient comfort and function.

9.5.3. *Cognitive rehabilitation*

Chemotherapy-induced cognitive impairment – commonly referred to as “chemo brain” – affects attention, memory and executive function, impairing patients’ daily activities and quality of life (Allegra et al. 2020). RNNs and other sequence-based models can analyze longitudinal cognitive performance data, identifying subtle trends in decline or recovery.

Neural networks can:

– monitor cognitive metrics collected through digital assessments and wearable EEG devices;

– predict trajectories of cognitive recovery;

– recommend targeted brain-training exercises, mindfulness practices or referral to neurorehabilitation services.

Such AI-guided cognitive rehabilitation programs can be seamlessly integrated into survivorship care plans, offering a structured and responsive path to cognitive recovery.

9.6. Challenges and limitations

9.6.1. *Data privacy and ethics*

The application of neural networks in healthcare necessitates the collection and analysis of sensitive patient information, including imaging, genetic data and electronic health records (Xiao et al. 2018). This raises critical ethical concerns related to data privacy, patient consent and the potential for misuse or unauthorized access. Regulatory frameworks such as the General Data Protection Regulation (GDPR) and the Health Insurance Portability and Accountability Act (HIPAA) mandate strict safeguards, yet many AI systems are developed across jurisdictions with varying compliance standards.

Moreover, ethical dilemmas arise around the secondary use of data, such as using patient records to train models without explicit consent. Ensuring transparency in how data is collected, processed and stored is paramount for maintaining public trust and protecting patient rights.

9.6.2. *Data quality and bias*

Neural networks rely heavily on large, diverse and accurately labeled datasets to perform effectively (Song et al. 2022). However, many available medical datasets are fragmented, incomplete or underrepresent minority populations. This lack of representation can introduce significant bias into model predictions, potentially leading to diagnostic inaccuracies and unequal treatment outcomes.

For instance, models trained predominantly on data from one demographic group may underperform in others, reinforcing existing health disparities. Addressing these issues requires concerted efforts to curate high-quality, inclusive datasets and apply bias mitigation techniques during model development and validation.

9.6.3. *Interpretability*

Despite their predictive power, many neural networks – particularly deep learning models – function as "black boxes", providing little insight into how decisions are made. This lack of interpretability poses a major challenge in clinical settings, where transparency and explainability are crucial for physician trust and patient safety (Albahri et al. 2023).

Clinicians are often reluctant to adopt AI tools they cannot understand or justify, especially in high-stake environments such as oncology. Furthermore, regulatory bodies require clear explanations of algorithmic decisions before granting approval for clinical deployment.

Techniques such as saliency mapping, SHAP (SHapley Additive exPlanations) and LIME (Local Interpretable Model-agnostic Explanations) are being developed to address this issue, but they are not yet universally accepted or standardized.

9.6.4. *Integration with clinical workflows*

Integrating neural network-based tools into the existing healthcare infrastructure is a complex, resource-intensive process (Saeed et al. 2025). It involves aligning new technologies with established clinical protocols, retraining staff, ensuring compatibility with electronic health systems and maintaining interoperability across platforms.

Resistance to change, lack of technical expertise among healthcare professionals and concerns about workflow disruption can delay or impede adoption. Additionally, real-time application of neural networks in clinical environments demands high computational power and robust data pipelines, which may not be readily available in all healthcare settings.

To overcome these barriers, interdisciplinary collaboration among clinicians, data scientists and system engineers is essential. Solutions must be user-centered, ensuring that AI tools augment – rather than replace – clinical judgment and integrate seamlessly into the care delivery process.

9.7. Future prospects

9.7.1. *Explainable AI (XAI)*

As neural networks become more prevalent in clinical decision-making, the need for transparency and trust is critical (Sendak et al. 2020). Explainable AI (XAI) aims to demystify complex models by providing interpretable outputs that clinicians can understand and validate. Techniques such as attention mechanisms, feature attribution methods (e.g. SHAP, LIME, etc.) and visual explanation tools (e.g. Grad-CAM for imaging) help reveal which inputs most influence the model's predictions.

XAI holds particular promise in domains such as oncology and physiotherapy, where life-altering decisions require a clear rationale. Increasing model interpretability will also ease regulatory approvals and promote ethical deployment, fostering greater confidence among healthcare providers and patients.

9.7.2. *Federated learning*

To mitigate data privacy risks, federated learning (FL) offers a decentralized approach to model training (Sendak et al. 2020). Instead of transferring sensitive patient data to a central server, FL allows neural networks to be trained locally at individual institutions, with only model parameters shared and aggregated. This approach preserves patient confidentiality while enabling access to broader, more diverse datasets.

Federated learning is particularly valuable in healthcare, where data silos exist across hospitals, research centers and geographic regions. Its implementation promises collaborative model development across institutions without compromising data ownership or violating privacy laws.

9.7.3. *Hybrid models*

Combining neural networks with traditional machine learning techniques or rule-based systems can lead to more robust and interpretable models. Hybrid models capitalize on the predictive strength of deep learning and the clarity of conventional approaches, enhancing both accuracy and explainability.

In oncology, for instance, a hybrid system could use a neural network to detect tumor regions in imaging data and then apply a rule-based classifier for stage assignment based on clinical guidelines (in oncology). In physiotherapy, time-series data from wearable sensors can be modeled using RNNs, while decision trees classify recovery patterns based on predefined thresholds. Such combinations are likely to become standard in clinical AI tools.

9.7.4. *Integration with robotics*

The convergence of neural networks and robotics is shaping the next generation of intelligent rehabilitation devices. These AI-powered systems enable adaptive physiotherapy interventions by learning from real-time patient performance and adjusting assistance levels dynamically.

For example, robotic exoskeletons and therapy arms guided by neural networks can personalize resistance, range of motion and feedback for each session. This responsiveness improves motor relearning and patient engagement. Future advancements may allow fully autonomous yet supervised rehabilitation sessions in home-based settings, extending care beyond clinical walls.

9.7.5. *Multimodal data fusion*

A key frontier in AI-driven healthcare is the integration of multimodal data – such as imaging, genomics, clinical history, lab results and sensor data – into unified patient models. Neural networks, especially deep learning architectures, are well-suited for processing this heterogeneous information to uncover nuanced patterns that single-source analysis may miss (Sirocchi 2025).

Multimodal fusion enhances diagnostic accuracy, prognostic modeling and treatment personalization. In oncology, combining radiomics with genomic and histopathological data can lead to more precise tumor characterization. In physiotherapy, sensor data, EMG signals and patient-reported outcomes can be fused to generate a holistic understanding of recovery progress. Such systems will enable more comprehensive, data-driven and patient-centered care pathways.

9.8. Conclusion

Neural networks are profoundly transforming the fields of physiotherapy and oncology by introducing data-driven precision, speed and personalization into traditionally manual and subjective domains. Their ability to process vast, complex datasets – including medical images, genomic profiles, sensor data and patient histories – enables earlier disease detection, more accurate diagnoses and individualized treatment strategies. In oncology, neural networks support critical tasks such as tumor detection, classification, staging and prognosis modeling, thereby assisting clinicians in making faster, more informed decisions. Similarly, in physiotherapy, these models enhance recovery by analyzing biomechanical data, personalizing rehabilitation protocols and predicting functional outcomes based on real-time feedback and historical trends. Furthermore, the integration of neural networks with technologies such as robotics, mobile health applications and wearable sensors is reshaping how care is delivered, making it more continuous, remote and patient-centric. These advancements not only improve treatment efficacy but also increase accessibility and convenience, particularly for patients in remote or underserved areas. However, despite their promise, neural networks face significant challenges that must be addressed to ensure their responsible and effective deployment. Issues such as data privacy, algorithmic bias, model interpretability and clinical integration require sustained attention and collaboration among data scientists, clinicians, ethicists and policymakers. Emerging approaches such as explainable AI, federated learning and hybrid modeling offer potential solutions that can help bridge these gaps. Looking forward, as neural networks become more transparent, secure and integrated into clinical workflows, they are expected to play a central role in the evolution of modern healthcare. With continued innovation and responsible implementation, these systems will not only augment human expertise but also contribute to more equitable, efficient and outcome-driven care across the global health landscape.

9.9. References

Alagappan, T.R., Bhansali, N., Bhansali, T. (2024). Navigating the integration of technology in physiotherapy: A balanced approach. *IP Journal of Surgery and Allied Sciences*, 6(1), 1–4.

Albahri, A.S., Duhaim, A.M., Fadhel, M.A., Alnoor, A., Baqer, N.S., Alzubaidi, L., Albahri, O.S., Alamoodi, A.H., Bai, J., Salhi, A. et al. (2023). A systematic review of trustworthy and explainable artificial intelligence in healthcare: Assessment of quality, bias risk, and data fusion. *Information Fusion*, 96(C), 156–191.

Allegra, A., Innao, V., Basile, G., Pugliese, M., Allegra, A.G., Pulvirenti, N., Musolino, C. (2020). Post-chemotherapy cognitive impairment in hematological patients: Current understanding of chemobrain in hematology. *Expert Review of Hematology*, 13(4), 393–404.

Beutler, L.E., Someah, K., Kimpara, S., Miller, K. (2016). Selecting the most appropriate treatment for each patient. *International Journal of Clinical and Health Psychology*, 16(1), 99–108.

Calderone, A., Latella, D., Bonanno, M., Quartarone, A., Mojdehdehbaher, S., Celesti, A., Calabrò, R.S. (2024). Towards transforming neurorehabilitation: The impact of artificial intelligence on diagnosis and treatment of neurological disorders. *Biomedicines*, 12(10), 2415.

Deshmukh, P. (2023). The integration of artificial intelligence and wearable technology in personalized physical therapy programs for enhanced recovery outcomes. *International Journal of Artificial Intelligence, Data Science, and Machine Learning*, 4(3), 1–9.

Garg, P.P., Jayashree, J., Vijayashree, J. (2025). Revolutionizing healthcare: The transformative role of artificial intelligence. In *Responsible and Explainable Artificial Intelligence in Healthcare*, Singh, A., Singh, K.K., Izonin, I. (eds). Academic Press, Cambridge, MA.

Harbourne, R.T. and Stergiou, N. (2009). Movement variability and the use of nonlinear tools: Principles to guide physical therapist practice. *Physical Therapy*, 89(3), 267–282.

Khalighi, S., Reddy, K., Midya, A., Pandav, K.B., Madabhushi, A., Abedalthagafi, M. (2024). Artificial intelligence in neuro-oncology: Advances and challenges in brain tumor diagnosis, prognosis, and precision treatment. *NPJ Precision Oncology*, 8(1), 80.

Kumar, A., Rathore, P.S., Ahuja, S., Lilhore, U.K. (eds) (2025). *Integrating Neurocomputing with Artificial Intelligence*. John Wiley & Sons, New York.

Li-wei, H.L., Adams, R.P., Mayaud, L., Moody, G.B., Malhotra, A., Mark, R.G., Nemati, S. (2014). A physiological time series dynamics-based approach to patient monitoring and outcome prediction. *IEEE Journal of Biomedical and Health Informatics*, 19(3), 1068–1076.

Maleki Varnosfaderani, S. and Forouzanfar, M. (2024). The role of AI in hospitals and clinics: Transforming healthcare in the 21st century. *Bioengineering*, 11(4), 337.

Mohapatra, A., Mohanty, P., Pattnaik, M., Padhan, S. (2024). Physiotherapy in the digital age: A narrative review of the paradigm shift driven by the integration of artificial intelligence and machine learning. *Physiotherapy – The Journal of Indian Association of Physiotherapists*, 18(2), 63–71.

Muscedere, J., Bagshaw, S.M., Kho, M., Mehta, S., Cook, D.J., Boyd, J.G., Sibley, S., Wang, H.T., Archambault, P.M., Albert, M. et al. (2024). Frailty, outcomes, recovery and care steps of critically ill patients (FORECAST): A prospective, multi-centre, cohort study. *Intensive Care Medicine*, 50(7), 1064–1074.

Nekhlyudov, L., Campbell, G.B., Schmitz, K.H., Brooks, G.A., Kumar, A.J., Ganz, P.A., Von, Ah D. (2021). Cancer-related impairments and functional limitations among long-term cancer survivors: Gaps and opportunities for clinical practice. *Cancer*, 128(2), 222.

Ortega-Batista, A., Jaén-Alvarado, Y., Moreno-Labrador, D., Gómez, N., García, G., Guerrero, E.N. (2025). Single-cell sequencing: Genomic and transcriptomic approaches in cancer cell biology. *International Journal of Molecular Sciences*, 26(5), 2074.

Parhi, K.K. and Unnikrishnan, N.K. (2020). Brain-inspired computing: Models and architectures. *IEEE Open Journal of Circuits and Systems*, 1, 185–204.

Saeed, A., Khan, M.A., Akram, U., Obidallah, W.J., Jawed, S., Ahmad, A. (2025). Deep learning based approaches for intelligent industrial machinery health management and fault diagnosis in resource-constrained environments. *Scientific Reports*, 15(1), 1114.

Sendak, M., Elish, M.C., Gao, M., Futoma, J., Ratliff, W., Nichols, M., Bedoya, A., Balu, S., O'Brien, C. (2020). "The human body is a black box" supporting clinical decision-making with deep learning. In *Proceedings of the 2020 Conference on Fairness, Accountability, and Transparency*. ACM, New York.

Singh, P.K., Gupta, P.K., Gupta, V., Singh, S., Akhtar, S. (2024). Making cancer imaging smarter: Emerging techniques and computational outlook to guide precision diagnostics. *The Open Bioinformatics Journal*, 17(1). doi: 10.2174/01187503 62344593241111114723.

Sirocchi, C. (2025). Infusing knowledge into data-driven modelling of complex systems for improved quality and interpretability. PhD Thesis, Università degli Studi di Urbino Carlo Bo, Urbino.

Song, H., Kim, M., Park, D., Shin, Y., Lee, J.G. (2022). Learning from noisy labels with deep neural networks: A survey. *IEEE Transactions on Neural Networks and Learning Systems*, 34(11), 8135–8153.

Srivastava, S. (2024). *Bio-Inspired Optimization for Medical Data Mining*. John Wiley & Sons, New York and Scrivener Publishing, Beverly, MA.

Steyaert, S., Pizurica, M., Nagaraj, D., Khandelwal, P., Hernandez-Boussard, T., Gentles, A.J., Gevaert, O. (2023). Multimodal data fusion for cancer biomarker discovery with deep learning. *Nature Machine Intelligence*, 5(4), 351–362.

Surya, N. and Someshwar, H.P. (2025). Low-cost telerehabilitation in low-and middle-income countries (LMICs): Overcoming barriers to access and improving healthcare delivery. *NeuroRehabilitation*, 56(1), 30–36. doi: 10538135241303349.

Thacharodi, A., Singh, P., Meenatchi, R., Tawfeeq Ahmed, Z.H., Kumar, R.R., Neha, V., Kavish, S., Maqbool, M., Hassan, S. (2024). Revolutionizing healthcare and medicine: The impact of modern technologies for a healthier future – A comprehensive review. *Health Care Science*, 3(5), 329–349.

Walsh, S., de Jong, E.E., van Timmeren, J.E., Ibrahim, A., Compter, I., Peerlings, J., Sanduleanu, S., Refaee, T., Keek, S., Larue, R.T. et al. (2019). Decision support systems in oncology. *JCO Clinical Cancer Informatics*, 3, 1–9. doi: 10.1200/CCI.18.00001.

Xiao, C., Choi, E., Sun, J. (2018). Opportunities and challenges in developing deep learning models using electronic health records data: A systematic review. *Journal of the American Medical Informatics Association*, 25(10), 1419–1428.

Yan, X., Wang, W., Xiao, M., Li, Y., Gao, M. (2024). Survival prediction across diverse cancer types using neural networks. In *Proceedings of the 2024 7th International Conference on Machine Vision and Applications*. ACM, New York.

10

AI-enabled Chatbots and Virtual Assistants in Oncology and Physiotherapy

Artificial intelligence (AI) has emerged as a transformative force in modern healthcare, revolutionizing how services are delivered, accessed and personalized. Among its most promising applications are AI-enabled chatbots and virtual assistants – intelligent conversational agents designed to simulate human interaction and support both patients and healthcare providers. This chapter investigates the growing integration of such technologies within oncology and physiotherapy, two domains that demand personalized, continuous and multidisciplinary care. In oncology, AI-driven virtual assistants facilitate symptom monitoring, medication adherence, appointment scheduling and emotional support, thereby easing the burden on clinical staff and empowering patients throughout their cancer journey. In physiotherapy, these tools assist with home-based rehabilitation by offering real-time feedback, exercise guidance and progress tracking, enhancing patient compliance and clinical outcomes. Moreover, AI chatbots are being increasingly used for patient education, mental health screening and triage, acting as accessible and always-available resources.

This chapter reviews state-of-the-art technologies and evidence supporting their use, including natural language processing (NLP), machine learning (ML) and voice recognition systems. It critically examines the benefits, such as scalability, cost-effectiveness and improved patient engagement, while also addressing pressing challenges, including ethical dilemmas, data privacy concerns, algorithmic biases and limitations in user trust and digital literacy. Adopting an interdisciplinary perspective, this chapter outlines future directions and emerging trends such as integration with electronic health records (EHRs), support for multiple languages and adaptive learning for personalized care. In doing so, it repositions AI-driven conversational tools from mere digital assistants to essential elements of a hybrid care model that helps bridge gaps in both oncology and physiotherapy.

Chapter written by Namrata KADAM, Rashmi GUDUR, Dhiraj Kumar MANE and Vivek DESHPANDE.

10.1. Introduction

10.1.1. *Background*

Healthcare systems worldwide are under mounting pressure due to aging populations, an increasing prevalence of chronic diseases and rising demands for personalized care. Specialties such as oncology and physiotherapy are particularly affected, as they require ongoing monitoring, multidisciplinary coordination and long-term patient engagement (Raj et al. 2020). Traditional healthcare delivery models often struggle to maintain consistent communication and support for patients, especially between clinical visits. In this context, artificial intelligence (AI)-enabled chatbots and virtual assistants have emerged as promising tools to extend healthcare services beyond conventional settings. These technologies can enhance patient engagement, streamline workflows and provide scalable support throughout the care continuum.

10.1.2. *Definitions and scope*

To frame the discussion, it is important to define the key technologies explored in this chapter:

– *Chatbots*: AI-powered software applications designed to simulate human conversation via text or voice (Karn et al. 2024). They typically handle predefined queries and offer information, reminders or basic interactions in a user-friendly manner.

– *Virtual assistants*: more advanced systems that use natural language processing (NLP), machine learning (ML) and contextual awareness to perform complex tasks, interact intelligently with users and provide personalized guidance. These assistants can analyze user input, adapt to individual needs and integrate with other digital health platforms (Jimenez et al. 2023).

These tools can be deployed across various interfaces – including smartphones, web-based platforms, wearable devices and electronic health records (EHRs) – to support functions such as symptom tracking, patient education, rehabilitation coaching, appointment scheduling and mental health support.

10.1.3. *Objectives*

This chapter aims to:

– analyze the applications of AI-enabled chatbots and virtual assistants in oncology and physiotherapy;

– evaluate their effectiveness in improving both clinical outcomes and patient experience;

– discuss current limitations, ethical and privacy concerns, and future prospects for the integration of these tools into standard care practices.

10.2. Technological foundation

10.2.1. *Natural language processing (NLP)*

Natural language processing (NLP) forms the backbone of conversational AI systems, enabling chatbots and virtual assistants to comprehend, process and generate human language (Babu and Akshara 2024). In healthcare applications, NLP allows these tools to interact meaningfully with patients and clinicians. Key advancements include:

– *Named entity recognition (NER)*: facilitates the identification of specific medical terms such as symptoms, medications, diagnoses and anatomical references from patient input. This is critical for symptom triage and clinical decision support.

– *Sentiment analysis*: detects emotional cues within a patient's responses, enabling the chatbot to adjust its tone, offer reassurance or escalate issues to human professionals when necessary – particularly important in oncology and mental health support (Lal et al. 2025).

– *Context-aware dialogue management*: maintains coherent, multi-turn conversations by recognizing context, tracking user intent and handling ambiguity. This feature allows for dynamic and patient-specific interaction, improving user engagement and the quality of care provided.

These NLP capabilities collectively enable chatbots to act not merely as scripted responders, but as adaptive conversational agents capable of delivering nuanced healthcare communication.

10.2.2. *Machine learning and knowledge bases*

Machine learning (ML) enhances the functionality and adaptability of AI-driven chatbots and virtual assistants (Inavolu 2024). Through supervised and reinforcement learning, these systems can refine their responses, personalize interactions and predict patient needs over time. Key elements include:

– *Predictive analytics*: ML models can assess risk factors, forecast treatment adherence or anticipate complications based on user behavior and input patterns.

– *Knowledge bases and medical ontologies*: integration with standardized resources such as SNOMED CT (Systematized Nomenclature of Medicine Clinical Terms), ICD-10 (International Classification of Diseases) and clinical guidelines ensures that chatbot responses are medically valid and aligned with evidence-based practice (Vuokko et al. 2023).

– *Continuous learning*: these systems can improve performance and expand functionality by analyzing data from user interactions, clinical feedback and real-world outcomes, making them increasingly reliable and relevant.

10.2.3. *Integration with EHRs and mHealth devices*

A critical enabler of personalized care is the integration of chatbots with electronic health records (EHRs) and mobile health (mHealth) devices. Such integration allows for:

– *Personalized responses*: access to patient history, comorbidities, medications and recent clinical data enables chatbots to tailor advice, reminders and interventions more precisely (Hindelang et al. 2024).

– *Real-time monitoring*: when linked with wearable devices or mobile sensors, chatbots can monitor vital signs, mobility metrics and rehabilitation progress. This data can inform feedback on exercise routines, detect anomalies or alert clinicians when thresholds are exceeded.

– *Closed-loop systems*: integration with EHRs facilitates bidirectional communication, allowing clinicians to input care plans and receive updates on patient progress from chatbot interactions, enhancing continuity and coordination of care (Chen et al. 2024).

10.3. Applications in oncology

10.3.1. *Patient education and decision support*

Oncology patients frequently face complex and emotionally charged information regarding their diagnosis, treatment options and potential outcomes. AI-enabled chatbots can play a critical role in improving health literacy and empowering patients to participate actively in their care (Anisha et al. 2024). These tools provide 24/7 access to understandable, consistent and personalized medical information:

– Cancer Chatbot offers structured, evidence-based responses to frequently asked questions on chemotherapy, radiation therapy, immunotherapy and prognostic expectations. Its interactive format allows patients to receive tailored information based on cancer type and stage.

– Virtual assistants such as IBM Watson for Oncology assist healthcare professionals by synthesizing massive datasets, including medical literature, clinical trial data and treatment guidelines, to provide evidence-based treatment recommendations. This aids oncologists in clinical decision-making while ensuring transparency and explainability in care delivery (Adeniran et al. 2024).

By improving both patient comprehension and clinician decision-making, these tools support shared decision-making and enhance the quality of oncology care.

10.3.2. *Symptom monitoring and reporting*

Managing symptoms during cancer treatment is crucial for patient safety and quality of life.

Conversational AI can facilitate real-time symptom tracking, prompt interventions and improve communication between patients and providers (Li et al. 2023):

– *Vivibot*: developed by Hopelab, this targets young adult cancer survivors and assesses emotional wellbeing using chatbot interactions. It also delivers coping strategies and resilience-building exercises.

– *Belong.life*: a comprehensive cancer navigation platform, this incorporates AI-driven chat features that allow users to report symptoms such as fatigue, nausea and pain. It enables timely alerts to healthcare teams and connects patients with community-based peer support.

Such systems reduce delays in addressing adverse effects and improve adherence to treatment regimens, contributing to better clinical outcomes and reduced hospital readmissions.

10.3.3. *Mental health and psychosocial support*

Cancer diagnosis and treatment are often accompanied by significant psychological distress, including anxiety, depression, isolation and fear of recurrence (Simonelli et al. 2017).

AI-based conversational agents are increasingly used to address these psychosocial needs:

– Chatbots can deliver cognitive–behavioral therapy (CBT) modules, guiding users through structured exercises to manage negative thoughts and emotions.

– They can also provide motivational dialogues, daily affirmations.

10.3.4. *Clinical trial navigation*

Enrolling in clinical trials is a critical yet often overlooked component of cancer care. Despite the availability of numerous studies, many patients either remain unaware of relevant trials or are unable to navigate the complex enrollment processes (Siembida et al. 2021).

AI-powered chatbots and virtual assistants are increasingly being leveraged to address these challenges, offering a more accessible and efficient pathway to trial participation.

– *Trial matching through eligibility algorithms*: one of the key functions of AI chatbots is the use of sophisticated algorithms to evaluate a patient's demographic details, diagnosis, stage of disease, prior treatments, and molecular or genetic markers. By processing this data against comprehensive clinical trial registries (e.g. ClinicalTrials.gov), the chatbot can identify studies that closely match the patient's profile (Chuan and Morgan 2020). This not only saves time for both patients and clinicians but also ensures a higher likelihood of eligibility and successful enrollment.

– *Clarification of informed consent procedures*: understanding clinical trial protocols and informed consent documents can be daunting, particularly for patients dealing with the emotional burden of a cancer diagnosis. Chatbots can break down complex medical language into layperson-friendly terms, guide users step-by-step through the consent process and answer frequently asked questions. This fosters transparency, reduces cognitive overload and helps patients make informed decisions about participation (Fossa et al. 2018).

– *Automated follow-up and compliance support*: once enrolled, chatbots serve as ongoing communication channels to remind patients about trial-related appointments, medication adherence, data submissions and protocol compliance. These reminders can be personalized based on treatment schedules and delivered via text, mobile apps or voice interfaces. Moreover, chatbots can log patient-reported outcomes and flag issues to study coordinators in real time.

– *Ethical and logistical benefits*: by reducing administrative workload and increasing patient awareness, chatbot-assisted trial navigation contributes to more equitable access to research opportunities, especially for underrepresented populations or those in remote areas (Laymouna et al. 2024). Additionally, it enhances the quality and reliability of trial data through improved adherence and timely feedback.

10.4. Applications in physiotherapy

10.4.1. *Remote exercise supervision*

Remote physiotherapy has become increasingly important due to the growing demand for home-based rehabilitation and the need to minimize hospital visits. AI-powered virtual assistants can effectively guide patients through prescribed exercise regimens using voice, video or text interfaces, ensuring continuity of care outside traditional settings.

– Kaia Health is a notable example, offering digital therapeutic programs for musculoskeletal conditions, particularly chronic back pain (Rohaj and Bulaj 2023). It uses AI-driven motion tracking via smartphone cameras to provide real-time feedback on exercise form and adherence.

– Other chatbot platforms are integrated with wearable sensors or mobile devices, enabling the detection of deviations in movement and the delivery of corrective prompts. These systems adapt exercises based on real-time performance metrics and patient-reported pain levels.

This level of interactivity not only enhances the accuracy and safety of home exercises but also empowers patients with autonomy in managing their recovery.

10.4.2. *Post-surgical rehabilitation*

Following surgeries such as joint replacements or spinal procedures, patients often require structured rehabilitation programs to restore mobility and function. AI-enabled tools assist in this critical phase by providing personalized support and monitoring:

– These tools track range of motion, pain intensity, swelling and overall progress using self-reported data, motion sensors or camera-based assessments.

– Virtual assistants deliver timely reminders to perform exercises or attend follow-up appointments and can automatically adjust exercise difficulty based on individual progress or setbacks (Maher et al. 2020).

– Importantly, they can recognize early warning signs – such as increased swelling, redness or fever – that may indicate infection or complications, triggering alerts to healthcare providers for timely intervention.

Such automation reduces the risk of post-operative complications and supports better outcomes through consistent, personalized rehabilitation.

10.4.3. *Patient motivation and adherence*

Adherence to physiotherapy is often hindered by a lack of motivation, perceived monotony of exercises, or logistical barriers.

Conversational AI tools play a key role in addressing these issues through behavioral reinforcement strategies:

– Daily check-ins help establish routine, encourage accountability and provide positive reinforcement.

– Gamification techniques, such as progress streaks, achievement badges and performance dashboards, make therapy more engaging and reward consistency (Petkevičius 2023).

– Some platforms include social encouragement features, such as peer comparison or community chat, which tap into the motivational benefits of social support and shared goals.

By integrating behavioral science into the user experience, these AI systems enhance long-term compliance and contribute to more effective physiotherapy outcomes.

10.4.4. *Appointment scheduling and communication*

Conversational AI significantly reduces the administrative burden associated with routine communication in physiotherapy clinics.

By automating scheduling-related tasks, chatbots and virtual assistants free up human staff to focus on clinical care:

– Patients can use chatbots to book, reschedule or cancel appointments in real time, often without waiting on hold or navigating complex phone systems.

– Virtual assistants can respond to frequently asked questions, such as clinic hours, billing inquiries or preparation instructions for sessions, enhancing accessibility and reducing delays (Benites et al. 2024).

– By providing automated follow-ups and reminders, these tools help maintain consistent engagement, minimize missed appointments and ensure continuity of care.

These efficiencies improve the overall patient experience and contribute to the scalability and responsiveness of physiotherapy services.

10.5. Case studies

10.5.1. *Wysa for oncology mental health*

Wysa is an AI-powered mental health chatbot that leverages cognitive–behavioral therapy (CBT) techniques, mindfulness exercises and empathetic conversation to support users dealing with psychological distress.

In the context of oncology, where emotional health is often compromised by diagnosis, treatment of side effects and uncertainty about the future, Wysa has emerged as a valuable complementary tool (Pavlopoulos et al. 2024):

– Clinical and observational studies have shown that Wysa contributes to reductions in anxiety, depression and stress symptoms, particularly among patients undergoing or recovering from cancer treatment.

– It facilitates the development of emotional coping strategies through guided self-help techniques, journaling prompts and structured behavioral exercises tailored to user input.

– Wysa's 24/7 availability ensures that patients can access support during periods of acute distress, including outside clinic hours, thus filling a critical gap in psycho-oncology services, especially where mental health professionals are scarce.

Importantly, Wysa does not replace professional counseling but functions as a scalable, cost-effective adjunct that empowers patients to manage their mental health with greater autonomy.

10.5.2. *Replika for cancer survivors*

Replika is an advanced, AI-driven chatbot designed to engage users in open-ended, emotionally intelligent conversation (Moylan and Doherty 2025).

Unlike clinical tools, Replika focuses on companionship and emotional expression, making it particularly appealing to individuals coping with the long-term psychosocial effects of cancer:

– For cancer survivors, Replika functions as a digital companion, offering an empathetic ear for discussing fears, hopes and existential concerns in a non-judgmental space.

– It has been observed to reduce feelings of loneliness and isolation, which are common among individuals who have completed treatment and may lack consistent social or emotional support.

– While not intended for diagnostic or clinical guidance, Replika excels at providing emotional validation and fostering self-reflection, which are essential for long-term psychological resilience.

The use of Replika underscores the value of non-clinical AI tools in addressing the nuanced emotional needs of patients navigating survivorship and recovery.

10.5.3. *Ada Health in early cancer symptom checking*

Ada Health is a globally recognized AI symptom checker that uses a probabilistic reasoning model, medical knowledge base and user-reported inputs to suggest potential conditions and recommend appropriate action (Liu et al. 2024):

– In oncology, Ada has shown promise in early detection by identifying symptoms that may warrant further medical investigation, such as persistent fatigue, unexplained weight loss or abnormal bleeding – often early indicators of malignancy.

– Ada provides structured guidance on whether a user should seek urgent care, consult a general practitioner or monitor symptoms over time, supporting timely diagnosis and reducing diagnostic delays.

– It enhances health literacy by offering clear explanations of potential conditions and related symptoms, empowering users to take proactive steps in managing their health.

10.6. Benefits and impact

AI-enabled chatbots and virtual assistants are transforming oncology and physiotherapy by delivering scalable, patient-centered care (Shaheen and Salim 2025). Their integration into clinical practice brings a host of benefits across accessibility, efficiency, personalization and cost-effectiveness.

10.6.1. *Increased accessibility*

AI chatbots bridge the gap for patients in rural or underserved areas who may have limited access to oncology or physiotherapy services. These virtual tools reduce the dependency on geographical proximity to specialists by offering remote consultations, exercise guidance, symptom tracking and emotional support. This is particularly valuable in regions where healthcare infrastructure is lacking or overburdened. Moreover, AI tools can be embedded in mobile applications, ensuring broad accessibility even via low-cost smartphones and basic Internet connectivity.

10.6.2. *24/7 availability*

Unlike human practitioners who have limited working hours, virtual assistants are always available, allowing for continuous monitoring, prompt responses and immediate interventions. This around-the-clock support is critical in oncology and rehabilitation, where timely management of symptoms, medication adherence and complication alerts can significantly influence outcomes (Crafoord 2023). Patients can interact with AI chatbots during non-clinical hours for reassurance or guidance, enhancing the sense of being continuously cared for.

10.6.3. *Personalized care*

By integrating electronic health records (EHRs), wearable data, imaging results and patient-reported outcomes, AI-driven assistants can deliver highly individualized care. These systems adapt recommendations based on patient preferences, cancer type and stage, physiotherapy progress and comorbidities. For example, an AI chatbot might modify exercise routines in real time based on pain levels or fatigue reports, or provide dietary tips aligned with a patient's treatment phase. This personalized approach improves adherence, satisfaction and overall therapeutic outcomes.

10.6.4. *Resource optimization*

Automating routine administrative and clinical tasks – such as appointment scheduling, follow-up reminders, basic triage, patient education and data collection – allows healthcare providers to focus on complex decision-making and patient interactions (Granger and Cavalheri 2022). This shift enhances workforce efficiency and enables more strategic allocation of medical resources. Hospitals and clinics can reduce operational costs by minimizing repetitive workload, avoiding unnecessary visits and improving workflow coordination. In oncology and physiotherapy settings, this results in faster service delivery, shorter waiting times and better patient throughput.

10.6.5. *Enhanced patient engagement and education*

AI-powered chatbots are effective tools for educating patients about their conditions, treatments and rehabilitation protocols. Through interactive conversations, visual aids and quizzes, they foster deeper understanding and empowerment. Educated patients are more likely to adhere to prescribed regimens, ask relevant questions during appointments and participate proactively in their care, ultimately leading to improved clinical outcomes and quality of life.

10.6.6. *Data-driven insights for continuous improvement*

The continuous interaction between patients and AI systems generates large volumes of data (Aminizadeh et al. 2024). This information can be analyzed to identify patterns, predict outcomes and inform personalized treatment strategies. In addition, it allows healthcare providers and administrators to monitor performance metrics, detect gaps in care delivery and refine service models. Over time, these insights contribute to a learning healthcare system that evolves with patient needs and emerging clinical evidence.

10.6.7. *Psychological support and companionship*

Virtual assistants equipped with natural language processing and sentiment analysis can offer emotional support, reduce anxiety and help combat the sense of isolation often experienced by cancer patients or individuals undergoing long-term physiotherapy. While they are not substitutes for mental health professionals, they can serve as a first line of support and alert clinicians when psychological intervention might be required.

10.7. Limitations and challenges

Despite the promise of AI-enabled chatbots and virtual assistants in oncology and physiotherapy, their integration into healthcare settings presents several significant challenges (Kaur 2024). These limitations must be carefully considered and addressed to ensure safe, ethical and equitable implementation.

10.7.1. *Data privacy and security*

Handling sensitive health information necessitates strict compliance with data protection regulations such as the General Data Protection Regulation (GDPR) in Europe and the Health Insurance Portability and Accountability Act (HIPAA) in the United States. AI chatbots collect and store a wide range of patient data – symptom logs, behavioral patterns, clinical histories – which can be vulnerable to breaches or unauthorized access. Mitigating these risks requires end-to-end encryption, robust authentication protocols, secure data storage systems and clearly defined data governance policies. Furthermore, transparency regarding what data is collected, how it is used and with whom it is shared is essential for building trust among users.

10.7.2. *Accuracy and reliability*

AI systems can only be as reliable as the data and algorithms they are built upon. Incorrect, outdated or biased information can lead to erroneous recommendations, potentially resulting in harm or mismanagement of care. In the context of oncology and physiotherapy, where treatment plans must be highly precise, even minor inaccuracies can have serious consequences (Ciardiello et al. 2014). To ensure reliability, several measures are necessary:

– *Regular updates of medical databases* to reflect the latest clinical guidelines, research findings and drug interactions.

– *Rigorous clinical validation* of algorithms through peer-reviewed studies, real-world testing and continuous performance audits.

– *Human oversight* to review and validate AI-generated outputs, especially in high-risk or ambiguous scenarios.

10.7.3. *Ethical concerns*

The deployment of AI in patient-facing roles introduces complex ethical issues that go beyond technical performance:

– *Transparency*: users must be clearly informed when they are interacting with an AI system rather than a human. Misrepresentation can undermine trust and informed consent.

– *Dependency*: over-reliance on chatbots may diminish valuable human interaction, potentially leading to emotional detachment, reduced empathy and missed psychosocial cues (Farbod 2024).

– *Consent and data usage*: explicit, informed consent must be obtained for data collection and usage. Patients should have control over their personal data, with options to opt out or delete their records.

– *Bias and equity*: AI systems trained on non-diverse datasets may perpetuate or exacerbate health disparities by providing less accurate or inappropriate guidance for certain populations.

10.7.4. *Technological barriers*

Not all patients have equal access to (or familiarity with) digital tools. Older adults, individuals with cognitive impairments, low digital literacy or physical disabilities may struggle to interact with AI chatbots. In oncology and physiotherapy, where users may already be dealing with fatigue, pain or emotional stress, usability becomes a critical factor. To overcome these barriers:

– *Interfaces should be designed with universal accessibility in mind*, including options for voice interaction, multilingual support, large text and intuitive navigation.

– *Training and support should be offered* to help patients and caregivers understand how to use these systems.

– *Offline and hybrid solutions* should be considered for areas with limited internet access.

10.7.5. *Integration with clinical workflows*

AI chatbots often operate in isolation from electronic health records (EHRs) or hospital management systems, leading to fragmented care (Sahithya et al. 2024). Seamless integration with existing healthcare infrastructure is essential for maintaining continuity of care and avoiding duplicated efforts. Achieving this integration can be complex due to compatibility issues, institutional resistance and regulatory barriers.

10.7.6. *Legal and regulatory hurdles*

There is currently a lack of uniform regulatory frameworks guiding the deployment and liability of AI tools in healthcare. In the event of an error or adverse event, accountability is often unclear: should it rest with the developer, the clinician or the institution? Legal uncertainties can hinder adoption and innovation unless clearly addressed by health authorities and lawmakers.

10.8. Future directions

10.8.1. *Multimodal interactions*

Combining voice, text, video and haptics will enhance engagement and inclusivity.

10.8.2. *Advanced emotional intelligence*

Emotionally aware AI that adapts tone, content and timing based on user sentiment will make interactions more human-like.

10.8.3. *Integration with AI imaging and diagnostics*

Future systems may:

– interpret imaging results (e.g. tumor size, bone density);

– provide feedback on rehabilitation posture using computer vision.

10.8.4. *Regulation and certification*

To build trust, regulatory bodies may need to establish frameworks to certify medical chatbots for specific clinical uses (Hauglid and Mahler 2023).

10.9. Conclusion

AI-enabled chatbots and virtual assistants are poised to reshape oncology and physiotherapy by enhancing patient engagement, supporting clinicians and improving outcomes. These intelligent systems can offer real-time responses, deliver personalized guidance, monitor patient progress and streamline administrative tasks – significantly alleviating the burden on healthcare professionals.

For patients, especially those managing complex and chronic conditions such as cancer, these tools can provide timely information, emotional support and adherence reminders, contributing to more consistent and empowered care experiences. However, the integration of these technologies into clinical practice must be approached with caution and responsibility. Their effectiveness depends not only on technical performance but also on rigorous clinical validation, adherence to privacy and data protection standards, and thoughtful consideration of ethical implications, such as bias mitigation and informed consent. Moreover, user-centric design is essential to ensure these tools are accessible, intuitive and culturally sensitive to meet the diverse needs of patients and practitioners alike.

Looking ahead, the continuous evolution of natural language processing, machine learning and multimodal AI promises even more capable and compassionate virtual assistants. With ongoing collaboration among technologists, clinicians, ethicists and patients, AI-enabled chatbots have the potential to become indispensable allies in delivering personalized, scalable and empathetic healthcare – bridging gaps in access, improving health equity and fostering a more connected care ecosystem in oncology and physiotherapy.

10.10. References

Adeniran, A.A., Onebunne, A.P., William, P. (2024). Explainable AI (XAI) in healthcare: Enhancing trust and transparency in critical decision-making. *World Journal of Advanced Research and Reviews*, 23, 2647–2658.

Aminizadeh, S., Heidari, A., Dehghan, M., Toumaj, S., Rezaei, M., Navimipour, N.J., Stroppa, F., Unal, M. (2024). Opportunities and challenges of artificial intelligence and distributed systems to improve the quality of healthcare service. *Artificial Intelligence in Medicine*, 149, 102779.

Anisha, S.A., Sen, A., Bain, C. (2024). Evaluating the potential and pitfalls of AI-powered conversational agents as humanlike virtual health carers in the remote management of noncommunicable diseases: Scoping review. *Journal of Medical Internet Research*, 26, e56114.

Babu, C.S. and Akshara, P.M. (2024). Revolutionizing conversational AI: Unleashing the power of ChatGPT-based applications in generative AI and natural language processing. In *Advanced Applications of Generative AI and Natural Language Processing Models*, Obaid, A.J., Bhushan, B., Muthmainnah, S., Rajest, S.S. (eds). IGI Global, Hershey, PA, 228–248.

Benites, J.A., Ramos, B.H., Trujillo, E.A. (2024). Optimization of appointment scheduling for hospitals based on a multilingual virtual assistant in an instant messaging application. In *IEEE 4th International Conference on Advanced Learning Technologies on Education & Research (ICALTER)*, Tarma.

Chen, Y., Lehmann, C.U., Malin, B. (2024). Digital information ecosystems in modern care coordination and patient care pathways and the challenges and opportunities for AI solutions. *Journal of Medical Internet Research*, 26, e60258.

Chuan, C.H. and Morgan, S. (2020). Creating and evaluating chatbots as eligibility assistants for clinical trials: An active deep learning approach towards user-centered classification. *ACM Transactions on Computing for Healthcare*, 2(1), 1–9.

Ciardiello, F., Arnold, D., Casali, P.G., Cervantes, A., Douillard, J.Y., Eggermont, A., Eniu, A., McGregor, K., Peters, S., Piccart, M. et al. (2014). Delivering precision medicine in oncology today and in future – the promise and challenges of personalised cancer medicine: A position paper by the European Society for Medical Oncology (ESMO). *Annals of Oncology*, 25(9), 1673–1678.

Crafoord, M.T. (2023). Individualised care for patients with breast or prostate cancer aided by an interactive app: A frame of process evaluation. Doctoral Thesis, Karolinska Institutet, Stockholm.

Farbod, S. (2024). Exploring the dark side of AI-enabled services: Impacts on customer experience and well-being. Master's Thesis, University of Twente, Twente.

Fossa, A.J., Bell, S.K., DesRoches, C. (2018). OpenNotes and shared decision making: A growing practice in clinical transparency and how it can support patient-centered care. *Journal of the American Medical Informatics Association*, 25(9), 1153–1159.

Granger, C. and Cavalheri, V. (2022). Preoperative exercise training for people with non-small cell lung cancer. *Cochrane Database of Systematic Reviews*, 9(9), CD012020.

Hauglid, M.K. and Mahler, T. (2023). Doctor chatbot: The EU's regulatory prescription for generative medical AI. *Oslo Law Review*, 10(1), 1–23.

Hindelang, M., Sitaru, S., Zink, A. (2024). Transforming healthcare through chatbots for medical history-taking and future directions: Comprehensive systematic review. *JMIR Medical Informatics*, 12(1), e56628.

Inavolu, S.M. (2024). Exploring AI-driven customer service: Evolution, architectures, opportunities, challenges and future directions. *International Journal of Engineering and Advanced Technology*, 13(3), 156–163.

Jimenez, J., Del Rio, A., Berman, A.N., Grande, M. (2023). Personalizing digital health: Adapting health technology systems to meet the needs of different older populations. *Healthcare*, 11(15), 2140.

Karn, A., Singh, P.K., Agarwal, C., Verma, A., Singh, D., Kumari, M. (2024). Unraveling the power of AI assistants. In *Advances in AI for Biomedical Instrumentation, Electronics and Computing*, Sachan, V., Malik, S., Gautam, R., Kumar, P. (eds). CRC Press, Boca Raton.

Kaur, J. (2024). FutureCare: AI robots revolutionizing health and healing. In *Revolutionizing the Healthcare Sector with AI*, Singla, B., Shalender, K., Stamer, K. (eds). IGI Global, London.

Lal, M., Neduncheliyan, S., Kaushik, A., Goswami, A. (2025). Healthcare chatbots using artificial intelligence and sentiment analysis. In *Human-Centric AI in Digital Transformation and Entrepreneurship*, Misra, S., Kaushik, M., Jain, A. (eds). IGI Global, London.

Laymouna, M., Ma, Y., Lessard, D., Schuster, T., Engler, K., Lebouché, B. (2024). Roles, users, benefits, and limitations of chatbots in health care: Rapid review. *Journal of Medical Internet Research*, 26, e56930.

Li, Y., Liang, S., Zhu, B., Liu, X., Li, J., Chen, D., Qin, J., Bressington, D. (2023). Feasibility and effectiveness of artificial intelligence-driven conversational agents in healthcare interventions: A systematic review of randomized controlled trials. *International Journal of Nursing Studies*, 143, 104494.

Liu, V., Kaila, M., Koskela, T. (2024). Triage accuracy and the safety of user-initiated symptom assessment with an electronic symptom checker in a real-life setting: Instrument validation study. *JMIR Human Factors*, 11, e55099.

Maher, C.A., Davis, C.R., Curtis, R.G., Short, C.E., Murphy, K.J. (2020). A physical activity and diet program delivered by artificially intelligent virtual health coach: Proof-of-concept study. *JMIR mHealth and uHealth*, 8(7), e17558.

Moylan, K. and Doherty, K. (2025). Expert and interdisciplinary analysis of AI-driven chatbots for mental health support: Mixed methods study. *Journal of Medical Internet Research*, 27, e67114.

Pavlopoulos, A., Rachiotis, T., Maglogiannis, I. (2024). An overview of tools and technologies for anxiety and depression management using AI. *Applied Sciences*. 14(19), 9068.

Petkevičius, M. (2023). The impact of gamification elements on the intention to use health related applications. Doctoral Dissertation, Vilnius University, Vilnius.

Raj, V.S., Pugh, T.M., Yaguda, S.I., Mitchell, C.H., Mullan, S.S., Garces, N.S. (2020). The who, what, why, when, where, and how of team-based interdisciplinary cancer rehabilitation. *Seminars in Oncology Nursing*, 36(1), 150974.

Rohaj, A. and Bulaj, G. (2023). Digital therapeutics (DTx) expand multimodal treatment options for chronic low back pain: The nexus of precision medicine, patient education, and public health. *Healthcare*, 11(10), 1469.

Sahithya, B., Guru Prasad, M.S., Sahithi, B., Madhu Krishna, K., Devarlla, A.C., Yashavanth, T.R. (2024). Empowering healthcare with AI: Advancements in medical image analysis, electronic health records analysis, and AI-driven chatbots. In *3rd International Conference for Innovation in Technology (INOCON)*. IEEE, Bengaluru.

Shaheen, R. and Salim, H. (2025). Role of artificial intelligence in healthcare. In *Intersection of Human Rights and AI in Healthcare*, Shaheen, R. (ed.). IGI Global, Hershey, PA, 173–200.

Siembida, E.J., Loomans-Kropp, H.A., Tami-Maury, I., Freyer, D.R., Sung, L., Crosswell, H.E., Pollock, B.H., Roth, M.E. (2021). Barriers and facilitators to adolescent and young adult cancer trial enrollment: NCORP site perspectives. *JNCI Cancer Spectrum*, 5(3), pkab027.

Simonelli, L.E., Siegel, S.D., Duffy, N.M. (2017). Fear of cancer recurrence: A theoretical review and its relevance for clinical presentation and management. *Psycho-Oncology*, 26(10), 1444–1454.

Vuokko, R., Vakkuri, A., Palojoki, S. (2023). Systematized nomenclature of medicine – Clinical terminology (SNOMED CT) clinical use cases in the context of electronic health record systems: Systematic literature review. *JMIR Medical Informatics*, 11, e43750.

11

AI-driven Sensor Technologies in Physiotherapy and Oncology: Transforming Rehabilitation Through Intelligent Biomechanical Monitoring

The integration of biomedical sensors into physiotherapy has revolutionized the assessment and rehabilitation of human movement. These sensors, both invasive and noninvasive, measure critical biomechanical parameters such as force, pressure and motion, enabling physiotherapists to conduct accurate diagnostics and monitor rehabilitation progress. Traditional silicon-based sensors such as accelerometers, gyroscopes and magnetometers form the backbone of wearable technologies, enabling real-time tracking of body movement. Complementing these are flexible sensors, including textile-based strain and pressure sensors, offering superior comfort and adaptability for patients.

The chapter also explores biophysical measurements, emphasizing the role of force and pressure sensors in evaluating muscle strength, joint stability and gait dynamics. Modern physiotherapy uses advanced tools such as digital goniometers, tilt sensors and inertial measurement units (IMUs) for comprehensive joint mobility assessments. Additionally, artificial intelligence (AI) is being integrated with sensor technologies to enhance rehabilitation outcomes. AI enables predictive analysis, personalized treatment planning and real-time biofeedback, thereby improving patient motivation and safety. The synergy between AI and sensor data facilitates early detection of abnormalities, risk prevention and continuous remote monitoring. Together, these innovations ensure precision-driven, patient-centered physiotherapy interventions.

This chapter underscores the importance of ongoing research and technological advancement in developing smarter, safer and more efficient rehabilitation tools tailored to individual patient needs.

Chapter written by G. VARADHARAJULU, Radhika CHINTAMANI and G. HIMASHREE.

For a color version of all of the figures in this chapter, see: www.iste.co.uk/kumar/ai-driven1.zip.

11.1. Introduction

Physiotherapy research frequently uses biomedical sensors to measure a variety of biomechanical factors. Often, they are called biomedical transducers and are the main building blocks of diagnostic medical instrumentation found in physiotherapists' and biomedical engineers' laboratories (Jones 2010).

The need for accurate medical diagnostic procedures places stringent requirements on the design and use of biomedical sensors (Jones 2010). Usually, the first step in developing a biomedical sensor is to access the accuracy, sensor operating range, response time, sensitivity, resolution and reproducibility of the sensor. Once the need of the device is met, it becomes easy to choose among various available sensors and ensure that the measurement remains sensitive, stable, safe and cost-effective (Enderle and Bronzino 2012).

Sensors also provide a non-hindrance assessment of human kinematic motion and motion status during exercise training or during rehabilitation. Nevertheless, there is a significant lack of studies on motion monitoring index systems, comprising rigid and/or flexible sensors, with a well-designed wearable system along with different angles of joint motion, especially for gross trunk mobility, complex joint movements and limb movements (Jones 2010; Enderle and Bronzino 2012).

11.1.1. *Sensor classification*

Depending on their particular applications, sensors are usually categorized as physical, electrical or chemical based on the quantity to be measured (Jones 2010). Biosensors are a special subclassification of sensors that have two distinct components:

1. a biomechanical recognition element that serves as a mediator and offers the selectivity required to detect the mechanical component of interest, such as movement, range of motion, depth or muscle strength;

2. a supporting structure that is in close proximity to the biomechanical component and serves as a transducer.

The transducer's function is to convert the mechanical reaction into visual, electrical or physical signals proportional to the concentration of particular biomechanics (Enderle and Bronzino 2012).

Both invasive and noninvasive sensors used by biomedical engineers will be covered in this chapter.

Physiotherapy and rehabilitation can make use of a variety of mechanical sensors. Every sensor serves a distinct purpose, either supporting or carrying out the biomechanical processes of the human body. Physiotherapy equipment, investigative or therapy related, is mostly noninvasive with the sensor being attached to the external body part.

Thus, sensors can be classified into several categories (Bowman and Meindl 1986; Mendelson 2005; Jones 2010; Enderle and Bronzino 2012):

1. On the basis of invasiveness:

   a. invasive sensors, which are inside the body;

   b. noninvasive sensors, which are outside the body.

2. On the basis of function:

   a. *active sensors* require an external power source to function, for example, accelerometers, pressure sensors, light sensors, temperature sensors and sound sensors;

   b. *passive sensors* are known to usually generate their own electric signal to function. Examples include vibrators, electric field sensors and infrared sensors;

   c. *analog sensors* produce a continuous output signal or measurement;

   d. *digital sensors* show the reading digitally on the device itself.

11.1.2. *Sensor packaging*

Whenever the sensor comes into contact with body fluids, the host itself may affect the sensor's function, or the sensor may affect the site in which it is implanted. Hence, the materials used in the construction of the sensor's outer body must be non-thrombogenic and nontoxic since they play a critical role in determining the overall performance and longevity of an implantable sensor (Bowman and Meindl 1986; Sani et al. 2021).

11.2. Sensors for monitoring human motion

According to Huang et al. (2023) motion monitoring sensors can be classified on the basis of:

1. a human motion gesture monitoring index system;
2. sensor characteristics;
3. system design.

Physiological signals, such as electromyography (EMG) and electrocardiography (ECG), can also identify muscle contractions and heart contractions, the latter of which is a muscle. They do, however, function as a less accurate indirect method for gross motor training, particularly when it comes to electromyography. Recent developments in EMG have suggested the use of needle electrodes, which provide a detailed analysis of the fiber recruitment method. However, it becomes challenging to use the voltage and frequency patterns of muscle activity to determine the precise motion amplitude or angle, particularly in a 3D gross motor assessment.

11.2.1. *Traditional silicon-based sensors*

– *Accelerometers*: devices that can measure the acceleration of the part to which it is attached. According to an accelerometer's measurement principle, there are two types:

1. linear accelerometers, which detect linear acceleration;
2. angular accelerometers, which detect angular acceleration.

Accelerometers use their electrical impulses to identify deflection or stress variances from external acceleration. Gait monitoring, human motion analysis, fall detection and energy consumption estimation are all common uses for these. Three-axial accelerometers, which have an accuracy of ± 2 g and a sampling rate of 50–100 Hz, are currently commonly used. These days, most wearable systems are made up of triaxial accelerometers, which are frequently positioned on the body's moving parts, such as joints and limbs (Huang et al. 2023).

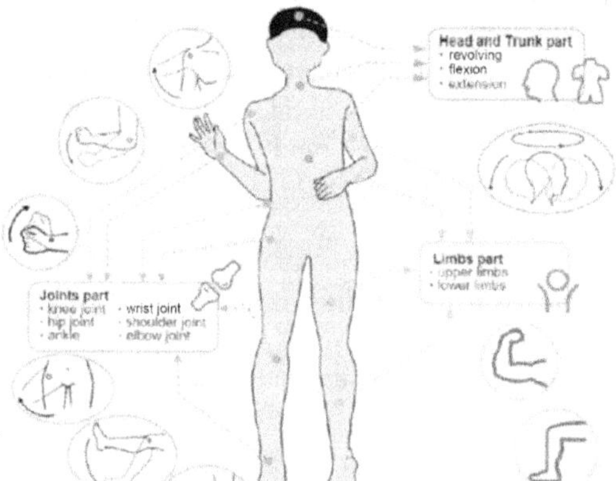

Figure 11.1. *Diagram of an accelerometer showing its components and basic working principle*

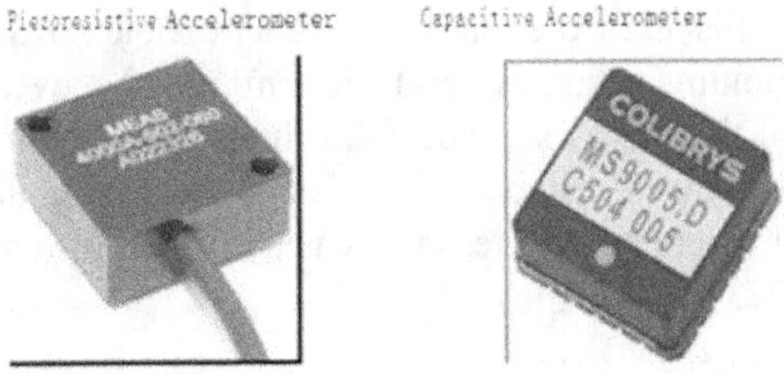

Figure 11.2. *Two basic types of accelerometers commonly used in prototype builds*

The above images are two basic types of accelerometers commonly used in prototype builds (Niu et al. 2018).

– *Gyroscope*: a sensor that uses the Coriolis force to measure the angular variance or angular rate. Accelerometers and gyroscopes are frequently used in tandem to detect the angular velocity of human motion in order to reduce error and increase monitoring accuracy. There are many gyroscopes available on the market, for example, L3G4200D, IMUZ, 3DM-GX3-25 and ISM330DHCX, which are multiaxial with the minimum axis being 3. The application area of what they are validated for differs variably from forearm to upper arms, waist and trunk (Schiefer et al. 2011).

Figure 11.3. *Example of a gyroscope sensor for measuring angular variance in human motion*

– *Magnetometer*: this uses the Earth's magnetic field to detect motion. Static motion yields maximum accuracy and determines the wearer's direction by measuring the flux of the local magnetic field. In order to rectify the heading angle offset brought about by gyroscope drift, accelerometers and gyroscopes are frequently combined to create an inertial measurement unit, or IMU (Huang et al. 2023; Zhang et al. 2023).

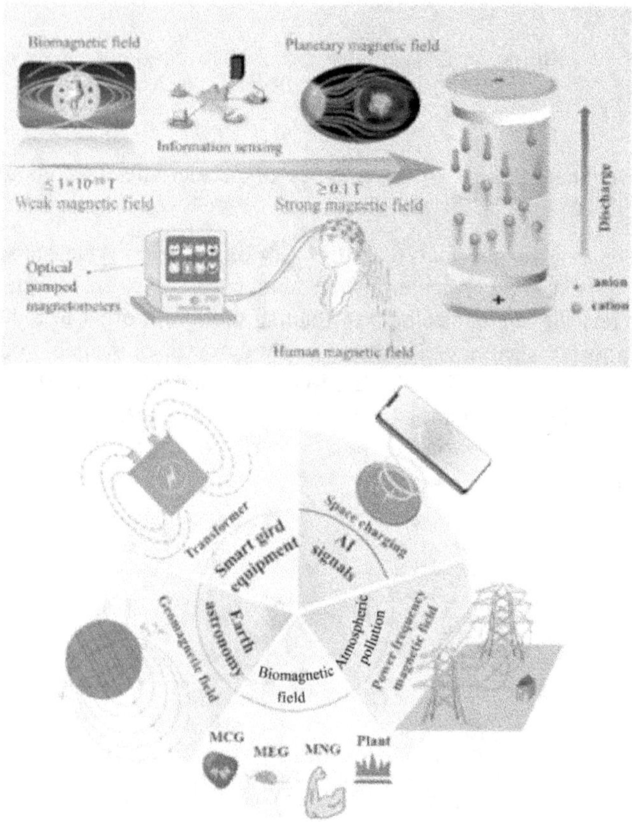

Figure 11.4. *Schematic of a magnetometer used for detecting orientation based on Earth's magnetic field*

– *Inertial sensors*: accelerometers, gyroscopes and magnetometers form an inertial measuring unit called an IMU. IMUs ensure accurate monitoring measurements, especially movements with three or more axes. The most popular inertial sensors for wearable technologies are made by InvenSense, Bosch Sensortec and Shimmer in Germany, and they are typically nine-axis (Ahmadi et al. 2015; Huang et al. 2023).

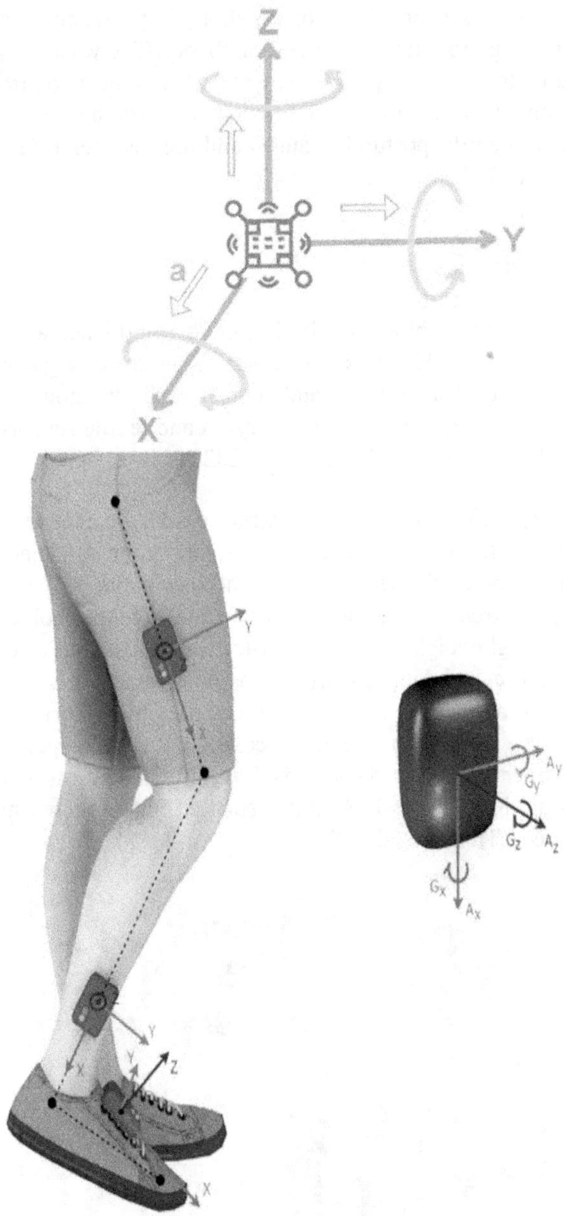

Figure 11.5. *An inertial measurement unit (IMU) combining accelerometer, gyroscope and magnetometer sensors*

– *Tilt sensors*: these sensors use a combination of gyroscopes and accelerometers and are based on the gravitational acceleration theory. The wearer's acceleration can be monitored using three-axial systems (x, y and z), and the angle between each axis and the corresponding acceleration of gravity can then be calculated. These are frequently used to identify postural changes and are also referred to as inclination sensors (Huang et al. 2023).

11.2.2. *Flexible sensors*

Excellent stretchability, biocompatibility, good compliance and high sensitivity are all included in them. The flexible sensor is used for augmented and virtual reality, human–computer interaction and human mobility monitoring and can be separated into two main groups: thin-film sensors and textile sensors, both of which have pressure and strain sensors (Huang et al. 2023).

– *Substrates of flexible sensors*: these textile sensors are composed of a variety of materials, including fabrics, yarns and fibers. In order to detect materials and maintain posture, the textile sensor is typically composed of materials that are omnidirectional, extremely elastic, permeable, hydrophobic, long-lasting, biocompatible and lightweight. At first, textiles were only used as carriers, with the rigid parts of electrical gadgets incorporated in or attached to them. Later, rigid microelectronic components combined with flexible fabric devices gained popularity, creating a heterogeneous technology that helped create flexible electronic components. These days, this hybrid technology is widely used to measure human motion. Figure 11.6 shows such an image representing the sensors used on a daily basis (Huang et al. 2023).

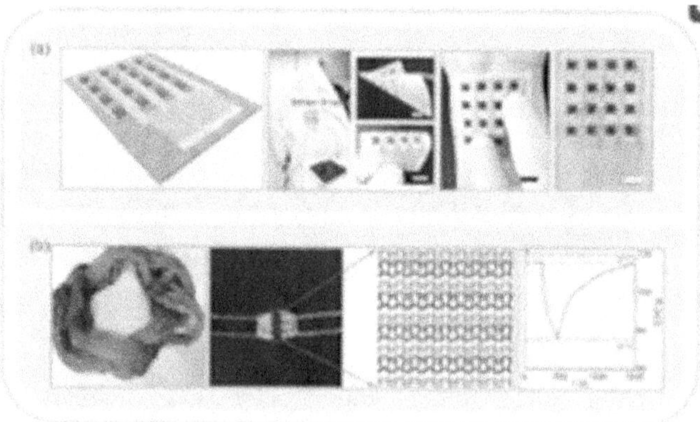

Figure 11.6. *Example of flexible textile-based sensors integrated into daily-use garments (Huang et al. 2023)*

– *Flexible strain and pressure sensors*: strain sensors translate a mechanical motion's degree of deformation – the alteration in the structure's shape upon the application of a load – into electrical signals. There are three categories in which these fall:

1. resistive;
2. capacitive;
3. voltage-based.

Current research suggests that strain sensors are commonly used to detect changes in the body's position and for kinesthetic monitoring (Huang et al. 2023).

Pressure sensors translate the pressure into an electrical signal, in contrast to strain sensors. They can be used extensively to detect finger motion, including pressure, twisting, extension and flexion force, and they have a detecting capacity of above 10 kPa. Joint motion recognition is another application for these sensors. These are now used for plantar pressure distribution during stance and gait monitoring. They are further illustrated in Figures 11.7 and 11.8 (Huang et al. 2023).

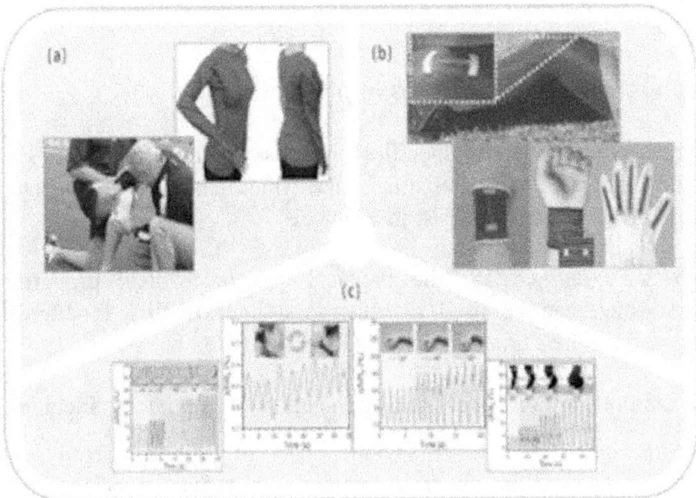

Figure 11.7. *Diagram of a flexible strain sensor translating mechanical deformation into electrical signals*

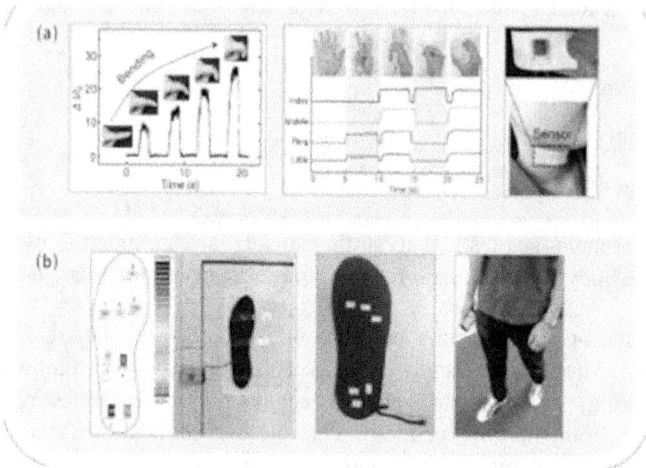

Figure 11.8. *Diagram of a flexible pressure sensor for detecting applied force and pressure distribution*

11.3. Biophysical measurements in physiotherapy

These measurements study specific functions which change over time. These changes can yield significant statistical data, which can be used to determine the effectiveness of the therapy or assist in diagnosis.

Biophysics is defined as "the physical and mechanical properties of the cells/tissues which can be used to either assess or modify functioning of the cells/tissues and can be studied in detail" (Jones 2010).

Some examples of biophysical measurements in physiotherapy include:

– *Electrotherapy*: these form a major part of adjunct therapy, where the electrotherapeutic modalities are used to treat pain, reduce swelling, inflammation and post-therapy soreness, assess the physiological characteristics of nerves by nerve conduction velocity, faradic curve and various other methods. The modalities preferred in electrotherapy are majorly noninvasive. The basic mechanism of electricity conduction in these modalities are listed in Table 11.1 (Bracciano 2024).

Parts	Function
Modality	Generates the current in required frequency.
Cables	Pass the current from the modality/system to the electrode (the contacting part of the machine to the body).
Electrodes	These are commonly made up of rubber or metal in order to conduct the electric current from the machine to the part in contact.
Aqua Gel	Decreases the resistance between the electrode and the skin so that the current passes smoothly.

Table 11.1. *Components of electrotherapy modalities and their functions in physiotherapy*

– *Physical therapists*: these can use electrical currents to stimulate muscles, which can help to reduce swelling, improve movement and maintain muscle tone and strength.

– *Ultra short-wave therapy*: research in this area has improved understanding of how high frequency energy is distributed and converted in the body. This research has led to the investigation of external conditions, such as the size, shape and arrangement of electrodes, as well as the distribution of energy within the body (Nazar and Tahir 2020).

Noninvasive sensors are most commonly used in physiotherapy. The biopotential measurements required in physiotherapy are as follows: force measurement (load or weight); muscular force measurement; joint movement measurement; fascia pliability measurement; response or reaction time measurement.

A few sensors along with their biomechanical measurements and uses are mentioned below. Yet, there are many biomechanical functions which can be developed or are under-development, necessitating further studies.

11.3.1. *Force measurement*

Force measurement is crucial for physiotherapists as it provides objective data to assess muscle strength, joint function and biomechanical performance during rehabilitation. By measuring the force exerted during exercises, physiotherapists can track a patient's progress, ensure safe load-bearing and adjust rehabilitation plans accordingly. Force measurement helps prevent re-injury by avoiding overloading or underloading the affected area, guiding exercise progression and ensuring the rehabilitation process is tailored to the patient's specific needs, promoting optimal recovery and functional outcomes.

Force measurement during gait is crucial in physiotherapy as it helps assess how weight is distributed across the joints and muscles while walking. This data enables physiotherapists to identify movement abnormalities, such as uneven load-bearing or muscle weakness, and tailor rehabilitation programs to address these issues. By monitoring gait forces, therapists can ensure more effective recovery and prevent re-injury, promoting better mobility and overall function.

Biomechanically, gait is classified into two major phases: swing and stance phases.

The stance phase is where the lower extremity touches the ground and the swing phase is where the lower extremity is off the ground. The main biomechanical analysis of gait in either phase is:

– *Stance phase*: area of weight bearing of the foot and angular motions of hip-knee and angle joints along with spine and upper extremities, in addition to muscle recruitment strategy for stability.

– *Swing phase*: angular motions of hip-knee and ankle joints along with spine and upper extremities, in addition to muscle recruitment strategy for mobility (Guerra et al. 2020).

Thus, as stated, the area of weight bearing is given more importance in stance phase than swing. The quantitative measurement of the amount of force induced by the muscle during stance and swing phases varies as the movements of the joint differ. Thus, quantitative assessment becomes equally important along with qualitative muscle recruitment strategy. Thus, muscular force measurement becomes important in gait.

An example of sensors which detect the muscular force during static posture or dynamic movements is electromyography sensors.

11.3.1.1. *Working principles (Zheng et al. 2022)*

Electromyography is a device which uses electrodes to pick up electrical signals from the muscle during its contraction and sends the impulse to the machine. The machine records the signals in detail and projects the information in graph form as an electromyograph. The signals are detected by the electrical pads placed on the area of contact through a velcro strap. Aqua Gel is placed between the electrode and the skin in order to reduce artifaction. Hence, these electric pads are called EMG sensors. Muscular contraction generates an electrical signal, or EMG signal, that can be detected by the EMG sensor.

The recordings vary according to the phases of EMG. Figure 11.9 shows the diagrammatic representation (Wang et al. 2021).

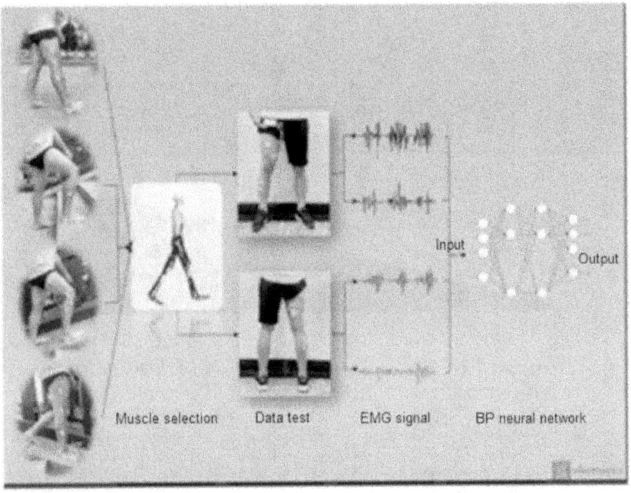

Figure 11.9. *Electromyography (EMG) signal patterns during different muscle contraction phases (Wang et al. 2021)*

Figures 11.10–11.12 demonstrate the position of the selected muscle and the reference electrodes (Wang et al. 2021).

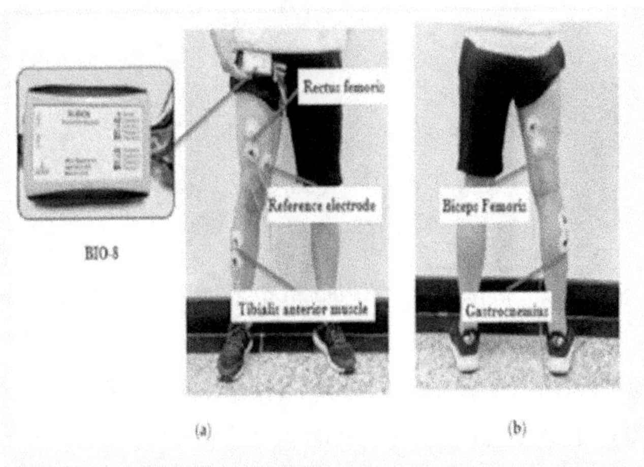

Figure 11.10. *Positioning of selected muscle and placement of reference electrodes for EMG measurement*

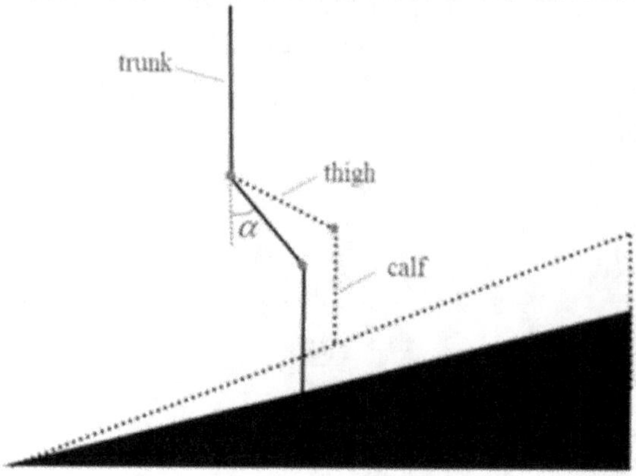

Figure 11.11. *Placement of EMG electrodes for upper or lower limb muscle activity detection*

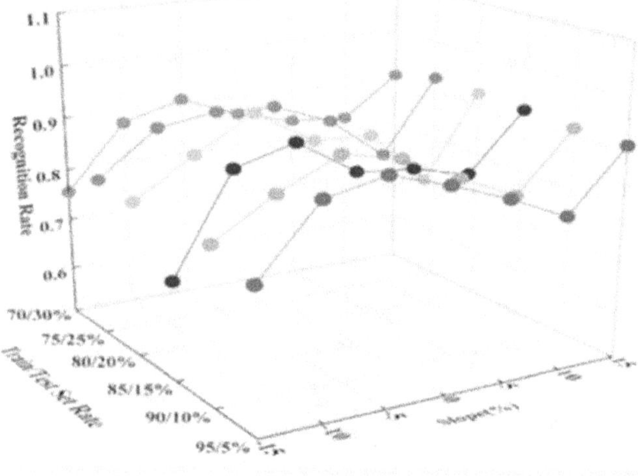

Figure 11.12. *Example of EMG electrode placement for gait analysis*

11.3.1.2. *Role of force measurement sensors in physiotherapy*

Because they provide precise, real-time data that aids physicians in evaluating, tracking and optimizing the recovery process, force measurement sensors are essential to physiotherapy. Both the patient and the physiotherapist can benefit from these sensors' ability to monitor the force or pressure used during particular movements or exercises. They support physiotherapy in the following ways:

1. *Objective assessment of force and strength*: force sensors enable physiotherapists to accurately quantify the strength or force a patient generates during exercises, joint movements or muscle contractions, offering a more objective and precise measurement compared to subjective evaluations such as verbal descriptions of pain or effort (Hollander et al. 2021). For instance, they can assess muscle strength by measuring the force exerted during exercises, helping identify weaknesses critical for conditions such as stroke rehabilitation, orthopedic recovery or sports injuries. Additionally, force sensors can evaluate joint stability by assessing the load a joint can bear, which is essential for tracking recovery after surgery or injury.

2. *Personalized treatment plans*: with detailed force data, physiotherapists can tailor treatment plans based on the individual's specific strengths and capabilities. Sensors can provide insights into which exercises the patient struggles with, or where they may be overexerting themselves (Rossi and Gatti 2020). This helps adjust the intensity or focus of rehabilitation programs for more effective outcomes.

3. *Monitoring progress*: force measurement sensors track a patient's progress over time by documenting force values during various movements or exercises, which is especially useful in rehabilitation after surgery or injury. For example, after knee surgery, sensors can monitor improvements in the patient's ability to bear weight or apply force through the leg (Ng and Lee 2018). They are also valuable in managing chronic conditions such as arthritis or neurological disorders, as they help assess changes in muscle strength and joint function. These data allow physiotherapists to make informed adjustments to the therapy plan based on the patient's progress or setbacks.

4. *Preventing overexertion and injury*: pushing the body too hard or too rapidly is one of the main hazards in rehabilitation. By preventing patients from going beyond their limit, force sensors reduce the chance of additional harm (Almeida and Vieira 2020). Therapists can help patients do exercises safely and within their physical limits by establishing force limits and monitoring real-time effort.

5. *Feedback for patient motivation and engagement*: real-time feedback from force sensors can be provided to patients, allowing them to understand how much force they are exerting and how they are progressing in their rehabilitation. This immediate feedback helps boost motivation, as patients can see tangible improvements, such as increased force or pressure tolerance during specific exercises (Kim and Lee 2019).

6. *Improved biomechanical analysis*: for more complex assessments, force sensors can be integrated into wearable devices or mats to track the distribution of force across the body during activities such as walking, standing or other

movements. This integration helps therapists analyze gait abnormalities by revealing how force is distributed, which can highlight posture or alignment issues that can be addressed through targeted therapy (Ding et al. 2020). Additionally, in balance training, sensors assess the force exerted by each limb or on different sides of the body, helping to develop a balanced rehabilitation approach and improve coordination.

7. *Evaluation of functional movements*: force sensors can be integrated into tools such as pressure mats, force plates or wearable devices to evaluate functional movements (e.g. squatting, standing up from a chair, etc.) (Molenaar et al. 2015). They help identify whether the patient is compensating during movements or engaging muscles inefficiently, which is useful for both rehabilitation and performance optimization.

8. *Posture and alignment assessment*: improper posture can lead to muscle imbalances and chronic pain. Force sensors can help measure how the body distributes weight while standing or sitting, identifying any asymmetries or improper alignments (Cavagna and Bianchi 2016). These data help create a treatment plan aimed at correcting posture and preventing future musculoskeletal issues.

11.3.1.3. *Artificial intelligence integration in force measurements*

– *Objective assessment of force and strength:*

- *AI integration*: machine learning models analyze real-time data from force sensors to identify specific patterns in force output, helping to quantify muscle strength and joint stability.

- *Basic AI algorithm*: linear regression or support vector machines (SVMs) can be used to model and predict muscle strength based on force data (Rana et al. 2020).

– *Personalized treatment plans:*

- *AI integration*: AI analyzes the force sensor data over time to understand the patient's capabilities, detecting weaknesses or areas of overexertion. This helps tailor the therapy to each individual's specific needs.

- *Basic AI algorithm*: decision trees or K-means clustering can group patients based on force output patterns, recommending personalized exercises or intensity levels for each group (Soni and Gupta 2020).

– *Monitoring progress:*

- *AI integration*: AI algorithms track force measurements over time, comparing them to baseline data to assess progress, improvements or setbacks in rehabilitation.

- *Basic AI algorithm*: time-series analysis (using algorithms such as ARIMA) for continuous tracking of force values and identifying trends over time (Vasudevan and Arjunan 2019).

– *Preventing overexertion and injury:*

- *AI integration*: AI predicts potential risks by analyzing the real-time force exertion data and comparing it to safe limits, adjusting intensity to prevent overexertion.

- *Basic AI algorithm*: anomaly detection algorithms such as isolation forests or K-means clustering can help detect when a patient is approaching unsafe force thresholds (Reddy and Suresh 2020).

– *Feedback for patient motivation and engagement:*

- *AI integration*: AI provides real-time feedback based on force sensor data, presenting patients with progress visualizations or motivational messages, and adjusting the difficulty of exercises based on performance.

- *Basic AI algorithm*: reinforcement learning (RL) could be used to adjust exercise intensity dynamically based on patient performance and feedback (Singh and Gupta 2021).

– *Biomechanical analysis:*

- *AI integration*: AI analyzes force distribution across different parts of the body, identifying abnormal gait patterns or imbalances that need attention.

- *Basic AI algorithm*: deep learning (convolutional neural networks, CNNs) can process sensor data from wearable devices to detect gait abnormalities or posture misalignments (Venkatesh and Sharma 2020).

– *Evaluation of functional movements:*

- *AI integration*: AI uses sensor data to assess whether a patient is compensating or performing movements inefficiently, such as during squats or standing exercises.

- *Basic AI algorithm*: recurrent neural networks (RNNs) or long short-term memory (LSTM) networks can analyze time series data from force sensors to evaluate functional movement patterns (Prakash and Ghosh 2021).

– *Posture and alignment assessment:*

- *AI integration*: AI algorithms detect weight distribution imbalances or misalignments in posture, helping to correct alignment and prevent musculoskeletal issues.

- *Basic AI algorithm*: random forests or SVMs can be used to classify posture alignment based on sensor data and recommend corrective actions (Kumar and Singh 2020).

11.3.2. *Pressure measurement*

Pressure measurement in physiotherapy is essential for assessing and optimizing a patient's biomechanical performance during rehabilitation. By measuring pressure distribution across the body, such as under the feet, back or other body parts, it means physiotherapists can analyze how forces are transmitted through the body during movements such as standing, walking or exercising. This helps identify areas of excessive pressure or insufficient load-bearing, which can indicate improper posture, misalignment or muscle imbalances. For example, an uneven distribution of pressure across the feet may suggest gait abnormalities, while concentrated pressure on certain body parts can indicate incorrect posture or compensation patterns. Real-time feedback from pressure sensors helps patients adjust their movement patterns, ensuring that forces are applied efficiently, minimizing the risk of injury and improving overall biomechanics (Patel and Patel 2020). Pressure data also assists in tracking the progress of joint mobility or strength after injury or surgery, providing insight into how the body is adapting to stress over time. In cases of conditions such as stroke, arthritis or paralysis, pressure measurement can help identify areas at risk for pressure ulcers or sores, prompting early interventions to prevent such injuries. By enabling precise, data-driven adjustments to rehabilitation programs, pressure measurement helps improve movement efficiency, stability and comfort, ultimately promoting safer and more effective recovery.

11.3.2.1. *Pressure sensors*

A sensor is an apparatus that converts a physical quantity into a signal. For example, the quantity may be temperature, length, force or, of course, pressure. Although it can also be optical, the signal is often electrical. It is a device that measures the real pressure applied to the sensor using a pressure-sensitive element and transforms the data into an output.

11.3.2.2. *Working principles (Homayounfar et al. 2021)*

There are a wide variety of different technologies used within pressure sensors to provide accurate results. The following section will highlight some of these:

– Additionally, strain gauge-based pressure sensors use a pressure-sensitive element on which thin-film gauges are sputter-applied or metal strain gauges are adhered. This measuring component can be a diaphragm or, in the case of metal foil gauges, measuring bodies of this kind. The capacity to assess the greatest pressures and enhanced rigidity are the main benefits of this monolithic can-type design.

– Capacitive pressure sensors can create a variable capacitor using a diaphragm and pressure cavity. Applying pressure causes the diaphragm to distort, which lowers capacitance. This electrically measurable change in capacity is then set in response to the pressure that is applied. Only low pressures can be detected by these sensors.

– In order to detect strain caused by applied pressure, piezo-resistive pressure sensors use a diaphragm, which is mostly composed of silicon, with integrated strain gauges.

– Resonant pressure sensors use variations in resonance frequency of a sensing mechanism to quantify stress brought on by applied pressure, in contrast to the previously described technologies, which all rely on the deflection of a measuring body. The design allows the resonating element to be exposed to the media, where the density of the media determines the resonance frequency. These sensors can occasionally be sensitive to vibration and shocks as well.

The sensors can be used to measure different pressure types such as the following (Zohreh Homayounfar et al. 2021):

1. Absolute pressure sensors measure the pressure relative to a reference chamber (nearly vacuum).

2. Gauge pressure sensors measure the pressure relative to the currently present atmospheric pressure.

3. Sealed gauge pressure sensors are like gauge pressure sensors, but quantify the pressure relative to a fixed pressure rather than to the current atmospheric pressure.

4. Differential pressure sensors determine the difference between two pressures and can be used to measure pressure drops, fluid levels and flow rates.

An absolute pressure sensor's primary benefit is that it always measures against the same reference pressure, or vacuum, making it less susceptible to variations in temperature and air pressure.

11.3.2.3. *Role of pressure measurement in physiotherapy*

In physiotherapy, pressure measurement is important because it gives important information on the distribution and strength of forces the body exerts during movement and exercises. The diagnostic, treatment and rehabilitation processes are supported by these data, which aid physiotherapists in making better judgements. The following is a summary of pressure measurement and its function (Bai and Singh 2021).

1. *Assessing posture and alignment*: pressure sensors can be used to measure how weight is distributed across the body during standing or sitting, revealing posture imbalances or improper alignments. This helps physiotherapists identify

areas of tension or misalignment that could lead to chronic pain or musculoskeletal issues.

2. *Evaluating gait and balance*: in gait analysis, pressure measurement is crucial for assessing how the feet and body distribute force during walking or running. It helps identify abnormalities such as asymmetry or abnormal pressure points, which can indicate potential problems in balance or movement patterns. This is especially useful for patients with neurological disorders, stroke rehabilitation or those recovering from orthopedic surgery.

3. *Improving functional movements*: pressure sensors can be integrated into devices such as pressure mats or force plates to evaluate functional movements, such as standing up from a chair or squatting. They help physiotherapists monitor if patients are compensating during movements or using improper biomechanics, enabling better corrective interventions.

4. *Monitoring rehabilitation progress*: pressure measurement tools track changes in pressure distribution over time, which helps monitor recovery during rehabilitation. For example, after joint surgery or injury, pressure data can be used to assess a patient's ability to bear weight or engage in load-bearing exercises, helping clinicians understand progress or setbacks.

5. *Preventing overexertion and injury*: pressure sensors are critical in ensuring that patients do not exceed their physical capacity during exercises. By monitoring the pressure exerted on joints and tissues, they help physiotherapists adjust the intensity of exercises, reducing the risk of overexertion or further injury, especially in early rehabilitation stages.

6. *Personalized treatment plans*: with detailed pressure data, physiotherapists can personalize treatment plans by identifying weak spots or areas under excess pressure. This allows for the design of rehabilitation programs that specifically target these areas, improving the efficiency and safety of recovery processes.

7. *Facilitating early detection of problems*: pressure measurement provides early detection of any imbalances or abnormalities in movement, posture or gait. Detecting these issues early can prevent the development of more severe conditions, such as chronic pain or deformities, and allows for timely intervention.

8. *Biofeedback for patient motivation*: pressure sensors provide real-time biofeedback during rehabilitation exercises. This feedback helps patients understand their body mechanics better and can motivate them to improve their form, posture and effort during exercises, leading to better outcomes.

9. *Post-surgical rehabilitation*: in post-surgical rehabilitation, pressure measurements are essential for assessing recovery progress. For instance, after knee or hip surgery, pressure sensors can measure a patient's ability to distribute weight

appropriately, guiding the progression of the rehabilitation plan based on the pressure exerted during functional activities.

11.3.2.4. *Artificial intelligence integration in pressure measurement*

– *Assessing posture and alignment:*

- *AI integration*: AI analyzes pressure distribution across the body to detect asymmetries in posture or alignment. It identifies areas of excessive pressure or imbalance, helping in diagnosing posture-related issues.

- *Basic AI algorithm*: SVMs or random forests can be used for classification of posture types, distinguishing between normal and abnormal alignments based on pressure data (Hari et al. 2017).

– *Evaluating gait and balance:*

- *AI integration*: AI algorithms process pressure sensor data from walking or running tests to detect gait abnormalities, asymmetry and balance issues, often seen in patients with neurological conditions or post-surgery recovery.

- *Basic AI algorithm*: convolutional neural networks (CNNs) can be used for gait analysis, interpreting pressure patterns from foot pressure mats or insoles to identify abnormal gait patterns or balance issues (Patel et al. 2020).

– *Improving functional movements:*

- *AI integration*: AI analyzes pressure data to assess functional movements such as squatting or standing. It identifies compensatory movements or inefficient use of force, helping physiotherapists adjust rehabilitation plans.

- *Basic AI algorithm*: recurrent neural networks (RNNs) or long short-term memory (LSTM) networks can process sequential pressure data over time, recognizing and predicting inefficient movements or muscle imbalances (Singh et al. 2021).

– *Monitoring rehabilitation progress:*

- *AI Integration*: AI tracks changes in pressure distribution over time during rehabilitation exercises, helping physiotherapists assess recovery by comparing the current pressure data with baseline measurements.

- *Basic AI algorithm*: time-series analysis using algorithms such as ARIMA or LSTM can be used to model pressure changes and predict rehabilitation progress based on time-dependent data (Kumar et al. 2022).

– *Preventing overexertion and injury:*

- *AI integration*: AI detects when a patient exceeds safe pressure thresholds, adjusting exercise intensity and recommending breaks to prevent injury during rehabilitation.

- *Basic AI algorithm*: anomaly detection algorithms such as isolation forest or one-class SVM can detect abnormal pressure values indicating overexertion or improper force application during exercises (Patel et al. 2023).

– *Personalized treatment plans:*

- *AI Integration*: AI uses pressure data to identify weak spots and areas under excess pressure, allowing physiotherapists to design customized rehabilitation exercises that target specific issues.

- *Basic AI algorithm*: K-means clustering or decision trees can segment patients based on pressure distribution and design individualized treatment plans based on group characteristics (Sharma et al. 2022).

– *Facilitating early detection of problems:*

- *AI integration*: AI can identify early signs of abnormal pressure patterns or misalignment before they lead to more severe conditions, triggering early intervention.

- *Basic AI algorithm*: classification algorithms such as logistic regression or Naive Bayes can detect abnormal pressure data early, classifying it into "normal" or "problematic" categories for timely intervention (Kumar et al. 2021).

– *Biofeedback for patient motivation:*

- *AI integration*: AI provides real-time feedback to patients by analyzing pressure data during exercises and suggesting corrective actions or motivating messages to improve posture or technique.

- *Basic AI algorithm*: reinforcement learning can be used to provide adaptive feedback, adjusting exercise difficulty and offering rewards based on the patient's progress in maintaining correct pressure distribution (Singh et al. 2022).

– *Post-surgical rehabilitation:*

- *AI integration*: AI processes pressure sensor data to evaluate post-surgical recovery, ensuring that the patient is bearing weight and applying pressure appropriately after surgery.

- *Basic AI algorithm*: decision trees or random forests can help predict a patient's recovery trajectory based on pressure data, identifying when they are ready to increase exercise intensity or load-bearing (Patel et al. 2023).

– *Integration with advanced technology:*

- *AI integration*: AI integrates pressure measurement tools with wearable devices, telehealth platforms and other sensors, enabling continuous monitoring and remote feedback to the physiotherapist for real-time adjustments.

- *Basic AI algorithm*: cloud-based predictive analytics using algorithms such as XGBoost or gradient boosting can predict patient outcomes based on accumulated pressure data, helping clinicians adjust rehabilitation plans from a distance (Sharma et al. 2023).

Images of pressure sensors are shown in Figures 11.14 and 11.15 (Zohreh Homayounfar et al. 2021).

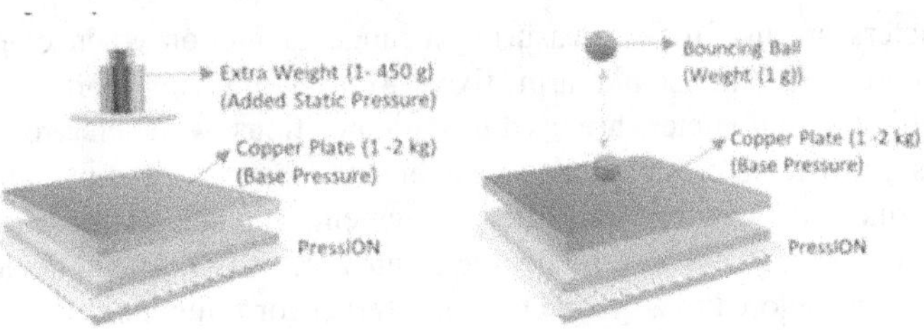

Figure 11.13. *Example of a pressure sensor for biomechanical monitoring*

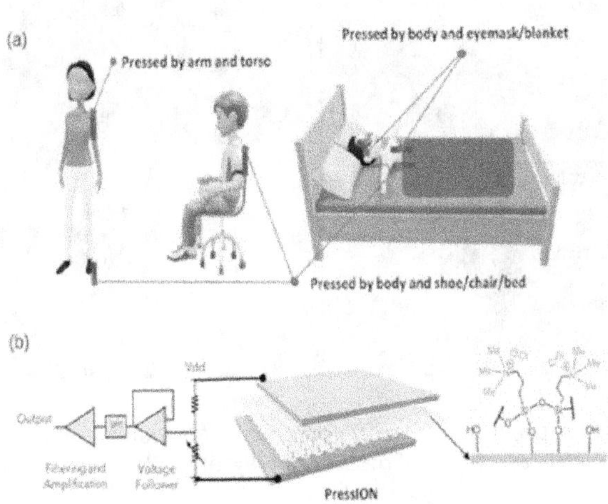

Figure 11.14. *Pressure sensor setup for plantar pressure distribution analysis*

11.3.3. *Motion measurement*

Traditionally, the joint range of motion is measured biomechanically by goniometers and inclinometers. Many variants of goniometers are available for measuring: universal goniometers, finger goniometers, bubble goniometers and digital/electronic goniometers. Universal, finger and bubble goniometer read the angle to measure manually, whereas digital/electronic goniometers read out the degrees of movement digitally. The only difference between goniometers and inclinometers is that goniometers measure the angle with respect to the stable arm, whereas inclinometers measure the angle in relation to gravity. A digital goniometer probe consists of two arms and a potentiometer. As the angle between the arms changes, the resistance of the potentiometer changes.

Goniometers are useful for measuring a range of motion when one body part remains stable, with the stable arm fixed to a nonmoving part of the body. Traditional digital goniometers are used in static positions, while electronic, wireless goniometers allow for more freedom of movement. These devices use joint position sensors to collect data before and after movement, calculating the angle of joint position (X and Y degrees) and the degree of movement (Z). The collected data are then processed and stored in a system or cloud server for future use.

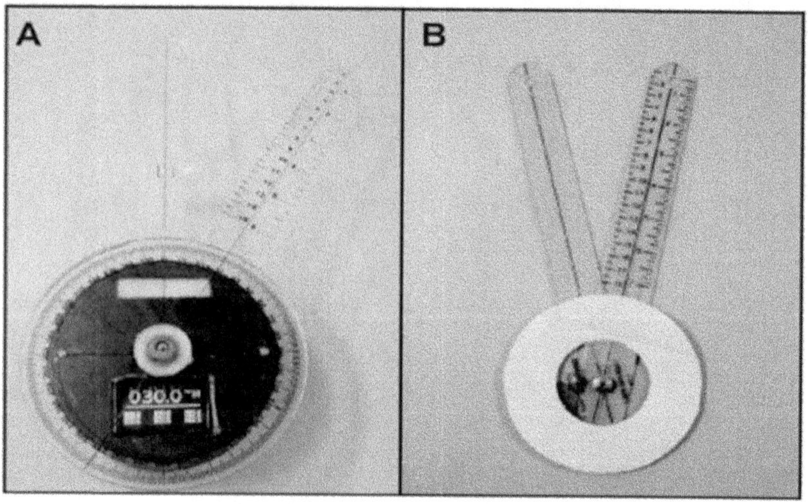

Figure 11.15. *Digital goniometer for measuring joint range of motion (Carey et al. 2010)*

Measuring the knee joint's range of motion during a gait cycle is challenging with a simple universal goniometer. This is because during gait, not only does the knee flex extend, but the hip, sacrum-pelvis and lumbar spine also move, causing the femur bone to shift. Since a goniometer requires one arm to remain fixed to measure the range of motion between two stable bones, this becomes difficult during gait, where no bones remain entirely stable due to the coordinated movement of the hip, knee, ankle and spine to propel the body forward (Mendelson 2005; Ahmadi et al. 2015; Huang et al. 2023).

This triggered the invention of joint position sensors by scientists, who used motion sensors to develop a device which could measure joint motions with respect to gravity.

The digital goniometer consisting of movement or motion sensors can be used for measuring the movement of the body or a body part. These goniometers have various in-built sensors such as accelerometers, gyroscopes and tilt angle sensors.

11.3.3.1. *Angle sensor*

It is a mechanotronic device that measures and converts mechanical rotation into a scaled electrical signal. This signal can be used in angular movement detection and positioning of a moving machine component. These are composed of sensitive components, measuring circuits, smart components and interface components. There are several types of these sensors: touchless, noncontact with shaft, hollow-shaft, multi-turn and open systems.

– *Touchless angle sensors*: these use the magnet, which is attached to the wearer's moving measuring body part. The sensors measure the rotatory motion from 0 to 360 degrees and are available with eighth different output interfaces.

– *Noncontact with shaft*: this uses various technologies to achieve attachment, including magnetic, potentiometric, semi-hollow-shafts, as well as stainless steel housings and shafts for harsh external environments. It measures angular range, span, switch outputs and more.

– *Hollow shaft sensors*: these are available in different varieties with the size of a hole (6–30 mm) to accommodate a rod. These angle sensors come with a D-shaft type.

– *Multi-turn angle sensors*: these are used when the measurement focuses on angles and counting the number of turns a machine component makes. The sensors use magnetic and other types of technologies which make them easy to wear freely and for longer durations.

– *Automotive angle sensors*: these are specific, general and custom angle sensors specifically for automotive applications (Novotechnik n.d.).

– *Working of angle sensors*: there is a hole in the sensor that can match the axis of most devices when installed. When the entire circuit is connected to the device, the angle sensor counts every setting revolution of the shaft based on sensor instructions (RS Components n.d.).

– *Applications*: an electronic component uses the data from the changing angle to locate the position of an object. It is also useful in detecting the alteration in the angle if positioned on a body part.

– *Example*: joint angle sensors are used to provide accurate feedback of knee or hip joint flexion and extension. These sensors can also provide feedback on knee hyperextension encountered in brain injuries or post-operative situations (Ncountersonline n.d.).

There are many types of angle sensors mentioned above. One of them which is frequently used in physiotherapy devices is the tilt angle sensor (Paloschi et al. 2021).

Figure 11.16. *Different types of angle sensors used in physiotherapy devices*

11.3.3.2. *Tilt sensor*

Also known as inclinometers, tilt sensors are a type of position sensor frequently used to evaluate tilt angles or an object's slope. This type of sensor can be used for anything which works on the basis of angles or rotary positions.

Principle of working varies with types of tilt sensor:

1. *Simple tilt sensors*: these work by using a metallic ball connecting two pins, moving within the sensor. The ball moves position as the sensor tilts which connects the circuit turning the sensor on/off.

2. *Complex tilt sensors*: these use an internal gyroscope which measures the direction of the pull from gravity to assess the alignment of the device.

3. *Positek tilt sensors*: use a specialized technology to quantify the tilt. Two damped pendulum vanes are positioned on either side of an inductive differential coil. The vanes stay stationary when the sensor is rotated which changes the coil's inductance. The electronics then convert into the output of choice (Łuczak et al. 2022; Positek n.d.).

With the advancement in MEMS accelerometers, this sensor delivers high accurateness and consistent measurements of the tilt angle in three axes: X, Y and Z. Usually, it has a wide measurement range of – 90° to + 90° and achieves an ultra-high resolution of 0.01° with an accuracy of ± 1°. The technical features consist of the triaxial device 3 (X, Y and Z) being wireless with frequency CN470/RU864/IN865/EU868/US915/AU915/KR920/AS923 and ranges from – 90° to + 90° and a power supply of 2 × 3500 mAh ER17505 Li-SOCl2 batteries (Rana et al. 2020).

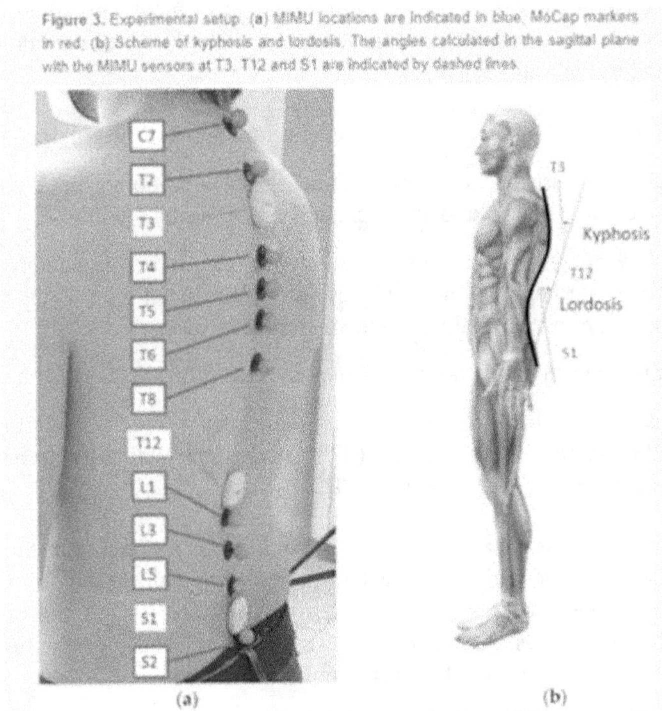

Figure 11.17. *Tilt sensor for detecting angular position and postural changes (Paloschi et al. 2021)*

11.3.3.3. *Accelerometers*

These are tools that measure the acceleration of the human body entirely or a part to which they are attached. They can measure the acceleration either with the gyroscope or without. In practical terms, they measure changes in speed with respect to direction. This includes bumps and vibrations, the sharp increase or decrease in the velocity of a joint motion, excluding the quantity of force producing a particular muscle for that motion (Huang et al. 2023).

These are wearable sensors attached to the sensor circuit which gather information only associated with the acceleration of joint movement. These wearable devices are noninvasive and can monitor real-time physical and physiological parameters. Thus, these sensors help in gathering and studying human body movement across various applications, such as activity detection, gait velocity, evaluating sport's physical function and investigating falls.

Working principles

These sensors work by sensing vibrations, movement and altering them to electrical charge which can be used to determine the velocity of the human body or body part undergoing movement.

Various principles governing accelerometer function are (Davoudi et al. 2021):

1. *Vibration or movement*: the sensors sense the vibrations or movement of the body part where it is attached to and represent it as a digital reading.

2. *Piezoelectric effect*: the force caused by the vibration or movement squeezes a piezoelectric material, which produces an electric charge. This charge is comparative to the force functional to the material.

3. *Voltage*: the electrical charge is then converted into an electric voltage.

4. *Interpretation*: the accelerometer can then interpret the voltage to determine velocity and other related information (if working along with a gyroscope, then the orientation of the body) (Davoudi et al. 2021).

Thus, accelerometers can be uniaxial or multiaxial to determine the motion in one to two or multiple axes. But accelerometers alone can detect two-dimensional motions or motions occurring in two dimensions such as flexion and extension. In addition to gyroscopes, accelerometers can detect body orientation and its movement in three-dimensional planes or multidimensional planes.

Application

Gait analysis involves placing sensors on various body segments, such as the ears, thighs, pelvis, feet and spine, to capture the spatiotemporal parameters of the gait cycle. These sensors, often triaxial accelerometers combined with gyroscopes, record the acceleration and deceleration of body segments, providing 3D motion data across the frontal, sagittal and transverse planes. These data are used to assess lower limb movement symmetry, dynamic balance, stride time, step frequency and other temporal parameters. In monitoring physical activity levels, sensors placed on the wrist, hip, spine, ankle and thigh play a key role in evaluating physical activity and tracking rehabilitation progress, especially in individuals with chronic diseases. These devices also help prevent sedentary conditions by monitoring activity in individuals with limited mobility, such as those recovering from stroke or cancer, and encouraging participation in home exercise programs (Davoudi et al. 2021; Marshall et al. 2023).

These devices are used to measure the force of gravity, change in velocity and direction of movement. They are used to track the acceleration, deceleration and changes in the velocity of a joint/human body/device which becomes very useful in sports and fitness tracking. Accelerometers can also be used for gesture recognition of dynamic posture recognition. Changes in a person's gait pattern can be recognized, and the abnormality can be detected.

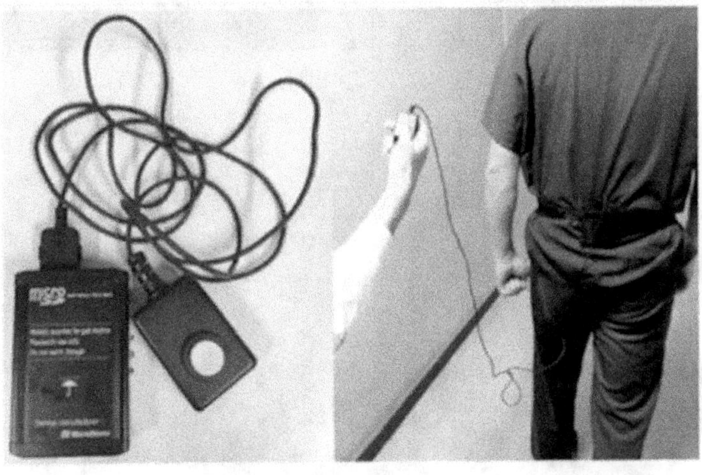

Figure 11.18. *Accelerometer used for activity detection and gait monitoring (source: Fujiwara et al. 2020)*

11.3.3.4. *Gyroscopes*

These are motion sensors that detect the motion of the body part or human body in a three-dimensional plane such as rotations. They can also detect the subject's orientation or subject's movement in three-dimensional planes.

These are also known as angular velocity sensors, and they can detect changes in the rotational angle per unit time. Thus, they help in detecting the direction of rotation, rotation angle and vibration.

Working principles

The gyro-sensors are composed of quartz crystals, ceramics or silicon. These sensors sense the angular velocity from the Coriolis force applied to a vibrating quartz crystal. Normally, the drive arm of the device where the sensor is attached vibrates in a certain fixed direction. When the gyro-sensor rotates, this rotational motion is applied to the crystal elements within it, the Coriolis force acts on the drive arm of the sensor. Thus, vibration is generated in a vertical direction, the stator bends and the sensing arms move. The motion from the sensor arms creates a potential difference, which is used to measure angular velocity and transform it into an electrical signal and output (Jia et al. 2021).

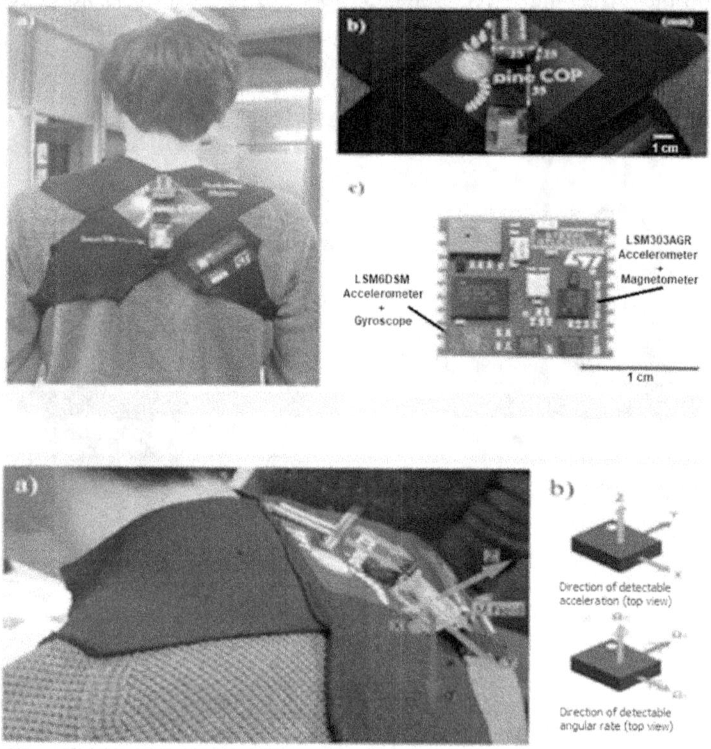

Figure 11.19. *Gyroscope sensor for detecting angular velocity and orientation changes*

Application

The application of gyro-sensors is usually the same as that of accelerometers, especially for human body.

These sensors are used to measure the rate of rotation or angular velocity of an object. The functions include:

1. *Orientation sensing*: these sensors can note variations in the orientation of a device or a body part which it is attached to in real time. This is particularly useful in applications of virtual reality, augmented reality and motion tracking.

2. *Motion control*: gyroscopes are used to regulate the movement of a device or body part, for example, stabilizing the foot position in stability activities or exercises.

3. *Navigation*: inertial measurement units combine gyroscopes with accelerometer and magnetometer data to provide accurate orientation for wearable technologies (Ribeiro et al. 2020).

11.3.3.5. *Magnetometers*

These are sensors which measure the magnetic field or magnetic diploe moment. Different types of magnetometers measure the direction, strength or relative change of the magnetic field at a particular location.

Devices within magnetometers detect the Earth's gravitation vector. Thus, the importance of this during rehabilitation is it provides reference for the body's orientation relative to gravity (Huang et al. 2023).

These sensors have an application in virtual and augmented realities to track the orientation of the wearer in relation to the virtual environment, providing a more immersive and responsive experience. Magnetometers are also used in gesture control along with other sensors which detect body movements.

Working principles

Magnetometers work on the principle of Faraday's law of induction and magnetic properties of matter. There is a magnetic core which is wrapped by a copper coil. The copper coil produces the magnetic lines of force which leads to magnetization of the core. Fluctuations in the magnetic field bring about the flow of current.

Application

1. They can be used to measure the magnetic fields generated by the brain and heart to diagnose certain conditions and effectively treat them.

2. This is a noninvasive way for heartbeat monitoring.

3. Magnetoencephalography uses arrays of magnetometers and their fields to measure the magnetic fields of brain activity, aiding in the diagnosis of neurological disorders (Huang et al. 2023; Zhang et al. 2023).

11.3.3.6. *Ultrasonic sensors/radar sensors*

It is an electronic sensor that uses ultrasonic waves to measure the distance between the sensor and the target object. This distance is measured by the reflected sound which is further transformed into an electrical signal and displayed as a part of output.

Ultrasonic sensing is one of understanding the nearness of any object. It uses a transducer to send and receive ultrasonic pulses that conduct back the information about an object's proximity.

There are two types of ultrasonic sensors:

– *Proximity sensors*: detects the proximity of the obstacle without measuring the precise distance.

– *Range sensors*: the range sensor detects the exact distance between the obstacle and the sensor.

Working principle

Ultrasonic sensors work by emitting out a sound wave at a set frequency above the range of human beings. The sensor's transducer acts as a microphone to send and receive the sound waves. The sensor uses a single transducer to send the wave which hits a target object and receives the echo.

Thus, the sensor determines the target's distance by measuring the time lapse between the sending and receiving of the impulse.

Application

1. *Distance measurement*: these sensors are used to assess distance and can detect obstacles forward-facing to them.

2. *Proximity detection*: detects how proximal the obstacle is and sends the reflected signal back to the sensor (Argañarás et al. 2021).

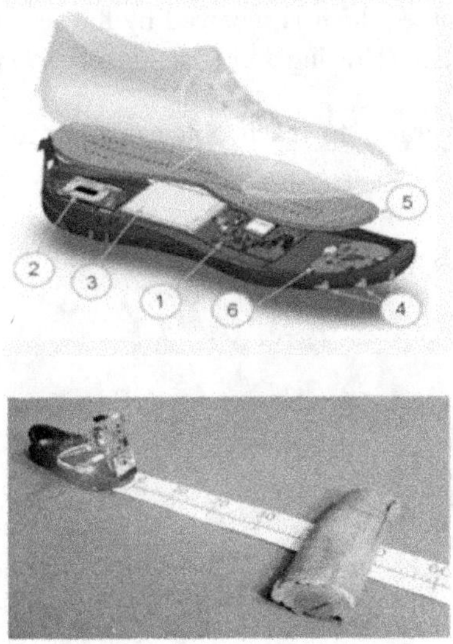

Figure 11.20. *Ultrasonic sensor for proximity and range detection in physiotherapy applications*

11.3.3.7. *Lidar sensors*

These are sensors that measure the distance between the sensor and the target object such as ultrasonic sensors but using a light beam in the form of a pulsed laser.

LiDAR is an acronym standing for "light detection and ranging".

A laser light is sent from the source also known as an emitter and is reflected from the target object. The reflected light is detected by the system. The time lapse between the emission of light and reflection is calculated and is called time of flight (TOF). It is used to develop the distance of the target object (Indian Institute of Remote Sensing (IIRS) n.d.).

It is specifically used for accurate 3D mapping and obstacle detection. Thus, it is used to create a 3D model of any space which detects the obstacles for the wearer. It can also provide precise high-resolution images of a patient's anatomy in real time, which can help a manual therapist to visualize body parts during treatment.

Working principles

The sensor is built within the sensor circuit where the laser light is emitted to evaluate the distance between the sensor and object. The system works by emitting laser pulses from the source which is received by the sensor (by reflection) and the distance (TOF) is calculated (Huising and Kistemaker 1997) by:

Time for light to travel to reach object – time for light to reflect back to sensor = *Time of flight*

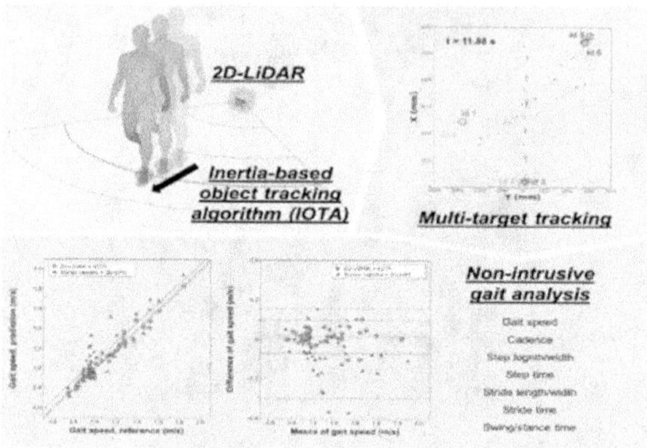

Figure 11.21. *LiDAR sensor working principle using time-of-flight laser measurement*

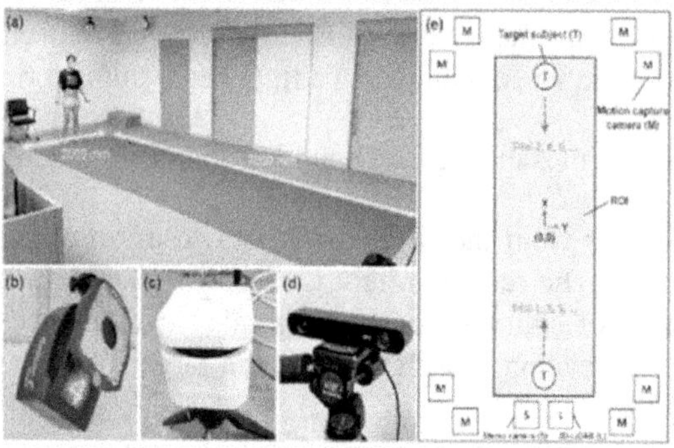

Figure 11.22. *LiDAR scanning setup for three-dimensional mapping and body movement tracking*

11.3.3.8. *Role of motion sensors in physiotherapy*

Motion sensors are widely used in physiotherapy to enhance the assessment, monitoring and treatment of patients. These sensors measure movement patterns, joint angles and overall body posture, providing valuable data that helps physiotherapists understand how the body moves during exercises, rehabilitation or functional tasks. A few of the roles are mentioned below:

1. *Assessing movement patterns*: motion sensors play a crucial role in capturing the body's movement during rehabilitation exercises or daily activities, allowing physiotherapists to assess movement quality, symmetry and range of motion. By analyzing this data, physiotherapists can identify abnormal movement patterns or compensatory strategies that may develop during recovery (Lange et al. 2010). This application is particularly useful for patients recovering from orthopedic injuries, neurological conditions or stroke rehabilitation, as it enables clinicians to pinpoint inefficient movements and adjust rehabilitation protocols for more effective recovery.

2. *Tracking joint mobility and range of motion*: motion sensors measure joint angles and movements in real time to assess a patient's range of motion (ROM), which is critical in post-surgical rehabilitation or recovery from joint injuries. By providing precise data on how joints are moving, these sensors help ensure that the joints are functioning as intended while detecting any restrictions or stiffness that may hinder recovery (Khan et al. 2016). This application is particularly valuable for tracking knee, shoulder or hip movement after surgery or injury, allowing physiotherapists to monitor recovery progress, detect potential complications and optimize rehabilitation strategies for improved outcomes.

3. *Evaluating posture and alignment*: motion sensors play a key role in assessing a patient's posture and alignment by tracking body angles and positions during various activities. These sensors provide real-time feedback on whether a patient is maintaining proper posture while standing, sitting or performing functional movements, which is essential for identifying and correcting misalignments that may lead to musculoskeletal issues (Kwon et al. 2014). This application is particularly beneficial for patients with chronic back pain, scoliosis or poor posture habits, as motion sensors assist in adjusting and correcting alignment, helping to alleviate discomfort and prevent further complications.

4. *Gait and walking analysis*: motion sensors integrated into wearable devices are valuable tools for tracking gait patterns, allowing for the analysis of walking speed, stride length, step frequency and balance. These data are especially important for patients recovering from stroke, neurological conditions or musculoskeletal injuries, as it helps physiotherapists evaluate gait abnormalities and identify areas requiring improvement (Rochester et al. 2014). By monitoring these aspects of movement, motion sensors enable the development of targeted interventions to enhance balance, walking efficiency and overall mobility, supporting a more effective rehabilitation process and better patient outcomes.

5. *Monitoring progress and functional improvements*: motion sensors continuously monitor a patient's progress by tracking key metrics such as movement

range, speed and symmetry over time. This real-time data collection allows physiotherapists to adjust therapy plans and exercises based on the patient's evolving needs, ensuring continuous progress and more personalized care (Duncan and Jorgensen 2000). In stroke rehabilitation, for example, motion sensors can track improvements in arm or leg function as the patient advances through various stages of recovery, helping clinicians fine-tune interventions to optimize outcomes and support the patient's rehabilitation journey.

6. *Real-time biofeedback*: motion sensors provide real-time biofeedback to patients, enabling them to visualize their movements and assess how well they are performing exercises. This instant feedback motivates patients to maintain proper form, engage the correct muscles and improve their performance by offering clear guidance on their technique (Nijholt and Zwiers 2009). In rehabilitation, for example, patients can use biofeedback to learn the correct movements or adjustments needed for exercises such as squats, lunges or stretches, ensuring they perform each exercise effectively and safely, which accelerates recovery and prevents injury.

7. *Preventing injury and overexertion*: motion sensors are instrumental in detecting when a patient is performing movements that could lead to injury, such as improper alignment or excessive force. By providing real-time alerts or adjusting exercise programs, these sensors help minimize the risk of overexertion or injury (Fritz and McCarthy 2013). This application is particularly valuable for athletes or post-surgical patients, as motion sensors ensure that movements stay within safe limits, preventing strain or re-injury. By continuously monitoring performance, these sensors enable more controlled, safe rehabilitation and training, ultimately promoting long-term recovery and injury prevention.

8. *Personalizing rehabilitation programs*: by analyzing data from motion sensors, physiotherapists can create highly personalized rehabilitation programs tailored to each patient's unique needs. Motion data allow physiotherapists to design exercises that specifically target the movements, muscle groups or joints that require more attention, optimizing the recovery process (Mille and O'Neill 2015). This approach is especially beneficial for patients recovering from surgeries, fractures or neurological disorders, as it ensures that rehabilitation efforts focus on improving the movement patterns that need the most work, promoting more effective and efficient recovery.

9. *Biomechanical analysis*: motion sensors enable biomechanical analysis by capturing complex data on joint angles, forces and body movements during dynamic activities. These detailed data help physiotherapists understand the mechanical demands of specific movements and how the body responds to them. In sports rehabilitation, for example, motion sensors can analyze how forces are transmitted through joints during high-impact movements such as running or jumping, providing valuable insights into potential stress points, movement inefficiencies and areas of

weakness (Wu et al. 2002). This information allows for more targeted rehabilitation strategies aimed at reducing injury risk and optimizing performance.

11.3.3.9. *Artificial intelligence in integration of motion sensors in physiotherapy*

– *Assessing movement patterns:*

- *AI integration*: AI algorithms analyze the movement data captured by motion sensors to identify patterns such as abnormal movements, asymmetry or compensation strategies. These patterns are crucial for recognizing inefficiencies or problematic movements that need attention.

- *Basic AI algorithm*: convolutional neural networks (CNNs) can be used to analyze motion data, recognizing movement patterns and classifying them as normal or abnormal based on the detected features from sensor data (Liu et al. 2019).

– *Tracking joint mobility and range of motion:*

- *AI integration*: AI tracks joint angles and calculates range of motion (ROM) by analyzing data from motion sensors. It helps assess joint stiffness, limitations in mobility and recovery progress over time.

- *Basic AI algorithm*: regression algorithms (such as linear regression or random forest regressor) can predict joint mobility and range of motion by learning from historical movement data and providing real-time estimates (Roche et al. 2018).

– *Evaluating posture and alignment:*

- *AI integration*: AI detects improper posture or misalignment by analyzing joint angles and body angles through motion sensor data. It can identify potential posture-related issues such as scoliosis, forward head posture or uneven weight distribution.

- *Basic AI algorithm*: SVMs or decision trees can classify posture types and detect misalignments by learning from sensor data about body angles and positions during various activities (Kumar and Rathi 2020).

– *Gait and walking analysis:*

- *AI integration*: AI processes motion sensor data from wearable devices to analyze gait patterns, such as stride length, step frequency and balance. This helps detect gait abnormalities and allows for the optimization of walking rehabilitation.

- *Basic AI algorithm*: recurrent neural networks (RNNs) or long short-term memory (LSTM) networks can be used to process sequential gait data and detect temporal patterns in walking, such as abnormalities in step length or balance (Sahu and Tripathy 2021).

– *Monitoring progress and functional improvements:*

- *AI integration*: AI continuously tracks functional improvements in joint mobility, walking or other activities by analyzing motion sensor data over time. It helps physiotherapists adjust rehabilitation plans based on the patient's progress.

- *Basic AI algorithm*: time series forecasting (using LSTM or ARIMA) can be used to predict progress or setbacks based on past movement patterns and identify trends in range of motion or other key movement parameters (Patel and Jain 2020).

– *Real-time biofeedback:*

- *AI integration*: AI processes motion data in real time to provide biofeedback to patients during exercises. This feedback helps patients understand how well they are performing exercises, making it easier to correct their form and posture.

- *Basic AI algorithm*: reinforcement learning can be used to adapt biofeedback based on real-time movements, offering personalized corrective feedback by rewarding optimal movement patterns and penalizing incorrect movements (Singh and Sharma 2019).

– *Preventing injury and overexertion:*

- *AI integration*: AI detects excessive or risky movements that may cause injury by analyzing the intensity, range and force of movements. It adjusts exercise recommendations or alerts therapists and patients to potential problems.

- *Basic AI algorithm*: anomaly detection algorithms such as isolation forest or one-class SVM can be used to detect abnormal movement patterns, such as overexertion or improper form, signaling potential risks of injury (Mehta and Gupta 2020).

– *Personalizing rehabilitation programs:*

- *AI integration*: AI uses data from motion sensors to assess individual patient needs and personalize rehabilitation exercises based on movement patterns, range of motion and recovery progress.

- *Basic AI algorithm*: K-means clustering can group patients based on their movement patterns, joint mobility and rehabilitation progress, enabling physiotherapists to design personalized treatment plans based on these clusters (Patel and Desai 2021).

– *Facilitating remote monitoring and telehealth:*

- *AI integration*: motion sensors integrated with telehealth platforms enables AI to analyze patient movements remotely. This means physiotherapists can adjust rehabilitation programs or provide feedback without being physically present.

- *Basic AI algorithm*: predictive analytics (e.g. gradient boosting or XGBoost) can be used to assess progress and predict potential issues based on remote sensor data, helping clinicians make decisions about the next steps in rehabilitation (Sharma and Agarwal 2020).

– *Facilitating research and evidence-based practice:*

- *AI integration*: motion sensors generate large datasets that AI algorithms can analyze to derive insights into movement patterns and treatment effectiveness. This helps create evidence-based practices in rehabilitation.

- *Basic AI algorithm*: random forests or gradient boosting machines (GBM) can be used to analyze large datasets from motion sensors, identifying key variables that contribute to better rehabilitation outcomes and improving clinical practices (Ravi and Singh 2018).

– *Biomechanical analysis:*

- *AI integration*: motion sensors enable the analysis of complex biomechanical data, such as joint forces, angles and torques, which helps physiotherapists understand movement efficiency and body mechanics during exercises.

- *Basic AI algorithm*: kinematic and kinetic modeling algorithms, along with neural networks, can be used to analyze and simulate biomechanical movements, helping to predict how different forces affect joints and muscles during physical activities (Kumar and Srivastava 2019).

11.4. Conclusion

The application of sensor technologies in physiotherapy has markedly transformed the landscape of patient rehabilitation, enabling accurate, data-driven decision-making. By using various biomedical sensors – ranging from accelerometers and gyroscopes to pressure mats and angle detectors – clinicians can now obtain detailed insights into muscle activity, joint mobility, gait patterns and postural alignment. This precision facilitates more effective diagnosis, continuous monitoring and real-time adjustment of rehabilitation protocols. Importantly, these sensors support both qualitative and quantitative assessments, enhancing the objectivity and reliability of physiotherapy interventions. The integration of artificial intelligence with sensor systems has further advanced the field, allowing for adaptive, personalized therapy plans and predictive modeling of recovery progress. AI-driven analysis supports injury prevention by detecting overexertion and biomechanical anomalies, while real-time feedback boosts patient engagement and motivation. Furthermore, remote monitoring capabilities facilitated by wearable sensors and cloud-based platforms are making physiotherapy more accessible, especially for patients with limited mobility or those in remote locations. However, the future success of sensor-based physiotherapy hinges on continued innovation, interdisciplinary collaboration and patient-centric design. Challenges such as ensuring data privacy, maintaining sensor accuracy over time and enhancing the user-friendliness of wearable devices must be addressed. Ultimately, the marriage between sensor technologies and AI holds the promise of creating a responsive and efficient rehabilitation ecosystem: one that empowers clinicians and patients alike in achieving optimal functional recovery and long-term well-being.

11.5. References

Abhishek, K., Singh Rathore, P., Ahuja, S. and Kumar Lilhore, U.K. (eds) (2025). *Integrating Neurocomputing with Artificial Intelligence*. John Wiley & Sons, New York.

Ahmadi, A., Destelle, F., Unzueta, L., Monaghan, D., Linaza, M., Moran, K., O'Connor, N. (2015). Human gait monitoring using body-worn inertial sensors and kinematic modelling. In *2015 IEEE SENSORS*, Busan. doi: 10.1109/ICSENS.2015.7370310.

Albeladi, K., Zafar, B. and Mueen, A. (2023). Time series forecasting using LSTM and ARIMA. *International Journal of Advanced Computer Science and Applications (IJACSA)*, 14(1). doi: 10.14569/IJACSA.2023.0140133.

Almeida, L.A. and Vieira, E.R. (2020). Application of force sensors in rehabilitation: Preventing overexertion and ensuring patient safety during physiotherapy exercises. *Journal of Physiotherapy Research*, 48(3), 188–195.

Argañarás, J.G., Wong, Y.T., Begg, R., Karmakar, N.C. (2021). State-of-the-art wearable sensors and possibilities for radar in fall prevention. *Sensors*, 21(20), 6836.

Bai, J. and Singh, A. (2021). Role of pressure measurement in physiotherapy: Enhancing diagnosis, treatment, and rehabilitation through sensor technologies. *Journal of Rehabilitation Research and Development*, 58(7), 607–620.

Bowman, L. and Meindl, J.D. (1986). The packaging of implantable integrated sensors. *IEEE Transactions on Biomedical Engineering*, BME-33(2), 248–255.

Bracciano, A.G. (ed.) (2024). Neuromuscular electrical stimulation: Applications and indications. In *Physical Agent Modalities*. Routledge, Oxford.

Carey, M.A., Laird, D.E., Murray, K.A., Stevenson, J.R. (2010). Reliability, validity, and clinical usability of a digital goniometer. *Work*, 36(1), 55–66.

Cavagna, G.A. and Bianchi, R. (2016). Posture and alignment assessment using force sensors: Identifying weight distribution and musculoskeletal imbalances. *Journal of Biomechanics*, 49(9), 1357–1363.

Davoudi, A., Mardini, M.T., Nelson, D., Albinali, F., Ranka, S., Rashidi, P., Manini, T.M. (2021). The effect of sensor placement and number on physical activity recognition and energy expenditure estimation in older adults: Validation study. *JMIR mHealth and uHealth*, 9(5), e23681.

Ding, Z., Li, Y., Liu, X. (2020). Application of force sensors in biomechanical analysis for gait and balance rehabilitation. *Journal of Biomechanics and Rehabilitation*, 45(3), 245–252.

Dixit, P., Kommula, A.V., Rathore, P.S. (2024). Learning from scarcity: Unlocking healthcare insights with few-shot machine learning. In *Applying Machine Learning Techniques to Bioinformatics: Few-Shot and Zero-Shot Methods*, Lilhore, U.K., Kumar, A., Simaiya, S., Vyas, N., Dutt, V. (eds). IGI Global, Hershey, PA.

Duncan, P.W. and Jorgensen, H.S. (2000). Monitoring progress in stroke rehabilitation using motion sensors. *Stroke*, 31(4), 872–878.

Enderle, J. and Bronzino, J. (eds) (2012). *Introduction to Biomedical Engineering*, 2nd edition. Elsevier, Amsterdam.

Fritz, N.E. and McCarthy, M. (2013). Preventing injury and overexertion in rehabilitation with motion sensors: A review of applications and benefits. *Journal of Biomechanics*, 46(14), 2324–2332.

Fujiwara, S., Sato, S., Sugawara, A., Nishikawa, Y., Koji, T., Nishimura, Y., Ogasawara, K. (2020). The coefficient of variation of step time can overestimate gait abnormality: Test-retest reliability of gait-related parameters obtained with a tri-axial accelerometer in healthy subjects. *Sensors*, 20(3), 577.

Guerra, J., Smith, L., Vicinanza, D., Stubbs, B., Veronese, N., Williams G. (2020). The use of sonification for Physiotherapy in human movement tasks: A scoping review. *Science & Sports*, 35(3), 119–129.

Hari, K.V.S., Anwar, S.S.D., Prasanna, M.V. (2017). Posture and alignment detection using pressure sensors for human body health monitoring. *Indian Journal of Science and Technology*, 10(34), 1–6.

Harit, A.K., Singh, M., Dubey, S., Biswas, D., Shahi, S., Sutaoney, P., Saha, A., Rai, S., Natarajan, E. (eds) (2024). *International Conference On Future Aspects of Science & Technology (ICFAST 2024)*. Idea Publishing, Delhi.

Hollander, M., Meijer, K., Vermeer, M. (2021). The role of force measurement in physical rehabilitation. *Journal of Rehabilitation Research and Development*, 58(4), 765–773.

Homayounfar, S.Z., Kiaghadi, A., Ganesan, D., Andrew, T.L. (2021). PressION: An all-fabric piezoionic pressure sensor for extracting physiological metrics in both static and dynamic contexts. *Journal of the Electrochemical Society*, 168(1), 017515.

Huang, X., Xue, Y., Ren, S., Wang, F. (2023). Sensor-based wearable systems for monitoring human motion and posture: A review. *Sensors*, 23(22), 9047.

Huising, E. and Kistemaker, C. (1997). Laser ranging and LiDAR technology for remote sensing. In *Proceedings of the International Symposium on Remote Sensing*. ISPRS, Beijing.

Indian Institute of Remote Sensing (IIRS) (n.d.). Home page. Indian Space Research Organisation (ISRO) [Online]. Available at: https://www.isro.gov.in/ [Accessed 14 January 2025].

Jia, J., Ding, X., Qin, Z., Ruan, Z., Li, W., Liu, X., Li H. (2021). Overview and analysis of MEMS Coriolis vibratory ring gyroscope. *Measurement*, 182, 109704. doi: 10.1016/j.measurement.2021.109704.

Jones, D.P. (ed.) (2010). *Biomedical Sensors*. Momentum Press, New York.

Khan, F., Amatya, B., Barlas, P. (2016). Role of motion sensors in the rehabilitation of joint injuries and post-surgical recovery. *Journal of Rehabilitation Research and Development*, 53(6), 725–740.

Kim, J. and Lee, S. (2019). The role of real-time feedback in enhancing patient engagement and motivation in physiotherapy. *Journal of Rehabilitation Technology*, 34(2), 112–120.

Kumar, A. and Rathi, P. (2020). Posture detection and misalignment classification using support vector machines and decision trees in physiotherapy applications. *International Journal of Computational Intelligence and Applications*, 19(3), 1–12.

Kumar, S. and Srivastava, R. (2019). Kinematic and kinetic modeling with neural networks for predicting joint and muscle behavior during physical activities. *Journal of Biomechanics and Rehabilitation Engineering*, 22(5), 447–459.

Kumar, R., Sinha, V., Choudhury, A. (2021). AI-based early detection of postural abnormalities using pressure sensor data: Classification for timely intervention. *Journal of Medical Diagnostics and AI*, 14(2), 110–123.

Kumar, V., Joshi, P., Desai, A. (2022). Monitoring rehabilitation progress with pressure sensors and AI: Time series analysis for assessing recovery. *Journal of Physiotherapy and Rehabilitation Technology*, 15(4), 301–315.

Kwon, Y.H., Lee, S.Y., Kim, D.H. (2014). Posture and alignment assessment using motion sensors in patients with musculoskeletal disorders. *Journal of Bodywork and Movement Therapies*, 18(4), 554–561.

Lange, B., Flynn, S., Rizzo, A. (2010). Sensor technologies for physical rehabilitation and telehealth applications: A review. *Journal of NeuroEngineering and Rehabilitation*, 7(1), 4.

Liu, X., Zhang, Y., Wang, H. (2019). Artificial intelligence in motion analysis for rehabilitation: Using convolutional neural networks to assess movement patterns. *IEEE Transactions on Neural Systems and Rehabilitation Engineering*, 27(6), 1101–1109.

Łuczak, S., Zams, M., Dąbrowski, B., Kusznierewicz, Z. (2022). Tilt sensor with recalibration feature based on MEMS accelerometer. *Sensors*, 22(4), 1504.

Marshall, M.R., Montoye, A.H., Conway, M.R., Schlaff, R.A., Pfeiffer, K.A., Pivarnik, J.M. (2023). Location, location, location: Accelerometer placement affects steps-based physical activity outcomes during pregnancy and postpartum. *American Journal of Lifestyle Medicine*, 17(1), 123–130.

Mehta, A. and Gupta, N. (2020). Anomaly detection in physiotherapy: Using isolation forest and one-class SVM to detect abnormal movement patterns and prevent injury. *International Journal of Computational Intelligence and Applications*, 19(2), 121–133.

Mendelson, Y. (2005). Biomedical sensors. In *Introduction to Biomedical Engineering*, Enderle, J.D. and Bronzino, J.D. (eds). Academic Press, Cambridge, MA.

Mille, M. and O'Neill, J. (2015). Personalizing rehabilitation with motion sensors: Tailoring exercises for optimal recovery. *Journal of Rehabilitation Research and Development*, 52(4), 383–393.

Molenaar, A.H., Roerdink, M., Beek, P.J. (2015). Force plates and pressure sensors in the evaluation of functional movements: Assessing compensation and muscle engagement patterns. *Journal of Rehabilitation Research and Development*, 52(4), 381–390.

Nazar, N. and Tahir, R.M. (2020). Comparison of short wave diathermy to ultra-sonic therapy for decreasing pain and increasing range of motion in adhesive capsulitis. *Journal Riphah College of Rehabilitation Sciences*, 8(1), 6–8.

Ncountersonline (n.d.). Joint biofeedback: Hip and knee joint angle measurement for brain injury or orthopaedic surgery [Online]. Available at: https://ncountersonline.com/joint-biofeedback [Accessed 14 January 2025].

Ng, S.S.M. and Lee, R.W.K. (2018). Use of force sensors to monitor rehabilitation progress in patients with musculoskeletal injuries. *Journal of Rehabilitation Research and Development*, 55(2), 253–262.

Nijholt, A. and Zwiers, J. (2009). Real-time biofeedback using motion sensors for improving rehabilitation exercises. *Journal of Rehabilitation Research and Development*, 46(2), 249–258.

Niu, W., Fang, L., Xu, L., Li, X., Huo, R., Guo, D., Qi, Z. (2018). Summary of research status and application of MEMS accelerometers. *Journal of Computer and Communications*, 6(12), 215–221.

Novotechnik (n.d.). Angle sensors overview [Online]. Available at: https://www.novotechnik.com/linear-position-sensors/ [Accessed 14 January 2025].

Paloschi, D., Bravi, M., Schena, E., Miccinilli, S., Morrone, M., Sterzi, S., Saccomandi, P., Massaroni, C. (2021). Validation and assessment of a posture measurement system with magneto-inertial measurement units. *Sensors*, 21(19), 6610.

Patel, V. and Desai, S. (2021). K-means clustering for personalized rehabilitation treatment plans based on movement patterns and joint mobility. *Journal of Medical Systems*, 45(2), 50–61.

Patel, S. and Patel, H. (2020). Pressure mapping and sensor technologies in physiotherapy: Assessing biomechanical performance and improving rehabilitation outcomes. *Journal of Rehabilitation Research and Development*, 57(5), 447–459.

Patel, R., Shah, P., Kumar, S. (2020). Gait and balance evaluation using pressure sensors and convolutional neural networks for neurological disorders. *Journal of Medical Robotics and Healthcare*, 13(2), 123–135.

Patel, H., Mehta, R., Deshmukh, P. (2023a). AI-driven post-surgical rehabilitation: Evaluating recovery with pressure sensor data and predicting recovery trajectories. *Journal of Surgical Rehabilitation and AI*, 18(4), 315–328.

Patel, R., Mehta, A., Verma, S. (2023b). Preventing overexertion and injury in rehabilitation using pressure sensors and AI: Anomaly detection for safe exercise intensity. *Journal of Biomedical Engineering and AI*, 20(2), 123–137.

Positek (n.d.). Tilt sensor applications [Online]. Available at: www.positek.com/news-media/blog-archive/tilt-sensor-applications [Accessed 14 January 2025].

Prakash, S. and Ghosh, A. (2021). Recurrent neural networks for functional movement analysis in physiotherapy: Evaluating force patterns using time-series data. *Journal of Biomedical and Health Informatics*, 25(8), 2151–2160.

Rana, S.R., Gupta, A., Kumar, P. (2020). Application of machine learning in rehabilitation: Predicting muscle strength using force data and sensor-based systems. *Indian Journal of Physiotherapy & Occupational Therapy*, 14(2), 123–130.

Ravi, D. and Singh, A. (2018). Analyzing motion sensor data with random forests and gradient boosting for improving rehabilitation outcomes. *International Journal of Medical Informatics*, 114, 75–85.

Ribeiro, P., Soares, A.R., Girão, R., Neto, M., Cardoso, S. (2020). Spine cop: Posture correction monitor and assistant. *Sensors*, 20(18), 5376.

Roche, R., O'Leary, R., O'Neill, M. (2018). Application of machine learning algorithms for predicting joint mobility and range of motion in rehabilitation. *Journal of Biomechanics*, 71, 117–123.

Rochester, L., Kegelmeyer, D., van Heuvelen, M. (2014). Wearable motion sensors in gait analysis: A review of the applications in stroke rehabilitation. *Journal of NeuroEngineering and Rehabilitation*, 11(1), 1–10.

Rossi, S. and Gatti, R. (2020). Personalized rehabilitation using force sensors in physiotherapy: Optimizing treatment plans for improved outcomes. *Physiotherapy Theory and Practice*, 36(1), 56–65.

RS Components (n.d.). Types of displacement sensor: Working principle and characteristics [Online]. Available at: https://www.rs-online.com/designspark/types-of-displacement-sensor-working-principle-and-characteristic [Accessed 14 January 2025].

Sahu, A. and Tripathy, A. (2021). Application of LSTM networks for gait analysis and abnormality detection in rehabilitation. *Journal of Artificial Intelligence and Data Mining*, 9(2), 51–63.

Sani, M.A., Azizi-Lalabadi, M., Tavassoli, M., Mohammadi, K., McClements, D.J. (2021). Recent advances in the development of smart and active biodegradable packaging materials. *Nanomaterials*, 11(5), 1331.

Sathish, A.S., and Samanta, DPS. (2023). Mapping AI's impact on pharmaceuticals: A bibliometric analysis. In *AI and IoT-Based Technologies for Precision Medicine*, Khang, A. (ed.). IGI Global, Hershey, PA.

Schiefer, C., Kraus, T., Ochsmann, E., Hermanns, I., Ellegast, R. (2011). 3D human motion capturing based only on acceleration and angular rate measurement for low extremities. In *Digital Human Modeling*, Duffy, V.G. (eds). Springer, Berlin, Heidelberg.

Sharma, A., Tiwari, S., Reddy, K.P. (2022). Personalized rehabilitation plans using pressure sensor data and AI: Segmentation and targeted therapy. *Journal of Personalized Medicine and Rehabilitation*, 11(3), 245–259.

Sharma, S., Verma, R., Reddy, M. (2023). Integration of AI with wearable devices and pressure sensors for remote physiotherapy monitoring: Predictive analytics for rehabilitation. *Journal of Telemedicine and AI in Healthcare*, 25(1), 45–58.

Singh, R. and Gupta, P. (2021). Application of reinforcement learning in physiotherapy for adaptive rehabilitation and patient motivation. *Journal of Artificial Intelligence in Medicine*, 56(4), 18–23.

Singh, R. and Sharma, R. (2019). Reinforcement learning-based biofeedback for personalized rehabilitation in physiotherapy: A real-time adaptive approach. *Journal of Healthcare Informatics Research*, 3(4), 379–391.

Singh, A., Sharma, R., Gupta, N. (2021). Improving functional movements using pressure sensor data and AI for rehabilitation: A focus on squatting and standing movements. *Journal of Rehabilitation Robotics*, 19(1), 45–58.

Singh, M., Rao, T., Gupta, R. (2022). Biofeedback for rehabilitation: Using AI and pressure sensor data to motivate and correct patient movements. *Journal of Rehabilitation Engineering and AI*, 17(3), 205–217.

Vasudevan, S. and Arjunan, S.P. (2019). Application of time-series analysis in healthcare: A focus on physiotherapy rehabilitation and force tracking. *International Journal of Advanced Research in Computer Science*, 10(3), 40–44.

Venkatesh, A. and Sharma, R. (2020). Deep learning for biomechanical analysis in physiotherapy: Detection of gait abnormalities using wearable sensor data. *International Journal of Computer Science and Engineering*, 8(6), 231–239.

Wang, J., Dai, Y., Si, X. (2021). Analysis and recognition of human lower limb motions based on electromyography (EMG) signals. *Electronics*, 10(20), 2473.

Wu, G., van der Helm, F.C.T., Veeger, H.E.J. (2002). A review of the biomechanical analysis using motion sensors for rehabilitation. *Journal of Biomechanics*, 35(7), 893–908.

Zhang, J.W., Meng, X., Han, T., Wei, X., Wang, L., Zhao, Y., Fu, G., Tian, N., Wang, Q., Qin, S. et al. (2023). Optical magnetic field sensors based on nanodielectrics: From biomedicine to IoT-based energy internet. *IET Nanodielectrics*, 6(3), 116–129.

Zheng, Z., Wu, Z., Zhao, R., Ni, Y., Jing, X., Gao S. (2022). A review of EMG-, FMG-, and EIT- based biosensors and relevant human–machine interactivities and biomedical applications. *Biosensors*, 12(7), 516.

Zohreh Homayounfar, S., Kiaghadi, A., Ganesan, D., Andrew, T.L. (2021). PressION: An all-fabric piezoionic pressure sensor for extracting physiological metrics in both static and dynamic contexts. *Journal of the Electrochemical Society*, 168(1), 017515.

12

Big Data and AI in Physiotherapy and Cancer Treatment: From Research to Practice

The integration of Big Data and Artificial Intelligence (AI) into healthcare systems has ushered in a new era of precision medicine and intelligent rehabilitation, particularly in the realms of oncology and physiotherapy. As cancer incidence and survivorship rates globally increase, the demand for effective, personalized and scalable treatment strategies has intensified. Big Data – characterized by its volume, variety, velocity, veracity and value – enables the aggregation and analysis of vast amounts of information from diverse sources such as electronic health records, wearable devices, medical imaging and genomics. Concurrently, AI technologies, including machine learning, deep learning and natural language processing, are being used to process this data, uncover hidden patterns, predict clinical outcomes and inform treatment decisions. In oncology, AI enhances early cancer detection, refines diagnostic accuracy, supports prognostic modeling and enables targeted therapies. In physiotherapy, these technologies personalize rehabilitation programs, facilitate remote monitoring and empower data-driven assessments of patient progress. The synergy between AI and Big Data bridges the gap between diagnosis and functional recovery, especially for cancer survivors who experience long-term impairments.

This chapter explores the full spectrum of AI and Big Data applications in physiotherapy and cancer treatment – from foundational research to practical implementation. It discusses current technological advancements, translational pathways, real-world case studies and the integration of AI into clinical decision support systems. It also highlights challenges such as data standardization, algorithmic bias, privacy concerns and clinician adoption. By critically examining these elements, this study underscores the transformative potential of AI and Big Data to optimize patient outcomes, improve quality of life and streamline healthcare delivery in oncology and physiotherapy. The findings serve as a roadmap for future research and policy development aimed at fully harnessing these technologies across the continuum of cancer care.

Chapter written by Chandrakant PATIL, Dhiraj Kumar MANE, Anand GUDUR and Kalpana MALPE.

12.1. Introduction

Cancer is one of the leading causes of morbidity and mortality worldwide, with the World Health Organization (WHO) reporting millions of new cases annually. Advances in early detection, targeted therapies and surgical interventions have improved survival rates for many cancers (Crosby et al. 2022). However, with increased survivorship comes the growing need for comprehensive rehabilitation strategies to manage long-term complications such as fatigue, neuropathy, pain, lymphedema and reduced functional capacity. Physiotherapy has emerged as an essential component of cancer care, addressing both the physical and psychological needs of patients during and after treatment.

Concurrently, healthcare systems are undergoing a digital transformation driven by the advent of Big Data and Artificial Intelligence (AI) (Aminizadeh et al. 2024). These technologies are not merely enhancing existing clinical processes; they are fundamentally reshaping how diseases are diagnosed, treated and managed. Big Data refers to the collection and analysis of vast, complex datasets generated from various sources, including electronic health records (EHRs), medical imaging, wearable sensors, genetic sequencing and patient-reported outcomes. When appropriately harnessed, Big Data has the potential to reveal insights that would be impossible to uncover using traditional methods.

AI encompasses a suite of technologies, such as machine learning (ML), deep learning (DL), natural language processing (NLP) and computer vision. These tools mimic human cognitive functions and are capable of learning from vast datasets, recognizing patterns, and making predictions or decisions with minimal human intervention. In oncology, AI algorithms have demonstrated remarkable capabilities in interpreting radiological and histopathological images, predicting disease progression and personalizing treatment protocols (Shmatko et al. 2022). In physiotherapy, AI is enabling more accurate movement analysis, personalized rehabilitation plans and remote patient monitoring through smart devices and sensors.

The intersection of Big Data and AI is particularly promising for cancer treatment and rehabilitation. In oncology, precision medicine has become increasingly reliant on the integration of genomic, imaging and clinical data to tailor treatment strategies to individual patients. AI algorithms can rapidly analyze this data to identify actionable mutations, predict treatment response and flag high-risk cases for early intervention. In physiotherapy, AI models are being trained on thousands of data points from patient histories, biomechanics and therapy outcomes to create adaptive exercise regimens and monitor patient progress in real time (Mikolajewska et al. 2024).

Moreover, cancer survivors often face chronic conditions that require long-term physiotherapy, which can be resource-intensive and difficult to standardize. AI-powered systems can address these challenges by enabling scalable, personalized and data-driven rehabilitation strategies. For example, motion capture technologies combined with AI can provide detailed assessments of patient mobility, flagging issues that may not be visible during traditional clinical evaluations. Similarly, virtual assistants and AI chatbots can support patients outside of clinical settings, offering real-time feedback, reminders and encouragement, thereby improving adherence to rehabilitation protocols (Aggarwal et al. 2023).

While the potential benefits of integrating Big Data and AI into cancer care and physiotherapy are substantial, the path from research to practical application is not without challenges. First, the quality and consistency of healthcare data are often variable. Clinical data may be unstructured, incomplete or stored across siloed systems that are not easily interoperable. Standardizing data formats and ensuring the integrity of information is essential for reliable AI training and deployment. Additionally, there are significant ethical concerns surrounding the use of patient data, including issues of privacy, consent and algorithmic transparency. Ensuring that AI systems are explainable, fair and secure is critical to their acceptance and sustainability in clinical environments.

Another barrier lies in the adoption and implementation of these technologies in clinical practice. Despite their potential, many AI and Big Data innovations remain confined to research settings or pilot programs (Pencheva et al. 2020). Clinicians may be hesitant to rely on algorithmic recommendations, particularly when the decision-making process is opaque. Furthermore, the integration of AI tools into existing healthcare workflows requires significant investment in infrastructure, training and interdisciplinary collaboration. Successful implementation depends not only on the robustness of the technology but also on the readiness of healthcare institutions and practitioners to adapt to new models of care.

In this context, this chapter seeks to explore how Big Data and AI are transforming physiotherapy and cancer treatment, tracing the journey from foundational research to clinical practice. By examining current applications, technological advancements, case studies and implementation challenges, this study aims to provide a comprehensive understanding of this evolving landscape (Singh 2024). Special attention is given to the translational aspect – how cutting-edge research findings are being operationalized into real-world solutions that improve patient outcomes, enhance clinical efficiency and reduce healthcare disparities.

This chapter begins by defining the fundamental concepts of Big Data and AI within a healthcare context, followed by an in-depth examination of their applications in oncology and physiotherapy. This includes the use of AI in cancer diagnostics, treatment planning and prognostic modeling, as well as in physiotherapy for movement analysis, personalized interventions and tele-rehabilitation (Vaniya et al. 2024).

The integration of these technologies into combined oncology–physiotherapy care models is also explored, highlighting their role in survivorship programs and quality-of-life improvements.

Subsequently, this chapter delves into the translational pathways that facilitate the shift from research to clinical implementation. This includes an overview of academic research trends, clinical trials incorporating AI and Big Data, and real-world examples of adoption in healthcare settings. Challenges such as data standardization, clinician–patient acceptance, regulatory compliance and ethical considerations are critically analyzed.

Finally, this chapter presents future directions for research, policy and practice, including the development of multi-omics integration, real-world evidence frameworks and collaborative approaches that bridge the gap between data science and clinical care (Mohr et al. 2024).

While the journey from research to practice is complex and multifaceted, the promise of these technologies to revolutionize patient care is undeniable. By understanding the mechanisms, opportunities and challenges of this integration, healthcare professionals, researchers and policymakers can work together to build a future in which intelligent, data-driven care is the standard for all patients – especially those navigating the complexities of cancer and rehabilitation.

12.2. Big Data and AI: definitions and scope

12.2.1. *What is Big Data?*

Big Data in healthcare refers to large volumes of complex health data derived from multiple sources, including EHRs, imaging, genomics, wearable devices and clinical trials. These data are characterized by the “5 Vs”: volume, velocity, variety, veracity and value.

12.2.2. *What is AI in healthcare?*

AI involves machine learning (ML), deep learning (DL), natural language processing (NLP) and computer vision technologies that simulate human intelligence to process data and support decision-making (Sarker 2022). AI systems can identify hidden patterns, predict outcomes and automate administrative and clinical tasks.

12.3. Big Data and AI in oncology

The application of Big Data and AI in oncology represents a paradigm shift in how cancer is diagnosed, treated, and monitored. As cancer care becomes increasingly complex, the ability to process and interpret massive volumes of data – ranging from genetic profiles to imaging scans – is essential for delivering personalized and effective treatment. AI algorithms, trained on Big Data from diverse populations and cancer types, are capable of uncovering patterns that elude conventional analysis. This enables oncologists to make more informed decisions, reduce uncertainty and optimize patient outcomes across the cancer care continuum.

12.3.1. *Precision oncology*

Precision oncology is one of the most impactful areas where Big Data and AI are being leveraged (Dlamini et al. 2020). The approach aims to tailor cancer treatment based on the individual genetic and molecular profile of each patient. AI algorithms analyze data from next-generation sequencing (NGS), whole-genome and exome sequencing, and transcriptomics to identify actionable mutations and molecular targets. For instance, machine learning models are used to predict patient responsiveness to specific therapies such as tyrosine kinase inhibitors, checkpoint inhibitors, or CAR-T (Chimeric Antigen Receptor) cell therapy. These predictions are based on integrated datasets that include genomic alterations, tumor microenvironment characteristics and historical treatment responses.

Pharmacogenomic databases such as The Cancer Genome Atlas (TCGA) and cBioPortal, coupled with AI-driven analytical platforms, facilitate the identification of biomarkers and potential therapeutic targets. NLP also plays a crucial role in mining unstructured clinical notes and scientific literature to match patients with the most appropriate clinical trials or experimental therapies (Chen et al. 2020). As a result, precision oncology not only enhances treatment efficacy but also minimizes adverse effects by avoiding ineffective interventions.

12.3.2. *Early detection and diagnosis*

Early detection is critical for improving cancer survival rates, and AI has emerged as a powerful ally in diagnostic radiology and pathology. Deep learning models trained on vast datasets of radiological images – such as mammograms, CT scans, PET scans and MRIs – can detect tumors with sensitivity and specificity often comparable to (or exceeding that of) experienced clinicians. For example, convolutional neural networks (CNNs) are widely used to analyze breast cancer mammograms, detecting microcalcifications or lesions that may be missed in routine screenings.

AI is also transforming the field of digital pathology. Whole-slide imaging (WSI) combined with AI algorithms can automate the classification, grading and staging of various cancers. These tools reduce diagnostic variability and speed up workflows, particularly in regions with limited access to trained pathologists. Additionally, AI-based decision support systems can flag suspicious findings for further review, facilitating earlier intervention.

In hematological malignancies, AI systems analyze flow cytometry data and genetic markers to detect leukemia subtypes at an early stage (Gedefaw et al. 2023). In lung cancer, AI-integrated liquid biopsy platforms can identify circulating tumor DNA (ctDNA) and other biomarkers that indicate early disease presence. These advancements collectively represent a move toward proactive oncology, where cancers are detected before symptoms emerge, significantly improving the chances of successful treatment.

12.3.3. *Prognostic modeling and risk stratification*

Accurate prognostic modeling is essential for tailoring cancer treatment to individual patient trajectories. By integrating heterogeneous data sources – including EHRs, laboratory results, genetic data, patient-reported outcomes and social determinants of health – AI can create robust models that predict disease progression, recurrence risk and overall survival.

Machine learning algorithms such as random forests, support vector machines (SVMs) and neural networks are commonly used to build these predictive models. These tools stratify patients into risk categories (e.g. low, intermediate, high), which inform clinical decisions such as the need for adjuvant therapy, intensity of surveillance and timing of palliative care. For example, in breast cancer, AI models predict the likelihood of recurrence based on histopathological features, hormone receptor status and gene expression signatures such as Oncotype DX (Ji et al. 2024).

In prostate cancer, AI-driven nomograms estimate the risk of biochemical recurrence after surgery or radiation, guiding the use of adjuvant therapies. Similarly, in colorectal cancer, models integrating clinical and molecular data help predict metastatic potential and guide treatment escalation or de-escalation.

AI also enables dynamic risk stratification – updating prognostic models as new data becomes available. This real-time approach is particularly useful in monitoring treatment response, identifying early signs of resistance and adapting care plans accordingly. The use of longitudinal data and reinforcement learning techniques enhances the ability to personalize care throughout the cancer journey.

12.4. Big Data and AI in physiotherapy

As cancer survivorship increases, the demand for long-term rehabilitation and supportive care also grows. Physiotherapy plays a vital role in mitigating the functional impairments, fatigue and pain that often follow cancer treatments such as surgery, chemotherapy and radiation. The integration of Big Data and AI into physiotherapy offers the opportunity to transform traditional rehabilitation models by providing continuous, personalized and data-driven care. These technologies are particularly beneficial for cancer survivors who require adaptive rehabilitation strategies to address complex, evolving needs.

Big Data enables the collection of comprehensive health and functional data across multiple time points and settings, including clinical encounters, home environments and digital health platforms (Hong et al. 2018). AI leverages this information to make intelligent predictions, recommend tailored interventions and provide real-time feedback, thereby enhancing the effectiveness and accessibility of physiotherapy. The sections below explore key areas where AI and Big Data are reshaping physiotherapy practice, particularly for cancer patients.

12.4.1. *Movement analysis and assessment*

Accurate assessment of movement is critical in physiotherapy to evaluate functional limitations, track progress and design appropriate interventions. Traditionally, movement assessments were subjective and dependent on the clinician's expertise and observation. However, the introduction of AI-powered wearable devices, smartphone-based sensors and markerless motion capture systems has revolutionized this domain.

These technologies gather biomechanical data such as joint angles, stride length, gait velocity, balance metrics and muscle activation patterns. AI algorithms analyze this high-dimensional data to detect movement abnormalities, quantify impairments and identify compensatory mechanisms. For instance, machine learning models can distinguish between normal and pathological gait patterns in cancer survivors experiencing chemotherapy-induced peripheral neuropathy or post-surgical mobility restrictions.

These insights allow for objective benchmarking and early identification of functional decline, enabling timely intervention. Moreover, the data can be continuously collected during daily activities outside of clinical settings, offering a more holistic view of the patient's real-world performance and recovery trajectory (Zhang et al. 2022). For clinicians, this means enhanced decision-making, while for patients, it translates into more responsive and targeted care.

12.4.2. *Personalized rehabilitation programs*

One of the most promising applications of AI in physiotherapy is the development of personalized rehabilitation programs. Cancer survivors often exhibit diverse and unpredictable responses to rehabilitation due to differences in treatment types, comorbidities, baseline fitness and psychosocial factors. A one-size-fits-all approach is thus inadequate for this population.

Machine learning algorithms trained on large datasets of patient outcomes can identify patterns in therapy effectiveness and recovery timelines. By incorporating real-time sensor inputs and patient-specific data – including age, cancer type, functional status and adherence rates – AI systems can tailor exercise programs to optimize effectiveness while minimizing the risk of overexertion or injury.

For example, reinforcement learning algorithms can continuously adjust exercise intensity and type based on feedback from wearable sensors and patient-reported outcomes. If a patient reports increased fatigue or demonstrates suboptimal range of motion, the AI system can automatically modify the next day's regimen to focus on rest or alternative therapeutic activities.

Additionally, predictive models can estimate a patient's likely trajectory, enabling clinicians to set realistic goals and intervene proactively (Terranova and Venkatakrishnan 2024). This dynamic, data-driven personalization not only improves physical recovery but also enhances patient engagement and motivation by aligning therapy with individual capacities and goals.

12.4.3. *Remote monitoring and telerehabilitation*

Accessibility to physiotherapy is a significant barrier for many cancer patients, particularly those living in rural areas, facing transportation challenges, or dealing with fatigue and immunosuppression. Telerehabilitation and remote monitoring systems powered by AI and Big Data offer a solution by enabling the delivery of care beyond traditional clinical settings.

These platforms use data from wearable devices, home-based sensors and patient-reported outcomes to continuously monitor health and functional status.

AI algorithms synthesize this data to detect deviations from expected recovery patterns, alert clinicians to emerging issues and recommend modifications to therapy plans.

For example, a patient recovering from breast cancer surgery may use a mobile application integrated with a shoulder motion sensor. The app tracks daily movement metrics and uses AI to determine whether the patient is regaining shoulder mobility as expected (Sassi et al. 2024). If the progress deviates from the predicted path, the system notifies the physiotherapist or provides corrective feedback to the patient.

Moreover, AI-powered virtual assistants and chatbots can conduct preliminary assessments, provide educational support, schedule therapy sessions and collect symptom reports in real time.

These tools enhance communication, promote adherence and ensure continuity of care, particularly when in-person visits are limited or impractical.

The combination of telerehabilitation and AI-driven analytics ensures that patients receive timely, appropriate and consistent support throughout their recovery journey. It also enables physiotherapists to manage larger caseloads more efficiently while maintaining personalized care.

12.5. Integration in cancer care: combined oncology and physiotherapy applications

12.5.1. *Multimodal data fusion*

Combining imaging, genomics, clinical records and physiotherapy outcomes offers a comprehensive view of patient status. AI can synthesize these modalities to provide integrated treatment recommendations.

12.5.2. *Survivorship and quality of life*

Cancer survivors often deal with long-term side effects such as fatigue, lymphedema and functional limitations (Lovelace et al. 2019).

AI tools analyze long-term data to identify patterns and predict late effects, guiding long-term physiotherapy interventions.

12.5.3. *AI-driven decision support systems*

Clinical Decision Support Systems (CDSS) powered by AI help oncologists and physiotherapists choose optimal interventions by processing patient history, current status and best-practice guidelines.

12.6. Translational pathways: from research to practice

12.6.1. *Academic research and clinical trials*

Recent trials incorporate AI and Big Data tools to analyze patient outcomes and enhance trial efficiency. For example, digital phenotyping is used to classify patient subgroups for tailored interventions.

12.6.2. *Clinical implementation*

Healthcare systems are beginning to adopt AI-powered tools in oncology and physiotherapy departments. Success depends on usability, clinician training and system integration.

12.6.3. *Regulatory and ethical considerations*

The transition from research to practice must address privacy concerns (e.g. HIPAA compliance), algorithmic bias, data security and accountability in decision-making (Blessing 2024).

12.7. Case studies

12.7.1. *IBM Watson for oncology*

IBM Watson supports oncologists by recommending treatment options based on evidence and patient data. It has been integrated into clinical workflows in hospitals in the US, China and India.

12.7.2. AI-powered rehabilitation platforms

Platforms such as Kaia Health and Reflexion Health use AI and motion tracking for home-based physiotherapy (Kakegawa and Matsuda 2024). These tools have shown positive outcomes in both general and cancer-specific rehabilitation programs.

12.7.3. Predictive analytics in breast cancer recovery

Studies using AI to analyze EHR and physiotherapy data have successfully predicted recovery times and recommended interventions, post-lumpectomy or mastectomy.

12.8. Challenges and limitations

12.8.1. *Data quality and standardization*

Inconsistent data entry, missing values and heterogeneous formats make it difficult to harness Big Data effectively (Rozony et al. 2024). Interoperability standards such as HL7 FHIR (Health Level 7 Fast Healthcare Interoperability Resources) are essential but not yet universally adopted.

12.8.2. *Algorithm bias and equity*

AI models trained on biased datasets may underperform for minority populations, exacerbating healthcare disparities. Transparent model training and validation across diverse datasets are crucial.

12.8.3. *Clinician and patient acceptance*

Resistance to AI due to fear of job displacement or mistrust in automation is a barrier. Educating both clinicians and patients about the supportive – not substitutive – role of AI is key to building trust and ensuring successful adoption.

12.9. Future directions

The integration of Big Data and AI into oncology and physiotherapy is still evolving, with immense untapped potential (Vaniya et al. 2024). As technological capabilities expand, the future of personalized, data-driven cancer care will rely on more sophisticated algorithms, richer data sources and seamless integration into clinical workflows. The following sections explore emerging frontiers and strategic priorities that can shape the next decade of innovation and implementation.

12.9.1. *Integration of multi-omics data*

One of the most promising frontiers is the incorporation of multi-omics data – such as genomics, transcriptomics, proteomics, metabolomics and microbiomics – into clinical and physiotherapy decision-making. Integrating these layers of biological information allows for a more comprehensive understanding of cancer behavior, treatment response and recovery trajectories.

For example, combining tumor genomic data with proteomic biomarkers and microbiome profiles may help identify subtypes of cancer that respond better to certain physiotherapy modalities, or are more prone to treatment-induced fatigue or sarcopenia. AI can uncover hidden patterns and causal relationships within these complex, high-dimensional datasets, ultimately informing highly personalized interventions.

Furthermore, linking multi-omics data with physical performance metrics (e.g. gait analysis, strength tests) and EHRs can enhance predictive models, enabling early detection of functional decline and adverse effects. This holistic view supports both precision oncology and precision rehabilitation.

12.9.2. *AI-driven adaptive interventions*

The future of cancer care lies in adaptive, responsive and patient-specific treatment strategies. Reinforcement learning (RL), a branch of machine learning, is uniquely positioned to power these interventions (Sivamayil et al. 2023).

RL algorithms can learn optimal treatment strategies through continuous interaction with patient data and feedback, adjusting therapy in real time to maximize outcomes.

In physiotherapy, RL models could dynamically modify exercise intensity, frequency and type based on real-time sensor inputs, such as range of motion, pain levels or fatigue. In oncology, adaptive systems might optimize chemotherapy dosing or radiation planning by continuously analyzing tumor response and patient tolerance (Lavrova et al. 2023).

These AI-driven adaptive systems offer a significant leap from static treatment protocols to intelligent, self-correcting pathways that evolve alongside the patient.

However, their implementation will require rigorous validation, robust clinical trials and transparent explainability mechanisms to gain widespread clinical trust.

12.9.3. *Cross-disciplinary collaboration*

The development, validation and implementation of AI tools in cancer care and physiotherapy demand close collaboration across traditionally siloed disciplines (Rahman et al. 2024). Oncologists and physiotherapists bring clinical context and patient-centered knowledge, while data scientists contribute algorithmic expertise and engineer support system integration. Ethicists and legal experts are also vital to address issues related to fairness, consent, privacy and accountability.

Creating successful AI-driven healthcare ecosystems will require shared governance structures, common data standards and interdisciplinary training programs (Rahman et al. 2024). Collaborative research platforms and consortia can facilitate data sharing and model generalizability across diverse populations and healthcare settings.

Moreover, involving patients and caregivers as co-creators of AI solutions ensures that the tools developed are relevant, usable and aligned with real-world needs (Weiler et al. 2024). Such inclusive innovation will help bridge the gap between technological possibility and practical benefit.

12.9.4. *Real-world evidence and longitudinal studies*

To assess the true impact of AI and Big Data in oncology and physiotherapy, there is a pressing need for real-world evidence (RWE) and longitudinal data collection (Kataria and Ravindran 2022). While randomized controlled trials (RCTs) remain the gold standard for evaluating efficacy, they are often limited in scale, duration and generalizability.

Real-world data from EHRs, wearable devices, telerehabilitation platforms and patient-reported outcomes can provide valuable insights into how AI tools perform across various healthcare settings and population groups (Kataria and Ravindran 2022).

Longitudinal studies capturing the long-term outcomes of AI-assisted interventions – such as improved survival, quality of life, functional independence and healthcare cost savings – are crucial for building confidence and guiding policy.

Furthermore, continuous learning health systems, where AI models are updated in real time based on new incoming data, represent the next step toward truly intelligent, adaptive care pathways (Aminizadeh et al. 2024). These systems require ethical oversight and technical infrastructure but promise more responsive and equitable care delivery.

12.10. Conclusion

The integration of Big Data and AI into the fields of oncology and physiotherapy is fundamentally reshaping modern healthcare. These technologies offer unprecedented capabilities in collecting, analyzing and interpreting large volumes of diverse health data, enabling more precise diagnostics, individualized treatment plans and dynamic rehabilitation programs. In oncology, AI models support early detection of malignancies, accurate staging, personalized therapeutic decision-making and predictive outcome modeling. In physiotherapy, AI enhances movement analysis, adapts rehabilitation protocols in real time and enables continuous remote monitoring through wearable technologies and virtual platforms. The synergistic application of Big Data and AI facilitates a more holistic and responsive approach to patient care, especially for cancer survivors who require long-term rehabilitation. Importantly, this technological advancement supports a shift toward patient-centered care, where interventions are tailored not only to disease characteristics but also to individual functional goals, preferences and recovery trajectories. Despite these significant advancements, the transition from research to routine clinical practice presents multiple challenges. Data quality, interoperability and standardization remain critical hurdles. Ethical issues such as data privacy, algorithmic transparency and bias mitigation must also be rigorously addressed to maintain trust in AI-driven healthcare. Furthermore, the successful adoption of these technologies hinges on the readiness of healthcare institutions, the availability of skilled personnel and the willingness of clinicians and patients to embrace data-driven tools.

12.11. References

Aggarwal, A., Tam, C.C., Wu, D., Li, X., Qiao, S. (2023). Artificial intelligence-based chatbots for promoting health behavioral changes: Systematic review. *Journal of Medical Internet Research*, 25, e40789.

Aminizadeh, S., Heidari, A., Dehghan, M., Toumaj, S., Rezaei, M., Navimipour, N.J., Stroppa, F., Unal, M. (2024). Opportunities and challenges of artificial intelligence and distributed systems to improve the quality of healthcare service. *Artificial Intelligence in Medicine*, 149, 102779.

Blessing, E. (2024). Regulatory compliance and ethical considerations: Compliance challenges and opportunities with the integration of Big Data and AI. Report, HAL Open Science. doi: 10.5281/zenodo.14926009.

Chen, X., Xie, H., Cheng, G., Poon, L.K., Leng, M., Wang, F.L. (2020). Trends and features of the applications of natural language processing techniques for clinical trials text analysis. *Applied Sciences*, 10(6), 2157.

Crosby, D., Bhatia, S., Brindle, K.M., Coussens, L.M., Dive, C., Emberton, M., Esener, S., Fitzgerald, R.C., Gambhir, S.S., Kuhn, P. et al. (2022). Early detection of cancer. *Science*, 375(6586), eaay9040.

Dlamini, Z., Francies, F.Z., Hull, R., Marima, R. (2020). Artificial intelligence (AI) and big data in cancer and precision oncology. *Computational and Structural Biotechnology Journal*, 18, 2300–2311.

Gedefaw, L., Liu, C.F., Ip, R.K., Tse, H.F., Yeung, M.H., Yip, S.P., Huang, C.L. (2023). Artificial intelligence-assisted diagnostic cytology and genomic testing for hematologic disorders. *Cells*, 12(13), 1755.

Hong, L., Luo, M., Wang, R., Lu, P., Lu, W., Lu, L. (2018). Big data in healthcare: Applications and challenges. *Data and Information Management*, 2(3), 175–197.

Ji, J.H., Ahn, S.G., Yoo, Y., Park, S.Y., Kim, J.H., Jeong, J.Y., Park, S., Lee, I. (2024). Prediction of a multi-gene assay (oncotype DX and Mammaprint) recurrence risk group using machine learning in estrogen receptor-positive, HER2-negative breast cancer – the BRAIN study. *Cancers*, 16(4), 774.

Kakegawa, K. and Matsuda, T. (2024). Challenges and prospects of sensing technology for the promotion of tele-physiotherapy: A narrative review. *Sensors*, 25(1), 16.

Kataria, S. and Ravindran, V. (2022). Harnessing of real-world data and real-world evidence using digital tools: Utility and potential models in rheumatology practice. *Rheumatology*, 61(2), 502–513.

Lavrova, E., Garrett, M.D., Wang, Y.F., Chin, C., Elliston, C., Savacool, M., Price, M., Kachnic, L.A., Horowitz, D.P. (2023). Adaptive radiation therapy: A review of CT-based techniques. *Radiology: Imaging Cancer*, 5(4), e230011.

Lovelace, D.L., McDaniel, L.R., Golden, D. (2019). Long-term effects of breast cancer surgery, treatment, and survivor care. *Journal of Midwifery & Women's Health*, 64(6), 713–724.

Mikolajewska, E., Masiak, J., Mikolajewski, D. (2024). Applications of artificial intelligence-based patient digital twins in decision support in rehabilitation and physical therapy. *Electronics*, 13(24), 4994.

Mohr, A.E., Ortega-Santos, C.P., Whisner, C.M., Klein-Seetharaman, J., Jasbi, P. (2024). Navigating challenges and opportunities in multi-omics integration for personalized healthcare. *Biomedicines*, 12(7), 1496.

Pencheva, I., Esteve, M., Mikhaylov, S.J. (2020). Big data and AI – A transformational shift for government: So, what next for research? *Public Policy and Administration*, 35(1), 24–44.

Rahman, M.H., Hossan, K.M., Uddin, M.K., Hossain, M.D. (2024). Improving collaborative interactions between humans and artificial intelligence to achieve optimal patient outcomes in the healthcare industry. Report, SSRN.

Rozony, F.Z., Aktar, M.N., Ashrafuzzaman, M., Islam, A. (2024). A systematic review of big data integration challenges and solutions for heterogeneous data sources. *Academic Journal on Business Administration, Innovation & Sustainability*, 4(4), 1–8.

Sarker, I.H. (2022). AI-based modeling: Techniques, applications and research issues towards automation, intelligent and smart systems. *SN Computer Science*, 3(2), 158.

Sassi, M., Villa Corta, M., Pisani, M.G., Nicodemi, G., Schena, E., Pecchia, L., Longo, U.G. (2024). Advanced home-based shoulder rehabilitation: A systematic review of remote monitoring devices and their therapeutic efficacy. *Sensors*, 24(9), 2936.

Shmatko, A., Ghaffari Laleh, N., Gerstung, M., Kather, J.N. (2022). Artificial intelligence in histopathology: Enhancing cancer research and clinical oncology. *Nature Cancer*, 3(9), 1026–1038.

Singh, M. (2024). Navigating the landscape: An in-depth exploration of modern application development methodologies and practices. In *International Conference on Innovations and Challenges in Emerging Technologies (ICICET)*. IEEE, Nagpur.

Sivamayil, K., Rajasekar, E., Aljafari, B., Nikolovski, S., Vairavasundaram, S., Vairavasundaram, I. (2023). A systematic study on reinforcement learning based applications. *Energies*, 16(3), 1512.

Terranova, N. and Venkatakrishnan, K. (2024). Machine learning in modeling disease trajectory and treatment outcomes: An emerging enabler for model-informed precision medicine. *Clinical Pharmacology & Therapeutics*, 115(4), 720–726.

Vaniya, J., Gandhi, N.V., Patel, K.P., Bhatt, C., Kaur, K.P. (2024). Role and scope of artificial intelligence in physiotherapy: A literature review. *Research Journal of Medical Sciences*, 18, 399–412.

Weiler, T., Müller, C., Struzek, D., Huldtgren, A., Klapperich, H., Grosskopp, S., Fischer, F., Osterheider, A., Gaertner, W. (2024). *2nd International Workshop on Co-Creation of Hybrid Interactive Systems for Healthcare. Mensch und Computer 2024 – Workshopband.* Gesellschaft für Informatik, Karlsruhe.

Zhang, J., Symons, J., Agapow, P., Teo, J.T., Paxton, C.A., Abdi, J., Mattie, H., Davie, C., Torres, A.Z., Folarin, A. et al. (2022). Best practices in the real-world data life cycle. *PLOS Digital Health*, 1(1), e0000003.

13

Emerging AI Technologies Shaping the Future of Cancer Care

The integration of Artificial Intelligence (AI) into oncology is transforming cancer care, offering a new frontier in early detection, diagnosis, treatment planning, patient monitoring and drug discovery. With the increasing complexity of cancer and the exponential growth of healthcare data, traditional approaches are often insufficient to provide timely and accurate decisions. Emerging AI technologies – ranging from deep learning and natural language processing to radiomics and AI-assisted genomics – are now enabling a more data-driven, precise and personalized approach to cancer management. This chapter explores how cutting-edge AI technologies are reshaping every stage of the cancer care continuum. In diagnostics, AI models demonstrate expert-level performance in interpreting medical imaging, histopathology slides and genomic data. In treatment, AI supports oncologists in developing personalized protocols, predicting treatment responses and minimizing adverse effects. AI is also revolutionizing drug discovery by identifying novel compounds and accelerating clinical trials through patient matching algorithms. Furthermore, AI-driven virtual assistants and wearable-integrated monitoring systems are empowering patients, enhancing communication and reducing hospital readmissions. Despite these advancements, challenges such as algorithmic bias, data privacy concerns, lack of transparency and the need for clinical validation hinder full-scale integration. The ethical implications of decision-making automation and the trustworthiness of AI systems remain key concerns. Nonetheless, the future of AI in oncology is promising, driven by continued innovation, interdisciplinary collaboration and regulatory evolution. This chapter presents a comprehensive overview of current and emerging AI applications in cancer care, critically analyzing their potential, limitations and the roadmap ahead.

13.1. Introduction

Cancer remains one of the most formidable health challenges of the 21st century, accounting for nearly 10 million deaths annually, according to the World Health

Chapter written by Omkar SOMADE, Rashmi GUDUR, Nitin K. CHAUDHARY and Mahendra ALATE.

Organization (WHO) (Ibraheem et al. 2025). Its rising incidence, coupled with the increasing complexity of diagnosis and treatment, demands continual innovation in medical research and clinical practices. While traditional diagnostic and therapeutic approaches have improved survival rates over the decades, they often struggle to accommodate the vast heterogeneity of cancer across different individuals. Factors such as tumor biology, genetic mutations and variable responses to treatment make one-size-fits-all strategies ineffective for many patients.

Recent years have witnessed a dramatic surge in the use of Artificial Intelligence (AI) to tackle these challenges (Bohr and Memarzadeh 2020). AI technologies – particularly those powered by machine learning (ML), deep learning (DL) and natural language processing (NLP) – are rapidly transforming the oncology landscape. These tools are designed to replicate human cognitive functions such as learning, reasoning, pattern recognition and decision-making. Crucially, AI systems can analyze vast and complex datasets – including imaging studies, histopathological slides, genomic sequences, clinical notes and electronic health records (EHRs) – with a level of speed, accuracy and scalability that exceeds traditional methods.

The convergence of AI and oncology is facilitating a shift toward precision medicine, where diagnostics and treatments are tailored to the individual characteristics of each patient (Sherani et al. 2024). AI models can now assist in the early detection of tumors, automate image analysis, predict therapeutic responses, monitor treatment outcomes and even support drug discovery efforts. These innovations promise to enhance clinical decision-making, reduce diagnostic errors and improve patient outcomes.

This chapter explores the cutting-edge AI technologies that are shaping the future of cancer care. It examines their applications across the entire cancer care continuum – from prevention and diagnosis to treatment and follow-up – while also discussing the associated challenges, ethical considerations and the roadmap for integration into mainstream clinical practice (Mao et al. 2022). Through this comprehensive review, this chapter underscores the transformative potential of AI in achieving more accurate, efficient and personalized oncology care.

13.2. AI in early detection and diagnosis

Early detection of cancer significantly improves treatment outcomes and survival rates. However, conventional diagnostic methods often rely heavily on subjective interpretation, which may lead to variability in assessments and delayed diagnoses (Pascoal et al. 2022). Emerging AI technologies, particularly those based on machine learning and deep learning, have demonstrated remarkable potential in enhancing the precision, speed and consistency of cancer detection. AI can process vast amounts of clinical and imaging data to identify subtle patterns indicative of malignancy – patterns often invisible to the human eye.

13.2.1. *Imaging-based diagnostics*

Medical imaging modalities such as computed tomography (CT), magnetic resonance imaging (MRI), positron emission tomography (PET) and mammography are foundational tools in oncology (Grueneisen et al. 2015) Traditionally, radiologists interpret these images manually, but AI has brought transformative advancements in this domain.

13.2.1.1. *Deep learning in radiology*

Deep learning, especially through convolutional neural networks (CNNs), has enabled the development of highly accurate image analysis systems. These models learn from thousands of annotated images to identify tumors, assess tumor size and shape, and even distinguish between cancerous and non-cancerous lesions.

One landmark achievement came from Google Health's AI model, which was trained on over 90,000 mammograms (Daraghmeh 2024). The system not only matched but, in some cases, surpassed expert radiologists in detecting breast cancer, reducing both false positives and false negatives. Similar approaches are being applied to lung, prostate and brain cancer screening.

These models continue to improve through federated learning and transfer learning strategies, which allow algorithms to learn from diverse, decentralized datasets without compromising patient privacy.

13.2.1.2. *Radiomics*

Radiomics is the process of extracting high-dimensional quantitative features from medical images, including texture, shape and intensity, which can be correlated with genetic markers and clinical outcomes. When combined with AI, radiomic analysis provides a non-invasive method to characterize tumors at a molecular level.

AI-driven radiomics enables:

- differentiation between benign and malignant lesions;

- prediction of tumor aggressiveness;

- forecasting of treatment response and progression;

- identification of radiogenomic biomarkers for personalized therapy.

These techniques are particularly valuable in monitoring tumor heterogeneity and evolution over time, offering a new layer of diagnostic insight beyond conventional radiology (Hussain et al. 2024).

13.2.2. *Pathology and histology*

Histopathological analysis remains the gold standard for cancer diagnosis, involving the microscopic examination of tissue samples. However, manual slide review is time-consuming and subject to intra- and inter-observer variability. AI solutions are now being developed to address these challenges by digitizing and analyzing pathology slides using deep learning algorithms.

13.2.3. *AI-powered histopathology*

AI can:

– accurately classify tissue samples;

– detect mitotic figures;

– quantify tumor-infiltrating lymphocytes (TILs);

– grade tumors based on morphology;

– identify rare cell types that may indicate early disease stages.

In comparative studies, AI algorithms have achieved sensitivity and specificity levels comparable to (or exceeding) human pathologists, particularly in detecting breast, prostate and skin cancers (Rai et al. 2024).

13.2.4. *Case study: Lunit INSIGHT*

One notable example is Lunit INSIGHT, an AI-based pathology solution that analyzes digital histopathology images. It can detect tumor-infiltrating lymphocytes in breast cancer samples – a critical marker for immune response and a predictor of response to immunotherapy. The system supports oncologists by offering quantifiable, reproducible metrics that inform prognosis and treatment planning.

Lunit's tools have been integrated into clinical trials and research studies, demonstrating significant time savings and enhanced diagnostic precision. Moreover, the explainability feature in its platform allows clinicians to understand and trust the AI-generated results, promoting clinical adoption.

13.3. AI in genomics and precision medicine

The emergence of genomics has revolutionized our understanding of cancer as a molecular disease rather than merely a histological anomaly (Bell 2010). Cancer is driven by complex genetic mutations that vary significantly across individuals. The ability to decode the human genome, especially through next-generation sequencing (NGS), has created a deluge of data – offering unprecedented opportunities for individualized cancer care. However, this wealth of data exceeds the analytic capabilities of traditional bioinformatics tools. AI technologies are now stepping in to bridge this gap by rapidly processing, interpreting and translating genomic information into actionable clinical insights.

13.3.1. *AI-driven genomic analysis*

Next-generation sequencing technologies have unlocked access to detailed genetic profiles of cancer patients, identifying single nucleotide polymorphisms (SNPs), copy number variations, gene fusions and other mutations (Krubaa et al. 2024). AI, particularly machine learning algorithms, plays a crucial role in sifting through these massive datasets to uncover meaningful patterns that may indicate cancer subtypes, disease progression or treatment responsiveness.

13.3.2. *Applications of AI in genomic oncology*

– *Identifying driver mutations*: AI can distinguish between driver mutations that initiate or sustain tumor growth and passenger mutations that are biologically irrelevant. This aids in selecting therapeutic targets.

– *Predicting genetic predisposition*: AI models can analyze inherited germline mutations and polygenic risk scores to identify individuals at a high risk of developing specific cancers, enabling preemptive surveillance and preventive interventions.

– *Forecasting disease progression*: by analyzing mutation trajectories and clonal evolution, AI can predict how a tumor is likely to evolve, which is particularly important for understanding drug resistance mechanisms and metastatic potential.

13.3.3. *Personalized medicine*

Personalized medicine – or precision oncology – is the tailoring of medical treatment to the individual characteristics of each patient (Jiang and Wang 2010). AI is instrumental in integrating diverse data types – such as genomic, transcriptomic, proteomic, radiologic and clinical data – to create a comprehensive patient profile.

13.3.4. *AI-powered therapeutic decision support*

AI platforms can evaluate the genomic alterations in a patient's tumor and cross-reference this information with vast biomedical literature, clinical trial data and treatment guidelines to recommend targeted therapies. For example:

– IBM Watson for Genomics integrates genomic data with curated medical literature to identify clinically actionable mutations and suggest treatment options, including experimental therapies and clinical trial eligibility (Saeed 2024).

– Tempus and Foundation Medicine offer AI-enhanced decision support systems that provide oncologists with real-time recommendations based on tumor sequencing results and phenotypic information.

These tools are improving treatment efficacy, minimizing adverse reactions and reducing the trial-and-error approach historically common in oncology.

13.3.5. *Integration with electronic health records*

Modern AI systems can also draw on EHRs to contextualize genomic data with a patient's history, co-morbidities and past responses to treatment. This creates a feedback loop where clinical outcomes continuously inform and refine AI models, enhancing predictive accuracy over time.

13.4. AI in treatment planning and optimization

Effective cancer treatment requires a highly individualized approach due to the diverse biological behavior of tumors and varied patient responses. Artificial Intelligence (AI) is increasingly being integrated into oncologic treatment planning to enhance precision, reduce clinician workload and improve clinical outcomes (Dlamini et al. 2020). From radiotherapy to chemotherapy and immunotherapy, AI tools are transforming how treatment protocols are designed, optimized and administered.

13.4.1. *Radiotherapy*

Radiation therapy is a cornerstone of cancer treatment, used in approximately 50% of all cancer cases. However, the planning process is complex, requiring accurate delineation of tumors and surrounding organs-at-risk (OARs), dose distribution calculations and consideration of tissue tolerance.

13.4.2. *AI-enhanced treatment planning*

AI-driven systems, particularly those using deep learning, are significantly improving radiotherapy workflows by:

– automating contouring of tumors and OARs, which is traditionally time-consuming and subject to inter-observer variability;

– predicting optimal dose distributions based on historical plans and patient anatomy, ensuring maximum tumor control while minimizing harm to healthy tissues;

– reducing planning time, thereby accelerating treatment initiation, which is critical for aggressive malignancies.

13.4.3. *Auto-segmentation*

One of the most notable AI contributions is auto-segmentation – the use of convolutional neural networks (CNNs) to delineate tumor boundaries and critical structures on imaging scans such as CT and MRI (Kalantar et al. 2021).

Studies have shown that AI-assisted contouring can match expert-level performance, reduce planning time by up to 50% and improve consistency across clinical settings.

Several commercial solutions, such as Varian's Eclipse and RaySearch's RayStation, now incorporate AI for auto-segmentation and adaptive radiotherapy planning, marking a shift toward more personalized and precise radiotherapy.

13.4.4. *Chemotherapy and immunotherapy optimization*

Chemotherapy remains a fundamental treatment modality, often administered in combination with other therapies. However, determining the most effective regimen – considering factors such as drug resistance, toxicity and individual patient response – is a persistent challenge (Nikolaou et al. 2018).

AI is now playing a key role in personalizing and optimizing chemotherapy protocols.

13.4.5. *AI in chemotherapy planning*

AI systems use patient-specific clinical, molecular and pharmacogenomic data to:

– predict treatment response and toxicity;

– recommend drug combinations and dosages tailored to the patient's genetic and phenotypic profile;

– monitor patient tolerance and dynamically adjust therapy over time.

Machine learning models have been applied to datasets encompassing thousands of chemotherapy patients, enabling the forecasting of side effects such as neutropenia, gastrointestinal toxicity and cardiotoxicity (Hassan et al. 2024). This allows for more informed treatment decisions and preemptive management of adverse effects.

13.4.6. *AI in immunotherapy*

Immunotherapy, particularly immune checkpoint inhibitors, has revolutionized the treatment of several cancers, including melanoma, lung cancer and bladder cancer. However, only a subset of patients responds favorably, making predictive biomarker identification crucial.

13.4.7. *Biomarker prediction and patient stratification*

AI excels in predicting key biomarkers for immunotherapy, such as:

– PD-L1 expression levels;

– tumor mutational burden (TMB);

– microsatellite instability (MSI).

By analyzing histological images, genomic data and multi-omics profiles, AI can stratify patients based on their likelihood of response to immunotherapy. This minimizes the risk of administering costly and potentially ineffective treatments.

Additionally, AI is being used to discover novel immune markers by integrating data from immunogenomics, tumor microenvironment analysis and single-cell sequencing, thereby expanding the scope of personalized immunotherapy (Chang et al. 2025).

13.5. AI in drug discovery and development

AI is revolutionizing the pharmaceutical landscape by streamlining the traditionally time-consuming and expensive process of drug discovery and development. AI-driven systems can analyze vast datasets, recognize patterns and generate predictive models that enhance the identification of novel drug candidates, repurpose existing medications and forecast drug–target interactions. These capabilities are particularly critical in oncology, where timely access to effective therapies can significantly influence patient outcomes.

13.5.1. *Virtual screening and molecular design*

AI-powered virtual screening tools facilitate the rapid evaluation of millions of chemical compounds against biological targets to identify promising candidates for cancer treatment (Odah 2025). These platforms integrate deep learning models, such as convolutional neural networks (CNNs) and graph neural networks (GNNs), to predict the binding affinity, toxicity and pharmacokinetic properties of molecules.

Prominent tools such as DeepChem, AtomNet and MoleculeNet have demonstrated success in uncovering novel anticancer agents by learning from large-scale chemical and biological datasets.

AI can also assist in de novo molecular design, enabling the creation of entirely new compounds optimized for efficacy, safety and bioavailability. This process not only accelerates the discovery phase but also reduces the need for exhaustive laboratory experimentation.

13.5.2. *Drug repurposing*

Drug repurposing – identifying new therapeutic uses for existing drugs – has gained traction as a cost-effective and time-efficient strategy (Majeed et al. 2022).

AI models trained on genomic, proteomic and clinical data can uncover hidden relationships between known drugs and new cancer targets. This approach is particularly useful in rare and aggressive cancers where traditional drug development may be financially unviable.

For example, machine learning algorithms have been used to repurpose metformin (a diabetes medication) and statins (cholesterol-lowering agents) for their potential anticancer properties. By analyzing molecular pathways and patient data, AI uncovers novel indications, thereby opening new treatment avenues without the need for extensive safety testing.

13.5.3. *Predictive modeling of drug–target interactions*

Understanding how a drug interacts with its target is fundamental to effective drug development. AI enhances this process by creating predictive models that simulate drug–target interactions at the molecular level. These models can incorporate three-dimensional structural data, protein–ligand docking scores and biochemical properties to predict efficacy and off-target effects.

Technologies such as AlphaFold, which predicts protein structures with unprecedented accuracy, further complement AI-based drug–target interaction models (Ozaybi et al. 2024). This integration allows researchers to virtually test thousands of compounds, narrowing down viable options for in vitro and in vivo studies. These predictive capabilities are particularly useful in targeting cancer-related proteins with high mutation rates.

13.5.4. *AI in clinical trial design and optimization*

AI not only assists in preclinical phases but also plays a pivotal role in optimizing clinical trials. Machine learning algorithms analyze electronic health records (EHRs), genomic profiles and patient registries to identify ideal candidates for specific trials. This results in faster patient recruitment and ensures a more diverse and representative cohort, improving the generalizability of trial outcomes.

Moreover, AI tools can simulate adaptive trial designs, enabling real-time adjustments based on interim results. These dynamic models enhance the efficiency of trials by reducing time and cost while increasing the probability of success. AI also facilitates real-time monitoring of participants by integrating wearable sensor data and electronic diaries, allowing for the early detection of adverse events and improving patient safety.

13.5.5. *Regulatory considerations and ethical implications*

While AI offers transformative potential, it also introduces challenges in regulatory compliance and ethical considerations (Blessing 2024). Drug approval agencies such as the FDA and EMA are gradually adapting to incorporate AI-driven methodologies, but concerns around data bias, transparency and reproducibility persist. Ensuring that AI models are trained on diverse datasets and undergo rigorous validation is crucial for maintaining public trust and scientific integrity.

Additionally, the use of patient data in AI algorithms raises questions about privacy, consent and data ownership. Ethical frameworks must evolve alongside technological advances to ensure equitable access and avoid unintended consequences, particularly in vulnerable populations.

13.6. AI in patient monitoring and support

As cancer care becomes increasingly complex and patient-centered, AI technologies are playing a vital role in improving patient monitoring and support (Marion et al. 2024). These innovations ensure continuous care beyond hospital settings, enhance patient engagement and enable timely interventions. AI-driven tools, such as virtual assistants, chatbots and wearable devices, are transforming how patients manage their treatment journeys and how healthcare professionals track health metrics and treatment responses.

13.6.1. *Virtual assistants and chatbots*

AI-powered virtual assistants and chatbots offer 24/7 support for cancer patients, facilitating communication, education and self-management. These tools use natural language processing (NLP) and machine learning to interpret patient queries, provide accurate information and guide patients through complex treatment regimens.

13.6.1.1. *Key functionalities*

– Answering questions about symptoms, medications, and treatment options;

– sending personalized medication and appointment reminders;

– recording symptoms or side effects for clinical review;

– offering emotional support and health literacy improvement.

13.6.1.2. *Example: IBM Watson assistant for oncology*

This AI-driven assistant supports both patients and oncologists by providing evidence-based treatment recommendations, derived from a vast corpus of clinical guidelines, research publications and patient records (Shapiro et al. 2024). It helps clinicians assess treatment options tailored to a patient's genetic and clinical profile while offering patients understandable explanations and psychosocial support. By doing so, it bridges the communication gap and enhances shared decision-making.

Emerging tools such as CareAngel, Ada Health and Youper are also contributing to personalized cancer care by offering mental health tracking, pain management suggestions and integration with digital health records.

13.6.2. *Wearables and remote monitoring*

The integration of AI with wearable technology and remote monitoring devices has revolutionized how clinicians oversee patient health outside traditional settings (Kanakaprabha et al. 2024). Smartwatches, biosensors and mobile health apps collect continuous physiological and behavioral data, including:

– heart rate, respiratory rate and oxygen saturation;

– physical activity, sleep patterns and fatigue levels;

– medication adherence and side effect tracking.

AI algorithms analyze this data in real time to detect anomalies, predict health deterioration and trigger alerts to healthcare providers or caregivers. This proactive approach enables early intervention, potentially reducing hospital readmissions and improving outcomes in cancer patients, especially those undergoing chemotherapy or recovering from surgery.

13.6.2.1. *Example applications*

– Apple Watch and Fitbit devices monitor vital signs and can be paired with AI apps for symptom tracking in cancer patients;

– BioIntelliSense's BioSticker offers continuous vital sign monitoring, helping detect early signs of infection or respiratory distress in immunocompromised individuals;

– Current Health and VitalConnect platforms integrate wearable data into clinical dashboards for remote patient management by oncology care teams.

Additionally, AI-enhanced remote monitoring supports teleoncology, allowing for virtual consultations, progress tracking and treatment adjustments without requiring frequent hospital visits (Busnatu et al. 2022). This is especially beneficial for rural or mobility-limited patients, improving access to care and reducing burden on healthcare infrastructure.

13.6.3. *Predictive health analytics and personalized support*

Beyond monitoring, AI can predict future health events by analyzing trends in collected data. Predictive models can forecast:

– risk of neutropenia or febrile episodes during chemotherapy;

– likelihood of treatment non-adherence or depression;

– probability of unplanned hospital admissions.

Such insights allow clinicians to personalize interventions and allocate resources effectively. AI can also tailor educational materials and self-care strategies based on individual needs, improving engagement and health outcomes.

13.7. Ethical and legal considerations

As AI becomes increasingly embedded in healthcare systems, particularly in oncology and physiotherapy, it brings with it a host of ethical, legal and regulatory challenges (Mennella et al. 2024). These challenges must be carefully addressed to ensure that AI-driven solutions are not only effective but also safe, equitable and trustworthy. This section explores critical concerns related to data privacy, bias, fairness, explainability and clinical accountability.

13.7.1. *Data privacy and security*

AI systems depend on large-scale datasets that often include sensitive personal health information (PHI), such as genomic sequences, medical histories and real-time physiological data from wearables (Khatiwada et al. 2024). The collection, storage and processing of such data raise substantial concerns about privacy and data security.

To address these issues, AI developers and healthcare institutions must ensure compliance with regulations such as:

– *HIPAA* (Health Insurance Portability and Accountability Act) in the United States;

– *GDPR* (General Data Protection Regulation) in the European Union;

– *PIPEDA* in Canada and similar frameworks globally.

Key strategies include:

– de-identification and anonymization of datasets;

– use of federated learning, where AI models are trained across decentralized data sources without transferring raw data;

– implementation of robust cybersecurity protocols, including encryption, secure APIs and role-based access control.

Despite these safeguards, the risk of data breaches and misuse persists, particularly when patient data is shared with third-party vendors or used for secondary purposes such as AI model training. Clear patient consent and transparency in data use are essential to maintain public trust.

13.7.2. *Bias and fairness*

AI models are only as unbiased as the data they are trained on. Unfortunately, many healthcare datasets reflect historical inequities, including racial, socioeconomic and gender-based disparities (Franklin et al. 2024).

As a result, AI systems may inadvertently perpetuate or exacerbate health disparities.

For example:

– an AI model trained primarily on data from White patients may perform poorly in diagnosing conditions in patients of other ethnicities;

– socioeconomic bias in datasets can lead to lower accuracy for underrepresented or marginalized communities.

To ensure fairness, AI developers must:

– use diverse, representative datasets;

– conduct bias audits and fairness evaluations throughout model development;

– apply algorithmic fairness techniques, such as reweighting, sampling or adversarial debiasing;

– include ethics oversight boards to review data practices and model behavior.

Failing to address these issues not only undermines the utility of AI tools but also poses serious risks to equity in cancer diagnosis, treatment and rehabilitation.

13.7.3. *Explainability and trust*

The clinical adoption of AI hinges on trust and interpretability. Many high-performing AI models, especially deep learning systems, operate as "black boxes", producing outputs without clear explanations of their decision-making processes (Tuan 2024).

This lack of transparency poses a barrier for clinicians who must justify and understand AI-assisted decisions, especially in high-stake domains such as oncology.

Explainable AI (XAI) aims to make these models more interpretable and transparent. Key techniques include:

– saliency maps in medical imaging, highlighting areas contributing to a diagnosis;

– SHAP (SHapley Additive exPlanations) and LIME (Local Interpretable Model-agnostic Explanations) for feature importance;

– rule-based or decision-tree models that offer traceable logic.

Explainability is not only important for clinicians but also for patients, who deserve clear communication about how AI influences their care.

Transparent AI fosters informed consent, shared decision-making and broader acceptance among stakeholders.

13.7.4. *Regulatory oversight and accountability*

As AI tools transition from research to clinical deployment, clear regulatory frameworks are necessary to ensure safety and accountability (Goktas and Grzybowski 2025). However, current regulations often lag behind technological advancements.

Regulatory agencies such as the FDA, EMA and MHRA are evolving guidelines for:

– AI as a medical device (AIaMD);

– continuous learning systems that update in real time;

– risk stratification based on clinical impact.

13.7.4.1. *Challenges include*

– Determining liability in the case of AI-induced harm;

– addressing intellectual property and ownership of AI-generated insights;

– establishing post-market surveillance mechanisms for ongoing monitoring of AI performance.

Ultimately, a collaborative approach involving developers, clinicians, regulators, ethicists and patients is essential for creating trustworthy and responsible AI systems in oncology and physiotherapy (Mennella et al. 2024).

13.8. Integration challenges and implementation

13.8.1. *Interoperability*

Effective AI deployment requires integration with existing healthcare IT systems. Lack of standardization across platforms and datasets remains a major hurdle.

13.8.2. *Clinical validation*

Many AI tools perform well in research but lack large-scale clinical validation. Robust, multi-institutional trials are needed to establish safety and efficacy.

13.8.3. *Workforce readiness*

Oncologists, radiologists and pathologists must be trained to understand and interact with AI tools. Curricula in medical education need to evolve accordingly.

13.9. Future outlook

The next decade will witness deeper integration of AI in oncology. Advancements in federated learning will allow models to be trained on decentralized data, preserving privacy while improving generalizability. Quantum computing may also enhance AI capabilities in simulating complex biological interactions for cancer modeling.

Other promising areas include:

– AI-driven robotic surgery;

– AI-powered psychosocial support systems;

– integration with blockchain for secure health data exchange;

– synergy between AI and synthetic biology for targeted therapies.

13.10. Conclusion

Emerging AI technologies are not just augmenting existing practices – they are fundamentally reshaping the landscape of cancer care and physiotherapy. From enhancing early detection through intelligent imaging systems to driving the discovery of new therapeutics and enabling personalized rehabilitation plans, AI has demonstrated its potential to revolutionize every phase of the cancer care continuum. AI-powered tools are improving diagnostic accuracy, optimizing treatment strategies, facilitating real-time patient monitoring and providing virtual support that extends care beyond hospital walls. These advancements collectively contribute to improved patient outcomes, reduced healthcare costs and greater access to high-quality care, particularly in underserved or remote populations. However, this transformation is not without its challenges. The implementation of AI in oncology and physiotherapy demands careful attention to ethical principles, data governance, model transparency and clinical validation. Issues such as data privacy, algorithmic bias, lack of explainability and unclear regulatory pathways must be addressed to ensure that these technologies are trusted, fair and safe. Realizing the full potential of AI in cancer care requires a multidisciplinary approach. Clinicians, data scientists, engineers, ethicists and policymakers must work collaboratively to design, validate and govern AI systems that are clinically relevant, socially responsible and technologically robust. Educational efforts are also essential to equip healthcare professionals with the skills needed to interpret and integrate AI tools into clinical practice. In summary, while AI is poised to become a cornerstone of modern oncology and physiotherapy, its success will depend on inclusive design, rigorous oversight and a shared commitment to patient-centered care.

13.11. References

Bell, D.W. (2010). Our changing view of the genomic landscape of cancer. *The Journal of Pathology: A Journal of the Pathological Society of Great Britain and Ireland*, 220(2), 231–243.

Blessing, E. (2024). Regulatory compliance and ethical considerations: Compliance challenges and opportunities with the integration of Big Data and AI. Report, HAL Open Science. doi: 10.5281/zenodo.14926009.

Bohr, A. and Memarzadeh, K. (eds) (2020). The rise of artificial intelligence in healthcare applications. In *Artificial Intelligence in Healthcare*. Academic Press, Cambridge, 25–60.

Busnatu, Ș.S., Niculescu, A.G., Bolocan, A., Andronic, O., Pantea Stoian, A.M., Scafa-Udriște, A., Stănescu, A.M., Păduraru, D.N., Nicolescu, M.I., Grumezescu, A.M. et al. (2022). A review of digital health and biotelemetry: Modern approaches towards personalized medicine and remote health assessment. *Journal of Personalized Medicine*, 12(10), 1656.

Chang, L., Liu, J., Zhu, J., Guo, S., Wang, Y., Zhou, Z., Wei, X. (2025). Advancing precision medicine: The transformative role of artificial intelligence in immunogenomics, radiomics, and pathomics for biomarker discovery and immunotherapy optimization. *Cancer Biology & Medicine*, 22(1), 33–47.

Daraghmeh, O.F. (2024). A hybrid artificial intelligence approach for early detection of breast cancer and classification from mammogram images in palestine. Thesis, Al-Quds University, Jerusalem.

Dlamini, Z., Francies, F.Z., Hull, R., Marima, R. (2020). Artificial intelligence (AI) and big data in cancer and precision oncology. *Computational and Structural Biotechnology Journal*, 18, 2300–2311.

Franklin, G., Stephens, R., Piracha, M., Tiosano, S., Lehouillier, F., Koppel, R., Elkin, P.L. (2024). The sociodemographic biases in machine learning algorithms: A biomedical informatics perspective. *Life*, 14(6), 652.

Goktas, P. and Grzybowski, A. (2025). Shaping the future of healthcare: Ethical clinical challenges and pathways to trustworthy AI. *Journal of Clinical Medicine*, 14(5), 1605.

Grueneisen, J., Nagarajah, J., Buchbender, C., Hoffmann, O., Schaarschmidt, B.M., Poeppel, T., Forsting, M., Quick, H.H., Umutlu, L., Kinner, S. (2015). Positron emission tomography/magnetic resonance imaging for local tumor staging in patients with primary breast cancer: A comparison with positron emission tomography/computed tomography and magnetic resonance imaging. *Investigative Radiology*, 50(8), 505–513.

Hassan, J., Saeed, S.M., Deka, L., Uddin, M.J., Das, D.B. (2024). Applications of machine learning (ML) and mathematical modeling (MM) in healthcare with special focus on cancer prognosis and anticancer therapy: Current status and challenges. *Pharmaceutics*, 16(2), 260.

Hussain, D., Al-Masni, M.A., Aslam, M., Sadeghi-Niaraki, A., Hussain, J., Gu, Y.H., Naqvi, R.A. (2024). Revolutionizing tumor detection and classification in multimodality imaging based on deep learning approaches: Methods, applications and limitations. *Journal of X-Ray Science and Technology*, 32(4), 857–911.

Ibraheem, A.S., Nwosu, C.B., Omowumi, R.K., Ayodapo, A.O. (2025). Managing population health through prevention and early detection of cancer and other non-communicable diseases: A call for action. *Discover Public Health*, 22(1), 143.

Jiang, Y. and Wang, M. (2010). Personalized medicine in oncology: Tailoring the right drug to the right patient. *Biomarkers in Medicine*, 4(4), 523–533.

Kalantar, R., Lin, G., Winfield, J.M., Messiou, C., Lalondrelle, S., Blackledge, M.D., Koh, D.M. (2021). Automatic segmentation of pelvic cancers using deep learning: State-of-the-art approaches and challenges. *Diagnostics*, 11(11), 1964.

Kanakaprabha, S., Kumar, G.G., Reddy, B.P., Raju, Y.R., Rai, P.C. (2024). Wearable devices and health monitoring: Big data and AI for remote patient care. In *Intelligent Data Analytics for Bioinformatics and Biomedical Systems*, Sharma, N., Cengiz, K., Chatterjee, P. (eds). Scrivener Publishing, Beverly and John Wiley & Sons, New York.

Khatiwada, P., Yang, B., Lin, J.C., Blobel, B. (2024). Patient-generated health data (PGHD): Understanding, requirements, challenges, and existing techniques for data security and privacy. *Journal of Personalized Medicine*, 14(3), 282.

Krubaa, P., Jha, A.M., Mahmood, A.A., Kumar, A., Abraham, J. (2024). Next-generation sequencing technology in cancer. *International Journal of Trends in OncoScience*, 2(3), 23–31.

Majeed, S., Mehraj, S., Ahmad, Z. (2022). Antituberculosis drug repurposing: A new hope for tackling multi-challenging TB in timely manner. In *Drug Repurposing: Molecular Aspects and Therapeutic Applications*, Badria, F.A. (ed.). BoD – Books on Demand, Norderstedt.

Mao, J.J., Pillai, G.G., Andrade, C.J., Ligibel, J.A., Basu, P., Cohen, L., Khan, I.A., Mustian, K.M., Puthiyedath, R., Dhiman, K.S. et al. (2022). Integrative oncology: Addressing the global challenges of cancer prevention and treatment. *CA: A Cancer Journal for Clinicians*, 72(2), 144–164.

Marion, S., Ghazal, L., Roth, T., Shanahan, K., Thom, B., Chino, F. (2024). Prioritizing patient-centered care in a world of increasingly advanced technologies and disconnected care. *Seminars in Radiation Oncology*, 34(4), 452–462.

Mennella, C., Maniscalco, U., De Pietro, G., Esposito, M. (2024). Ethical and regulatory challenges of AI technologies in healthcare: A narrative review. *Heliyon*, 10(4), e26297.

Nikolaou, M., Pavlopoulou, A., Georgakilas, A.G., Kyrodimos, E. (2018). The challenge of drug resistance in cancer treatment: A current overview. *Clinical & Experimental Metastasis*, 35, 309–318.

Odah, M. (2025). Artificial intelligence meets drug discovery: A systematic review on AI-powered target identification and molecular design. *Preprints*. doi: 10.20944/preprints202503.0912.v1.

Ozaybi, M.Q., Madkhali, A.N., Alhazmi, M.A., Faqihi, H.M., Alanazi, M.M., Siraj, W.H., Zalah, A.H., Khormi, M.M., Al Salem, A.M., Mashragi, T.Q. et al. (2024). The role of artificial intelligence in drug discovery and development. *Egyptian Journal of Chemistry*, 67(13), 1541–1547.

Pascoal, E., Wessels, J.M., Aas-Eng, M.K., Abrao, M.S., Condous, G., Jurkovic, D., Espada, M., Exacoustos, C., Ferrero, S., Guerriero, S. et al. (2022). Strengths and limitations of diagnostic tools for endometriosis and relevance in diagnostic test accuracy research. *Ultrasound in Obstetrics & Gynecology*, 60(3), 309–327.

Rai, H.M., Yoo, J., Razaque, A. (2024). Comparative analysis of machine learning and deep learning models for improved cancer detection: A comprehensive review of recent advancements in diagnostic techniques. *Expert Systems with Applications*, 124838.

Saeed, M. (2024). Integrating genomic data into clinical practice: Implications for precision medicine. *Emerging Trends in Medicine*, 1(2), 57–62.

Shapiro, M.A., Stuhlmiller, T.J., Wasserman, A., Kramer, G.A., Federowicz, B., Hoos, W., Kaufman, Z., Chuyka, D., Mahoney, W., Newton, M.E. et al. (2024). AI-augmented clinical decision support in a patient-centric precision oncology registry. *AI in Precision Oncology*, 1(1), 58–68.

Sherani, A.M., Khan, M., Qayyum, M.U., Hussain, H.K. (2024). Synergizing AI and healthcare: Pioneering advances in cancer medicine for personalized treatment. *International Journal of Multidisciplinary Sciences and Arts*, 3(2), 270–277.

Tuan, D.A. (2024). Bridging the gap between black box AI and clinical practice: Advancing explainable AI for trust, ethics, and personalized healthcare diagnostics. *Preprints*. doi: 10.20944/preprints202409.1974.v2.

14

AI-driven Motion Analysis for Physiotherapy in Cancer Survivors

Cancer survivorship presents a multifaceted challenge involving the management of long-term physical impairments caused by aggressive oncological treatments such as chemotherapy, radiation and surgery. Physiotherapy plays a crucial role in restoring functionality, mobility and quality of life. However, traditional physiotherapy methods often fall short due to limitations in objective assessment, personalized treatment planning and continuous monitoring. The advent of Artificial Intelligence (AI) in healthcare, particularly through motion analysis technologies, is revolutionizing post-cancer rehabilitation by introducing a new era of intelligent, patient-centered care. AI-driven motion analysis uses wearable sensors, computer vision and machine learning algorithms to quantify movement patterns, detect functional impairments and provide real-time feedback. These technologies enable clinicians to make data-driven decisions, design customized rehabilitation protocols and monitor patient progress with high precision. Importantly, AI allows for remote and home-based physiotherapy by supporting tele-rehabilitation platforms that increase accessibility and adherence, especially for patients with mobility or geographic constraints. This chapter explores the integration of AI in motion analysis for cancer rehabilitation, discussing its core technologies, clinical applications and emerging innovations. It also presents case studies and trials demonstrating improved outcomes in range of motion, balance and adherence to therapy. Furthermore, the chapter addresses challenges such as data privacy, algorithmic bias, cost barriers and the need for clinical training. Ultimately, the research suggests that AI-driven motion analysis holds transformative potential to enhance the effectiveness, efficiency and equity of physiotherapy for cancer survivors.

14.1. Introduction

14.1.1. *Background*

Over the past few decades, advancements in early detection, targeted therapies, immunotherapy and multidisciplinary cancer care have led to a significant increase

Chapter written by Trupti YADAV, Anand GUDUR and Vibha VYAS.

in cancer survival rates globally. While this is a remarkable achievement, survivorship comes with its own set of complex and long-term challenges (Mennella et al. 2024). Cancer survivors frequently contend with a spectrum of physical, neurological and psychological impairments that result from aggressive treatments such as chemotherapy, radiation therapy and surgical interventions. These may include cancer-related fatigue, chemotherapy-induced peripheral neuropathy, joint stiffness, muscular atrophy and restricted mobility. Such impairments can severely affect the survivor's ability to perform daily activities and diminish the overall quality of life.

Physiotherapy is a cornerstone of post-treatment cancer care, aiming to restore functional independence, improve strength and endurance, and reduce the risk of secondary complications such as falls and muscle wasting (Bezzina 2023). However, traditional physiotherapy practices rely heavily on subjective observation and periodic assessments conducted during clinical visits. These methods are often insufficient for tracking subtle functional changes or adapting to the evolving needs of patients.

Additionally, logistical issues such as limited access to specialized care, patient nonadherence to home exercises and variability in clinical expertise present further barriers to effective rehabilitation.

The integration of AI, particularly through motion analysis technologies, offers a promising solution to these limitations (Phatak et al. 2021). AI-driven systems use computer vision, machine learning algorithms and sensor-based data to continuously and objectively monitor movement patterns. This allows for highly accurate analysis of joint kinematics, gait mechanics, posture and muscular coordination. The result is a more precise, personalized and adaptive rehabilitation approach that can respond in real time to a patient's progress or regression.

14.1.2. *Aim and scope*

The aim of this research chapter is to explore the transformative role of AI-driven motion analysis in enhancing physiotherapy for cancer survivors (Al-Akayleh et al. 2024). Specifically, it investigates how AI technologies are being applied to overcome conventional physiotherapy challenges by providing continuous, quantitative and personalized monitoring of patient movement. The chapter systematically examines the core technologies underpinning AI-based motion analysis – including wearable sensors, computer vision platforms and deep learning algorithms – and their applications in both clinical and home-based rehabilitation settings.

Furthermore, the scope includes an evaluation of real-world use cases, pilot studies and clinical trials that demonstrate the clinical efficacy and usability of AI-powered physiotherapy tools (Shinde et al. 2025). The chapter also addresses critical challenges such as data privacy, integration into existing healthcare workflows, cost and accessibility concerns and the risk of algorithmic bias. Finally, it outlines future directions and emerging innovations, such as virtual reality integration and federated learning, which could further enhance the potential of AI in oncological rehabilitation.

By providing a comprehensive analysis, this chapter aims to inform healthcare professionals, technologists and policymakers about the opportunities and considerations involved in implementing AI-driven motion analysis for the rehabilitation of cancer survivors. It underscores the necessity for interdisciplinary collaboration to ensure that these technological advancements translate into meaningful improvements in patient outcomes and health equity (Shinde et al. 2025).

14.2. The role of physiotherapy in cancer survivorship

14.2.1. *Physical impairments in cancer survivors*

Cancer treatments, while targeting the eradication of malignant cells, often have a profound impact on healthy tissues and organ systems (Kaur et al. 2023). These side effects can manifest as long-term or even permanent physical impairments that hinder survivors' ability to engage in normal daily activities, maintain independence or return to work.

Some of the most commonly observed physical complications include:

– *Lymphedema*: a frequent consequence of lymph node dissection during breast, prostate or gynecologic cancer surgeries. This condition results in fluid accumulation, swelling and discomfort, typically in the limbs, which can significantly impair mobility and function.

– *Peripheral neuropathy*: caused by neurotoxic chemotherapeutic agents such as taxanes and platinum-based compounds (Was et al. 2022). Patients experience numbness, tingling, pain or weakness, primarily in the hands and feet, which affects balance, gait and fine motor skills.

– *Reduced cardiovascular endurance*: chemotherapy and radiation can weaken the heart muscle and vascular structures, contributing to fatigue and decreased exercise tolerance. Survivors often struggle to meet basic physical activity recommendations.

– *Muscle atrophy and skeletal muscle loss (sarcopenia)*: prolonged bed rest, inactivity and catabolic effects of treatment lead to significant reductions in muscle mass and strength. This can increase the risk of falls, frailty and functional dependency.

– *Joint stiffness and limited mobility*: radiation fibrosis, scar formation after surgery or prolonged immobilization can lead to a decreased range of motion and joint contractures, affecting the ability to perform tasks such as reaching, bending or walking (Gomez et al. 2023).

Addressing these impairments requires a highly individualized rehabilitation approach that evolves with the patient's recovery status. Traditional one-size-fits-all programs often fail to capture this dynamic need. Physiotherapists must monitor progress continuously, adjust therapy intensity and detect complications early; all of which are areas where AI-driven motion analysis systems can play a pivotal role.

By enabling accurate, real-time assessment of joint angles, muscle activation and gait mechanics, AI technologies can assist clinicians in fine-tuning interventions and detecting subtle deteriorations before they become clinically significant.

14.2.2. *Current limitations in traditional physiotherapy*

While physiotherapy is widely recognized as essential in cancer rehabilitation, conventional approaches are constrained by structural, methodological and logistical barriers (Navntoft et al. 2024).

Key limitations include:

– *Reliance on manual goniometry and visual observation*: in many settings, the assessment of joint range of motion, posture and balance is still performed manually. These methods are prone to human error, inter-observer variability and lack precision, especially when evaluating small changes in mobility or movement quality.

– *Infrequent assessments due to limited clinician time*: most patients attend physiotherapy sessions once or twice a week. Between sessions, their progress or decline may go unmonitored. This episodic care model fails to capture the continuous nature of rehabilitation, potentially delaying important therapy modifications.

– *Patient noncompliance with home exercise programs*: adherence to prescribed exercises is crucial for recovery but often compromised due to factors such as pain, lack of motivation, forgetfulness or misunderstanding of instructions (Room et al. 2021). Without effective monitoring, therapists cannot know whether patients are performing exercises correctly, or at all.

– *Lack of data-driven personalization*: rehabilitation plans are frequently based on standardized guidelines rather than real-time, patient-specific data. This can result in under-treatment or overexertion, both of which negatively impact recovery and patient satisfaction. Moreover, traditional physiotherapy lacks the predictive analytics to foresee complications or adjust for comorbidities dynamically.

– *Limited access and resource constraints*: in rural or underserved areas, access to specialized oncology physiotherapy services is scarce. Transportation issues, financial burdens and a shortage of trained professionals can further restrict continuity of care.

These constraints highlight a growing need for intelligent, scalable and patient-centered physiotherapy solutions. AI-driven motion analysis offers a compelling pathway to overcome many of these barriers (Reddy 2025). By enabling automated, accurate assessments and supporting personalized exercise plans, AI systems can empower both clinicians and patients to engage more effectively in the rehabilitation process, whether in clinics, homes or telehealth settings.

14.3. AI and motion analysis: technologies and techniques

AI-driven motion analysis is revolutionizing the way physiotherapists assess, monitor and guide cancer survivors through rehabilitation. By capturing complex movement patterns and analyzing them using intelligent algorithms, AI systems offer unprecedented precision and personalization. This section explores the core technologies that make AI-powered motion analysis possible, from motion capture systems to machine learning models and real-time feedback integration.

14.3.1. *Motion capture systems*

At the heart of AI-based physiotherapy systems lies the ability to accurately record and interpret human movement (Nambi et al. 2024). Several motion capture modalities are used to extract biomechanical data relevant to rehabilitation:

– *Wearable sensors*: devices such as inertial measurement units (IMUs), accelerometers, gyroscopes and magnetometers are affixed to key anatomical locations (e.g. knees, hips, spine) to measure linear acceleration, angular velocity and orientation. These systems offer mobility and can be used in both clinical and home settings.

– *Computer vision systems*: cameras – ranging from standard RGB to depth-sensing cameras such as Microsoft Kinect or LiDAR – track movements without the need for physical sensors. These systems can analyze posture, balance and dynamic motions in 3D, offering noninvasive, contactless monitoring.

– *Markerless motion capture*: recent advances in computer vision have enabled motion capture without traditional reflective markers. AI frameworks such as OpenPose, MediaPipe and PoseNet can estimate body keypoints from video footage, allowing for scalable and user-friendly analysis (Roggio et al. 2024). These technologies are particularly useful for tele-rehabilitation and remote assessments.

Together, these systems measure joint angles, movement velocity, stride length, limb symmetry and postural alignment. These are critical parameters for identifying impairments and tailoring therapy.

14.3.2. *Machine learning and deep learning*

Motion data collected from sensors and cameras is highly complex and multidimensional. To interpret this data meaningfully, AI systems use advanced machine learning and deep learning models. These models not only analyze current performance but also predict future trends and personalize interventions.

Key functionalities of AI models in this context include:

– *Anomaly detection*: AI can identify subtle abnormalities such as gait asymmetry, reduced joint flexion or poor balance that may be overlooked by the human eye (Harris et al. 2022). These insights are crucial for early intervention and progress tracking.

– *Outcome prediction*: machine learning algorithms can forecast patient outcomes based on historical and real-time data. For example, they can predict recovery times, risk of re-injury or adherence levels, helping clinicians make informed decisions.

– *Adaptive therapy optimization*: AI systems dynamically update rehabilitation protocols in response to patient progress. For instance, exercises may become more challenging as strength and coordination improve or be adjusted if regressions are detected.

Common models used include:

– *Convolutional neural networks (CNNs)*: these are ideal for pose estimation and image-based motion tracking. CNNs process visual data to extract spatial patterns, such as joint locations or body alignment.

– *Recurrent neural networks (RNNs) and long short-term memory (LSTM) networks*: these are effective for analyzing sequential and time-series data from sensors. These models can learn temporal dependencies, making them well suited for understanding movement trajectories over time.

– *Support vector machines (SVMs)*: often used for classification tasks, such as categorizing movement patterns into normal or impaired. SVMs are particularly useful in clinical decision support systems due to their high accuracy in structured datasets (Zikos and DeLellis 2018).

These algorithms turn raw motion signals into actionable insights, enabling tailored rehabilitation approaches and facilitating continuous improvement.

14.3.3. *Data integration and feedback systems*

For AI-driven motion analysis to be clinically effective, it must integrate diverse data sources and provide meaningful feedback to both patients and clinicians. A comprehensive AI physiotherapy platform typically unifies the following:

– *Kinematic data from sensors*: real-time data on joint angles, stride length, body acceleration and range of motion are fed into AI models to detect trends and deviations.

– *Video-based motion patterns*: computer vision extracts visual markers such as posture or alignment, which are cross-referenced with sensor data to validate movement quality.

– *Patient-reported outcomes (PROs)*: subjective data on pain, fatigue or perceived difficulty is integrated into the analysis, offering a holistic view of the patient's status and therapy impact (Wideman et al. 2019).

Based on this multimodal data fusion, intelligent systems provide:

– *Real-time feedback*: patients receive immediate cues – via audio, visual or haptic signals – on whether their movements align with prescribed exercises. This helps correct poor form and reinforces correct execution, enhancing motor learning.

– *Performance dashboards*: clinicians can access visual summaries of patient progress, including adherence, improvement trends and detected issues. These insights support more informed, data-driven decisions in therapy planning.

– *Tele-rehabilitation support*: integrated platforms enable remote monitoring, where therapists can assess performance asynchronously and adjust plans accordingly (Schickler et al. 2017). Patients benefit from virtual coaching, video demonstrations and interactive guidance.

By bridging data analytics with user-centered feedback mechanisms, AI systems significantly improve engagement, accountability and outcomes in physiotherapy.

14.4. Applications in physiotherapy for cancer survivors

14.4.1. *Post-surgical rehabilitation*

AI can aid recovery after mastectomy, thoracotomy or limb-preserving surgeries by:

– monitoring shoulder range of motion;

– detecting compensatory movement patterns;

– guiding progressive resistance training.

14.4.2. *Management of chemotherapy-induced neuropathy*

AI-driven gait analysis can:

– quantify balance impairments and fall risk;

– guide proprioceptive training;

– offer biofeedback to correct irregular gait.

14.4.3. *Lymphedema and edema management*

Computer vision tools can:

– track limb volume changes over time;

– detect motion restrictions due to swelling;

– prompt early intervention strategies.

14.4.4. *Fatigue and activity monitoring*

Wearables embedded with AI algorithms can:

– track daily activity and energy expenditure;

– correlate movement patterns with reported fatigue;

– recommend activity pacing strategies.

14.5. AI in home-based rehabilitation

As cancer survivorship becomes increasingly prevalent, the demand for long-term, accessible rehabilitation solutions has grown substantially (Stout et al. 2019). Many survivors face barriers to traditional in-clinic physiotherapy, including limited mobility, transportation challenges, immune system vulnerabilities and geographic isolation. AI-driven home-based rehabilitation platforms address these barriers by offering personalized, consistent and remotely accessible physiotherapy support. These technologies not only promote convenience and adherence but also enhance clinical effectiveness through continuous monitoring and intelligent feedback.

14.5.1. *Remote physiotherapy platforms*

AI-powered rehabilitation platforms are reshaping the home physiotherapy experience by combining guided exercise programs with intelligent motion tracking. Some of the leading platforms in this space include Kaia Health, Sword Health and Reflexion, each offering advanced capabilities that improve patient outcomes and clinician oversight.

Key features of these platforms include:

– *Guided exercise routines*: using motion tracking from smartphone cameras, depth sensors or wearable devices, these platforms guide patients through personalized rehabilitation exercises (Nascimento et al. 2020). The exercises are tailored to the patient's condition, progress and limitations, often using adaptive algorithms that modify the intensity or sequence in real time.

– *Real-time corrective feedback*: through visual and audio prompts, patients receive immediate feedback on posture, range of motion and alignment. This helps reduce the risk of improper technique, which is especially important for patients dealing with post-surgical sensitivity or treatment-induced fatigue.

– *Progress analytics for clinicians*: these platforms automatically collect and analyze data on exercise frequency, movement quality, joint angles, and recovery trends. Clinicians can access dashboards summarizing each patient's progress, pain scores (if integrated with PROs) and overall adherence, enabling data-driven decision-making.

– *Accessibility and engagement*: the user-friendly interfaces of these systems allow cancer survivors of all ages to engage in physiotherapy independently. Instructional videos, interactive avatars and gamified elements improve motivation and consistency, which are often lacking in traditional home exercise programs (Mouatt et al. 2020).

The ability to deliver comprehensive, AI-assisted therapy from the comfort of home has proven particularly valuable for cancer survivors who are immunocompromised, elderly or reside in rural or underserved areas.

14.5.2. *Tele-rehabilitation with AI support*

Tele-rehabilitation integrates digital communication technologies with AI to extend physiotherapy services beyond clinical settings. When coupled with AI-driven motion analysis, tele-rehabilitation enables clinicians to monitor, evaluate and guide patients remotely with a level of precision previously only possible in person.

Key benefits of AI-enhanced tele-rehabilitation include:

– *Patient adherence monitoring*: AI continuously tracks whether patients are performing exercises correctly and consistently (Griefahn et al. 2024). Nonadherence triggers alerts for clinicians, who can then follow up with motivational support or program adjustments.

– *Dynamic therapy adjustments*: based on real-time performance data and patient feedback, AI algorithms can recommend modifications to exercise plans, such as adjusting intensity, frequency or types of exercises. These recommendations help ensure the therapy remains aligned with the patient's current capabilities and recovery goals.

– *Consistency in assessment*: traditional rehabilitation suffers from variability in assessments across different therapists or sessions. AI-driven analysis standardizes motion evaluations, ensuring objective, reproducible measurements of progress regardless of who is administering care or where the patient is located.

– *Trend analysis and longitudinal insights*: AI platforms maintain a comprehensive digital record of each patient's rehabilitation journey (Afridi and Khan 2024). Over time, these records can be analyzed to identify patterns, predict recovery timelines and flag signs of regression or complications. This longitudinal perspective supports proactive, preventative care rather than reactive interventions.

– *Enhanced communication*: clinicians and patients can interact through secure video consultations, where therapists overlay real-time motion data to assess form and provide visual cues for correction. This creates a hybrid model of care: one that is remote yet deeply interactive.

The integration of AI into tele-rehabilitation not only optimizes clinical outcomes but also democratizes access to high-quality physiotherapy. It empowers patients to actively participate in their recovery journey and ensures that geographical, physical or logistical constraints do not hinder rehabilitation success.

14.6. Case studies and clinical trials

Real-world applications of AI-driven motion analysis are emerging through clinical research and pilot implementations across various cancer populations (El Arab et al. 2025). These studies highlight the tangible benefits of integrating artificial intelligence into rehabilitation practices, demonstrating improvements in recovery speed, safety and patient satisfaction.

This section presents three representative examples that underline the clinical relevance and effectiveness of AI-based physiotherapy interventions.

14.6.1. *Case study: breast cancer survivors*

A clinical study conducted at the University of Toronto investigated the use of AI-driven motion tracking for shoulder rehabilitation in breast cancer survivors post-mastectomy. Traditional shoulder mobility assessment requires in-person goniometric measurements, which can be subjective and infrequent. To address this, researchers implemented a computer vision system using an AI-powered camera and pose estimation software to track shoulder range of motion (ROM) remotely.

Key features and outcomes of the study included:

– *Automated monitoring*: patients performed prescribed exercises at home while being recorded via a depth-sensing camera. AI algorithms analyzed shoulder joint angles in real time.

– *Remote feedback*: the system provided visual and auditory cues to help patients adjust posture and motion range during exercise sessions (Maskeliūnas et al. 2023).

– *Quantitative gains*:

- 35% faster recovery of ROM compared to the control group receiving standard physiotherapy;

- higher patient satisfaction, as measured by Likert-scale surveys, particularly with the convenience and clarity of the instructions;

- fewer in-person clinic visits, reducing the burden on both patients and the healthcare system.

This case illustrates the potential of AI in delivering precise, effective and patient-friendly rehabilitation for upper limb impairments following breast cancer surgery.

14.6.2. *Randomized controlled trial: AI-guided gait training*

A 2022 randomized controlled trial (RCT) published in Supportive Care in Cancer evaluated the efficacy of AI-assisted gait training among 120 patients undergoing chemotherapy (Jafleh et al. 2024).

Chemotherapy-induced peripheral neuropathy often impairs balance and increases the risk of falls. The intervention group used an AI-powered motion analysis platform that provided adaptive balance exercises and real-time gait correction.

Trial design and findings included:

– *Participants*: divided into control (standard physical therapy) and intervention (AI-guided) groups.

– *Intervention features*:

- gait analysis through wearable sensors and depth cameras;
- personalized balance exercises updated weekly using AI predictions;
- real-time feedback and alerts for unsafe gait patterns.

– *Clinical outcomes*:

- 42% reduction in fall incidence over a twelve-week period;
- significant improvement in timed up and go (TUG) test scores, indicating better functional mobility and lower fall risk (Barry et al. 2014);
- increased adherence to therapy, with > 90% completion rate in the intervention group compared to 68% in the control group.

These findings demonstrate how AI-enhanced gait training can effectively reduce risks and enhance functional outcomes during vulnerable periods of cancer treatment.

14.6.3. *Pilot study: AI in pediatric oncology rehabilitation*

In a pioneering pilot study conducted at a pediatric oncology center in Germany, researchers explored the feasibility and outcomes of AI-based motion monitoring for children recovering from bone tumors, such as osteosarcoma or Ewing sarcoma. These patients often face long-term mobility issues following limb-salvage surgery or chemotherapy-related muscle weakness (Wacker et al. 2017).

AI tools were deployed to support home-based physiotherapy through:

– *Computer vision and pose estimation*: using tools such as OpenPose, the system monitored children's movement patterns and joint functionality during exercise.

– *Early detection algorithms*: AI models were trained to detect subtle declines in mobility, such as reduced stride length or abnormal limb loading patterns, which could indicate relapse or complications.

– *Therapist alerts*: clinicians received automated notifications when regression was detected, allowing for proactive intervention.

14.6.3.1. *Outcomes and observations*

– Early interventions prevented significant deterioration in 78% of flagged cases.

– Functional mobility outcomes improved compared to historical controls, with patients achieving age-appropriate movement milestones more quickly.

– Children and parents reported high levels of engagement due to the interactive, game-such as interface of the AI platform (Alrusaini and Beyari 2022).

14.7. Benefits of AI-driven motion analysis

14.7.1. *Objective and quantifiable assessment*

AI eliminates subjectivity in motion analysis. Joint angles, gait symmetry and movement quality are measured precisely, aiding in more accurate evaluations.

14.7.2. *Personalized rehabilitation programs*

Machine learning models adapt exercises based on patient progress, enabling:

– individualized difficulty levels;

– timely progression or regression;

– consideration of comorbidities.

14.7.3. *Increased patient engagement and adherence*

Gamified feedback and real-time corrections improve patient motivation. Visual dashboards enhance awareness of progress.

14.7.4. *Enhanced clinical decision-making*

AI-derived insights support evidence-based decisions, identify subtle deterioration and optimize therapy pathways.

14.8. Challenges and limitations

14.8.1. *Data privacy and security*

Handling sensitive health and movement data requires strict compliance with HIPAA and GDPR regulations (Herath et al. 2024). Cloud storage and AI platforms must ensure encryption and controlled access.

14.8.2. *Algorithm bias and generalizability*

Most AI models are trained on healthy populations. There is a need for more training on diverse cancer survivor datasets to ensure accuracy and avoid false inferences.

14.8.3. *Accessibility and cost*

High-quality motion capture devices and software can be expensive, limiting adoption in resource-constrained settings. Scalable, affordable alternatives are needed.

14.8.4. *Integration into clinical workflows*

Clinicians may resist adopting unfamiliar technologies. Successful implementation demands clinician training, interdisciplinary collaboration and streamlined interfaces (Sieja et al. 2021).

14.9. Future directions

14.9.1. *Federated learning for privacy-preserving AI*

Federated learning allows models to be trained on decentralized data without transferring raw data to central servers. This could facilitate collaborative model training across hospitals while preserving patient privacy.

14.9.2. *Integration with wearable health ecosystems*

Future systems may combine motion analysis with:

– heart rate monitors;

– smart scales;

– EMG sensors;

– sleep trackers.

This will provide a holistic view of recovery and allow AI to model comprehensive patient health states.

14.9.3. *Augmented reality and virtual reality integration*

AI-powered AR/VR platforms can create immersive rehabilitation experiences, offering guided therapy, enhanced engagement and reduced isolation (Singha and Singha 2025).

14.9.4. *AI coaches and conversational agents*

Voice-enabled AI coaches can offer real-time instructions, encourage activity and answer rehabilitation-related queries, simulating therapist interaction.

14.10. Conclusion

AI-driven motion analysis marks a significant milestone in the evolution of physiotherapy for cancer survivors. Traditional rehabilitation practices, though effective, are often limited by subjectivity in assessment, irregular patient monitoring and poor adherence to home-based exercise regimens. By integrating advanced technologies such as computer vision, wearable sensors and machine learning, AI offers an innovative solution that enhances the precision, consistency and scalability of physiotherapeutic care. Through real-time tracking and analysis of biomechanical parameters, AI systems provide objective insights into patient movements, enabling early detection of functional decline and fine-tuning of rehabilitation protocols. This personalized, data-driven approach not only optimizes recovery outcomes but also empowers patients to engage more actively in their rehabilitation journey. Moreover, AI's capacity for remote monitoring and feedback bridges geographical and logistical barriers, making high-quality physiotherapy accessible to cancer survivors in rural, underserved or mobility-limited settings.

Evidence from case studies and clinical trials demonstrates clear benefits, including faster recovery of range of motion, reduced fall risk and improved patient satisfaction. These findings support the clinical validity of AI-based systems and underscore their potential to complement and augment the expertise of human therapists. Nonetheless, the path to widespread implementation is not without challenges. Concerns around data privacy, algorithmic bias, regulatory approval and the cost of AI infrastructure must be carefully navigated. Additionally, equipping clinicians with the knowledge and tools to effectively use AI technologies will be essential to ensure seamless integration into existing care frameworks.

14.11. References

Afridi, A. and Khan, S.N. (2024). Digital transformation in healthcare rehabilitation: A narrative review. *Journal of Process Management and New Technologies*, 12(1–2), 16–30.

Al-Akayleh, F., Al-Remawi, M., Agha, A.S. (2024). AI-driven physical rehabilitation strategies in post-cancer care. In *2024 2nd International Conference on Cyber Resilience (ICCR)*. IEEE, Dubai.

Alrusaini, O. and Beyari, H. (2022). The sustainable effect of artificial intelligence and parental control on children's behavior while using smart devices' apps: The case of Saudi Arabia. *Sustainability*, 14(15), 9388.

Barry, E., Galvin, R., Keogh, C., Horgan, F., Fahey, T. (2014). Is the Timed Up and Go test a useful predictor of risk of falls in community dwelling older adults: A systematic review and meta-analysis. *BMC Geriatrics*, 14, 1–4.

Bezzina B. (2023). Pulmonary rehabilitation in lung cancer management: An investigative study on health-related quality of life of patients and the perceived views of healthcare professionals. Master's Thesis, University of Malta, Malta.

Brown, A.G., Campo, R., Freemer, M., Ilias, M., Desvigne-Nickens, P., Redmond, N., Vitalis, D., Pratt, C.A. (2024). Perspective on advancing health equity: Enhancing impact through collaboration. *JACC: Advances*, 3(7 Part 2), 100964.

El Arab, R.A., Abu-Mahfouz, M.S., Abuadas, F.H., Alzghoul, H., Almari, M., Ghannam, A., Seweid, M.M. (2025). Bridging the gap: From AI success in clinical trials to real-world healthcare implementation – A narrative review. *Healthcare*, 13(7), 701.

Gomez, J., Wilson, D., Black, P., Friedman, L., Roche, A.M. (2023). Physical and occupational therapy. In *Complex Head and Neck Microvascular Surgery: Comprehensive Management and Perioperative Care*, Quimby, A., Parmar, S., Fernandes, R. (eds). Springer, Cham.

Griefahn, A., Zalpour, C., Luedtke, K. (2024). Identifying the risk of exercises, recommended by an artificial intelligence for patients with musculoskeletal disorders. *Scientific Reports*, 14(1), 14472.

Harris, E.J., Khoo, I.H., Demircan, E. (2022). A survey of human gait-based artificial intelligence applications. *Frontiers in Robotics and AI*, 8, 749274.

Herath, H.M., Herath, H.M., Madhusanka, B.G., Guruge, L.G. (2024). Data protection challenges in the processing of sensitive data. In *Data Protection: The Wake of AI and Machine Learning*, Hewage, C., Yasakethu, L., Jayakody, D.N.K. (eds). Springer, Cham.

Jafleh, E.A., Alnaqbi, F.A., Almaeeni, H.A., Faqeeh, S., Alzaabi, M.A., Al Zaman, K., Alnaqbi, F., Almaeeni, H., Alzaabi, M. (2024). The role of wearable devices in chronic disease monitoring and patient care: A comprehensive review. *Cureus*, 16(9), e68921.

Kaur, R., Bhardwaj, A., Gupta, S. (2023). Cancer treatment therapies: Traditional to modern approaches to combat cancers. *Molecular Biology Reports*, 50(11), 9663–9676.

Maskeliūnas, R., Damaševičius, R., Blažauskas, T., Canbulut, C., Adomavičienė, A., Griškevičius, J. (2023). BiomacVR: A virtual reality-based system for precise human posture and motion analysis in rehabilitation exercises using depth sensors. *Electronics*, 12(2), 339.

Mouatt, B., Smith, A.E., Mellow, M.L., Parfitt, G., Smith, R.T., Stanton, T.R. (2020). The use of virtual reality to influence motivation, affect, enjoyment, and engagement during exercise: A scoping review. *Frontiers in Virtual Reality*, 1, 564664.

Nambi, G., Alghadier, M., Mohamed, S.H., Aldhafian, O.R., Alshahrani, N.A., Albarakati, A.J. (2024). Clinical usefulness of artificial intelligence in physiotherapy – A practice-based review. *SBV Journal of Basic, Clinical and Applied Health Science*, 7(4), 184–188.

Nascimento, L.M., Bonfati, L.V., Freitas, M.L., Mendes Junior, J.J., Siqueira, H.V., Stevan Jr, S.L. (2020). Sensors and systems for physical rehabilitation and health monitoring – A review. *Sensors*, 20(15), 4063.

Navntoft, S., Andreasen, J., Petersen, K.S., Rossau, H.K., Jørgensen, L. (2024). Barriers and facilitators to cancer rehabilitation for patients with head and neck or lung cancer – A scoping review mapping structural and healthcare professionals' perspectives. *Disability and Rehabilitation*, 46(20), 4617–4629.

Phatak, A.A., Wieland, F.G., Vempala, K., Volkmar, F., Memmert, D. (2021). Artificial intelligence based body sensor network framework – Narrative review: Proposing an end-to-end framework using wearable sensors, real-time location systems and artificial intelligence/machine learning algorithms for data collection, data mining and knowledge discovery in sports and healthcare. *Sports Medicine-Open*, 7(1), 79.

Reddy, K.J. (ed.) (2025). Patient perspectives and outcomes. In *Innovations in Neurocognitive Rehabilitation: Harnessing Technology for Effective Therapy*. Springer, Cham.

Roggio, F., Trovato, B., Sortino, M., Musumeci, G. (2024). A comprehensive analysis of the machine learning pose estimation models used in human movement and posture analyses: A narrative review. *Heliyon*, 10(21), e39977.

Room, J., Boulton, M., Dawes, H., Archer, K., Barker, K. (2021). Physiotherapists' perceptions of how patient adherence and non-adherence to recommended exercise for musculoskeletal conditions affects their practice: A qualitative study. *Physiotherapy*, 113, 107–115.

Rowland, J.H. and Bellizzi, K.M. (2008). Cancer survivors and survivorship research: A reflection on today's successes and tomorrow's challenges. *Hematology/Oncology Clinics of North America*, 22(2), 181–200.

Schickler, M., Pryss, R., Stach, M., Schobel, J., Schlee, W., Probst, T., Langguth, B., Reichert, M. (2017). An it platform enabling remote therapeutic interventions. In *2017 IEEE 30th International Symposium on Computer-Based Medical Systems (CBMS)*, Thessaloniki.

Shinde, S., Baraskar, S., Dhumale, A., Jonnawar, S., Aphale, S.R. (2025). The scope of artificial intelligence in facilitating disability rehabilitation for rural population. *South Eastern European Journal of Public Health*, XXXVI(S2), 2392–2410. doi: 10.70135/seejph.vi.5420.

Sieja, A., Kim, E., Holmstrom, H., Rotholz, S., Lin, C.T., Gonzalez, C., Arellano, C., Hutchings, S., Henderson, D., Markley, K. (2021). Multidisciplinary sprint program achieved specialty-specific EHR optimization in 20 clinics. *Applied Clinical Informatics*, 12(02), 329–339.

Singha, R. and Singha, S. (2025) Use of virtual reality (VR) and AI in therapeutic settings. In *Transforming Neuropsychology and Cognitive Psychology With AI and Machine Learning*, Bansal, R., Maqableh, T., Shuklaa, G., Rabby, F., Lathabhavan, R. (eds). IGI Global, Hershey, PA.

Stout, N.L., Silver, J.K., Alfano, C.M., Ness, K.K., Gilchrist, L.S. (2019). Long-term survivorship care after cancer treatment: A new emphasis on the role of rehabilitation services. *Physical Therapy*, 99(1), 10–13.

Wacker, K., Tanner, L., Ovans, J., Mason, J., Gilchrist, L. (2017). Improving functional mobility in children and adolescents undergoing treatment for non–central nervous system cancers: A systematic review. *PM&R*, 9(9), S385–S397.

Was, H., Borkowska, A., Bagues, A., Tu, L., Liu, J.Y., Lu, Z., Rudd, J.A., Nurgali, K., Abalo, R. (2022). Mechanisms of chemotherapy-induced neurotoxicity. *Frontiers in Pharmacology*, 13, 750507.

Wideman, T.H., Edwards, R.R., Walton, D.M., Martel, M.O., Hudon, A., Seminowicz, D.A. (2019). The multimodal assessment model of pain: A novel framework for further integrating the subjective pain experience within research and practice. *The Clinical Journal of Pain*, 35(3), 212–221.

Zikos, D. and DeLellis, N. (2018). CDSS-RM: A clinical decision support system reference model. *BMC Medical Research Methodology*, 18(1), 137.

15

AI-driven Innovations in Physiotherapy and Oncology: Advancing Postural Assessment, Rehabilitation and Patient-centered Care

This chapter comprehensively explores the concept of posture, its significance in musculoskeletal health and the technological advancements used to detect and monitor it. Central to the discussion is the McKenzie method, a widely adopted physiotherapeutic approach that classifies posture-related musculoskeletal disorders into postural, dysfunction and derangement syndromes. The chapter then transitions into posture detection systems, highlighting how sensor-based wearable devices and vision-based tools use real-time data to assess posture deviations and provide corrective feedback. Deep learning algorithms, such as convolutional neural networks (CNNs), and machine learning (ML) models play a pivotal role in identifying skeletal misalignments and suggesting ergonomic interventions. The integration of artificial intelligence (AI) in smart systems – particularly smart chairs – is emphasized for their ability to autonomously detect poor posture, recommend corrections and monitor health parameters such as heart rate and muscle tension.

The chapter also details the applications of force plates in balance and gait assessment, discussing how AI enhances data interpretation, predictive modeling and feedback mechanisms. Through a literature review, the chapter presents evidence supporting the effectiveness of posture detectors and force plates in rehabilitation, sports performance and injury prevention. The AI-driven analysis of kinetic and kinematic data empowers clinicians and therapists to design personalized, data-informed rehabilitation protocols. Overall, the integration of posture detection systems with AI offers a transformative approach in promoting musculoskeletal health, workplace ergonomics and neurological rehabilitation.

Chapter written by G. HIMASHREE, Radhika CHINTAMANI and G. VARADHARAJULU.
For a color version of all of the figures in this chapter, see www.iste.co.uk/kumar/ai-driven1.zip.

15.1. Posture detector

15.1.1. *Introduction of posture*

Posture is commonly defined as a position that an individual adopts while performing any functional activities such as walking, sitting, standing, sleeping or at rest. The term postural hygiene is the set of rules that aims to maintain the correct body position, either in static or in dynamic position around the head–spine–pelvis axis with total weight distributed equally on both feet (Conroy et al. 2022). Poor posture resembles the opposite sense, for example, unequal distribution of weight, the head–spine–pelvis axis is not straight or is tilted and is not aligned (Rybski and Juckett 2024).

The McKenzie method, developed by physiotherapist Robin McKenzie in the 1960s, is a comprehensive approach for diagnosing and treating musculoskeletal conditions, particularly those related to the spine. One of the core concepts of the McKenzie method is posture and its significant impact on spinal health. McKenzie emphasized that poor posture, often due to prolonged sitting, improper alignment or repetitive strain, can lead to spinal dysfunction, pain and discomfort. He introduced the idea of centralization, where movement of the spine or extremities is used to reduce pain and promote healing by shifting discomfort from peripheral areas (such as the limbs) back towards the center of the body, where it can be more effectively managed. The McKenzie approach advocates self-treatment, empowering patients to learn correct postural alignment and perform exercises designed to reduce pain, improve mobility and strengthen the muscles surrounding the spine. By focusing on posture and spinal alignment, the McKenzie method aims to not only relieve pain but also to prevent recurrence and improve overall functionality.

The McKenzie method classifies musculoskeletal disorders primarily into three syndromes: postural syndrome, dysfunction syndrome and derangement syndrome. These classifications are based on the patterns of symptoms, the mechanisms causing them and the appropriate treatment approaches.

15.1.1.1. *Postural syndrome*

– *Definition*: postural syndrome is caused by prolonged poor posture or sustained positions that place abnormal stress on tissues, leading to discomfort or pain. It is typically the result of mechanical irritation of normal, healthy tissues, such as muscles, ligaments and discs, due to static loading or poor body alignment. Common causes include slouching while sitting, poor ergonomic habits or abnormal spinal alignment during daily activities.

– *Symptoms*: pain is often localized and typically occurs only after maintaining a poor posture for a prolonged period. The pain generally resolves quickly with movement or a change in posture. There is typically no structural damage, and the pain is usually intermittent, appearing when the body is held in certain positions.

– *Treatment and correctability*: the good news about postural syndrome is that it is easily correctable. Treatment often involves education on proper posture, ergonomic adjustments (such as sitting properly, standing with correct spinal alignment) and periodic breaks to avoid prolonged static positions. Additionally, strengthening exercises to improve posture, flexibility training and awareness of body mechanics can prevent recurrence. Postural syndrome is highly treatable and reversible with simple interventions.

15.1.1.2. *Dysfunction syndrome*

– *Definition*: dysfunction syndrome occurs when there is a loss of normal range of motion due to shortening or scarring of tissues, such as muscles, ligaments or joint capsules. It often follows a period of injury or immobilization where tissues become fibrotic and stiff. This syndrome is associated with pain due to tissue restriction when a joint or muscle is stretched beyond its normal range of motion.

– *Symptoms*: pain is usually localized and occurs when a restricted area of the body is moved, especially beyond its limited range of motion. The pain is more consistent and often persists over time unless treated. Unlike postural syndrome, the tissues involved are typically damaged or altered due to injury or immobilization.

– *Treatment*: treatment for dysfunction syndrome focuses on restoring movement and flexibility through stretching, mobilizations and specific exercises aimed at lengthening the shortened tissues. Although the syndrome is treatable, the process may take more time and consistent effort compared to postural syndrome.

15.1.1.3. *Derangement syndrome*

– *Definition*: derangement syndrome is the most complex of the McKenzie syndromes. It involves a mechanical disruption of the spinal joints or discs, typically resulting in a displacement of a disc or other structures within the spine. This can lead to pain, swelling, nerve compression or even radicular pain (pain that radiates down an arm or leg).

– *Symptoms*: the pain in derangement syndrome can be intense and is often associated with significant movement restrictions. Patients may experience pain that radiates to the limbs or a limited range of motion in the spine. The symptoms may fluctuate, with periods of increased pain and decreased function, particularly with certain movements or postures.

– *Treatment*: treatment for derangement syndrome is more complex and often requires more advanced techniques, including specific exercises to centralize the pain (moving it away from the limbs back to the spine), spinal mobilization and sometimes pain management strategies. Derangement may take longer to treat, and in some cases, if the disc or structures involved are severely damaged, surgical intervention may be necessary.

Thus, postural syndrome is easier to manage and treat compared to derangement syndrome, making early intervention for poor posture highly beneficial in preventing the development of more serious spinal conditions (Chan et al. 2021).

15.1.2. *Introduction to posture detectors*

A posture detector is a device that detects human posture either in static or dynamic movement. This machine is composed of several sensor components, for example, wearable cameras, joint position sensors, angle sensors and tilt sensors.

These sensors can be placed on various surface landmarks of the human body. It can be noted if there is movement or an abnormal position by the computer reading against a software created grid. The software within the computer detects the stable or moving sensor and forms a picture of only sensors. These sensors can be either left individual to assess each joint and its motion or can be connected with lines to assess the coordinated movement pattern of the human body.

Below is one such image to show the representation of the behind-the-lab posture assessment and the in-software sensor attachment with lines (Piñero-Fuentes et al. 2021).

There have been many devices made to assess change in posture and give feedback to the wearer.

These devices usually have angle sensors which detect the sense of change in angle from set neutral position to either increasing or decreasing. The use of machine learning (ML) and deep learning techniques has become vital (Piñero-Fuentes et al. 2021).

The recently developed deep learning-based processing techniques process the position of various joints with convolutional neural networks (CNN) to approximate the posture and detect if any changes are sensed. This work is feasible for skeletal detection and tracking systems, aiming for proper posture guidance. Thus, posture detectors are beneficial in avoiding or preventing the musculoskeletal disorders caused by poor posture (Piñero-Fuentes et al. 2021).

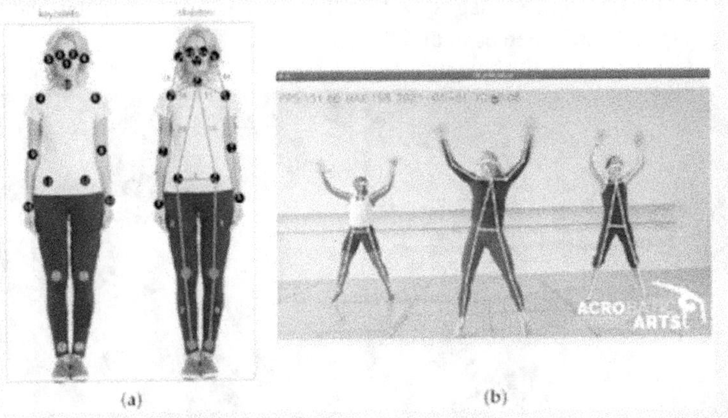

Figure 15.1. *Posture detector*

Posture detectors are beneficial for various applications: identifying musculoskeletal disorders caused by prolonged static loading on intervertebral discs; aiding muscle rehabilitation for patients with motor dysfunction; and promoting workplace safety by preventing injury through posture monitoring in factories and offices. They are also useful for the treatment, prevention and monitoring of spinal disorders, such as scoliosis, where irregular posture leads to misalignment in the shoulders, waist and hips, and kyphosis, which causes a rounded or stooped back.

These detectors use wearable technology to collect movement and posture data, as well as vision cameras to monitor multiple individuals simultaneously. However, sensor-based posture detectors are contraindicated for people with metal implants, open wounds, ulcers or burns, as these can distort sensor readings and affect data accuracy (Nascimento et al. 2020; Piñero-Fuentes et al. 2021; Huang et al. 2023).

15.1.3. *Working principle*

– *Posture detection system*: a real-time pose estimation software called TRT_Pose was used by the authors. It was designed to run on NVIDIA devices, whose main purpose was to obtain the performance matrix and determine the best platform of work for ergonomics. The posture detection also works offline with the MSCOCO dataset specifically on the key point task (see: https://cocodataset.org/keypoints-2020 (accessed on June 20, 2021). It is a public dataset that contains both gender details with 66,808 images, with people from different racial backgrounds and age groups (Piñero-Fuentes et al. 2021).

Figure 15.2. *Real time posture detection*

– *Skeletal processing*: TRT_Pose software was used to detect a human being's real-time skeleton. The processing framework was based on CNN model which was also implemented in Pytorch (name: RestNet). RestNet uses an inertial connection between layers which are not contiguous to communicate information from earlier or later moments in time with respect to moment. Thus, information could be stored and reused for posture system classification. Finally, by using a deep learning model, the authors were able to detect the wearer's joint position (defined by 18 points) (Piñero-Fuentes et al. 2021).

– *Posture recognition*: posture recognition by a posture detector uses wearable or sensor-based technology to monitor and assess an individual's posture in real time. These devices, often in the form of sensors attached to clothing or worn on the body, track key indicators such as spinal alignment, body positioning and movement patterns. By collecting data from the joints and muscles, the posture detector can identify poor posture, such as slouching or misalignment, and provide immediate feedback through alerts or notifications. The aim of posture recognition, through the use of these devices, is to help users correct their posture, reduce the risk of musculoskeletal issues and promote better ergonomics, especially in environments such as workplaces or during rehabilitation. The system may also offer personalized guidance or reminders to maintain optimal posture throughout daily activities.

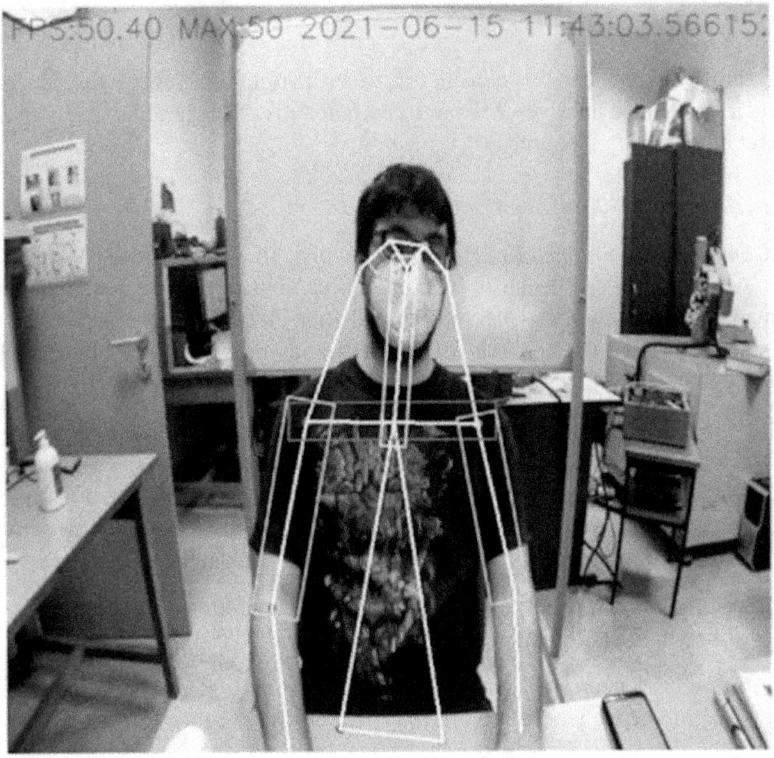

Figure 15.3. *Posture recognition*

– *Posture evaluation*: posture evaluation using posture detectors has been the focus of several studies, which demonstrate the effectiveness of these devices in monitoring and improving body alignment. These studies typically use wearable sensors, pressure mats or vision-based systems to assess posture in real time, providing valuable insights into the body's alignment during various activities. Researchers have found that posture detectors are particularly useful in identifying improper posture in both sedentary and dynamic movements, helping reduce musculoskeletal strain and discomfort. For instance, studies in workplace settings have used posture detectors to monitor employees' sitting or standing positions, while clinical studies have applied them in rehabilitation for musculoskeletal disorders or after surgeries. The data collected from these devices allow for targeted interventions, such as posture correction techniques or ergonomic adjustments, significantly contributing to better posture awareness, prevention of posture-related injuries and overall health improvement (Piñero-Fuentes et al. 2021).

15.1.4. *AI integration in posture detectors*

Posture detectors can be considered to be part of artificial intelligence (AI), particularly when it involves ML and computer vision techniques to analyze and interpret human body posture.

15.1.4.1. *How posture detection uses AI*

– *Computer vision*: posture detection typically relies on computer vision, which is an AI subfield that enables computers to interpret and understand visual information from the world (Chakraborty and Saha 2023). Using cameras or sensors, AI systems can track the positions of different body parts, such as joints and limbs, to determine the body's posture.

– *ML*: in many posture detection systems, AI models, particularly those based on ML, are trained on large datasets containing examples of human poses in various situations (Sundararajan and Sharma 2021). The AI uses these data to learn how to recognize and classify postures (e.g. sitting, standing or different physical activities). Over time, the AI improves its accuracy in detecting and analyzing human posture.

– *Real-time analysis*: AI-powered posture detectors can process the data in real time, enabling applications such as (Singh and Kapoor 2022):

- *Ergonomic monitoring*: identifying poor posture in real time to prevent strain or injury (e.g. in office workers, athletes or rehabilitation patients).

- *Fitness and exercise tracking*: AI can assess body alignment and movements to guide users through exercises with proper form.

- *Behavior analysis*: AI can be used for posture analysis in research or behavioral studies, providing insights into body language and communication.

– *Examples of AI in posture detection*:

- *Pose estimation algorithms*: these AI models use deep learning (a subset of ML) to analyze images or video frames, detecting the positions of key body joints and inferring posture (Zhang and Liu 2020). Open-source tools such as OpenPose or MediaPipe by Google are examples of such algorithms used for posture detection.

- *Wearable sensors with AI*: in some cases, AI is combined with wearable sensors, such as accelerometers or gyroscopes, that measure body movements and angles (Patel and Kumar 2021). AI models analyze the sensor data to detect changes in posture or track physical activity.

15.1.5. *Smart chairs*

Smart sensing chairs were first explored by Tan et al. in 2001, integrating sensors into the seating surface and backrest. The sensors embedded in the chair give real-time monitoring of posture, thus demonstrating a noncontact assessment of the sitting posture. Tan et al. (2001), being the first explorers of the smart sensing chair, developed the sitting posture classification system (Odesola et al. 2024).

The authors of the review studied 39 relevant papers according to the inclusion criteria.

Figure 15.4. *Smart sensing chairs*

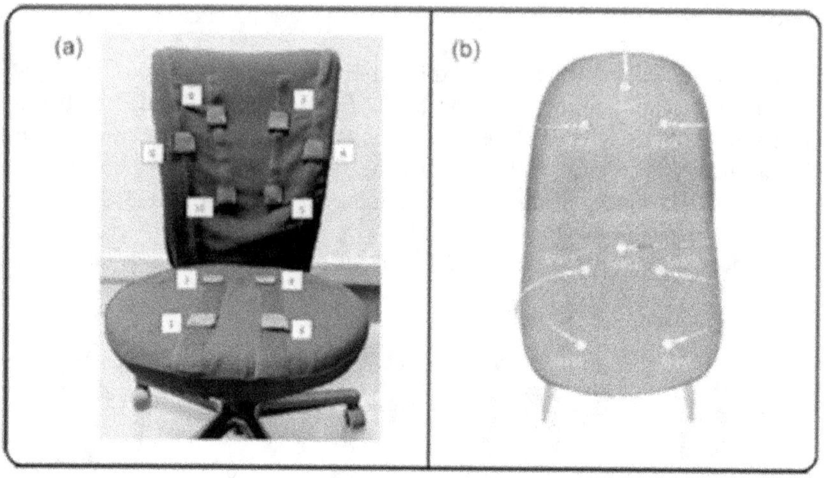

Figure 15.5. *The chair*

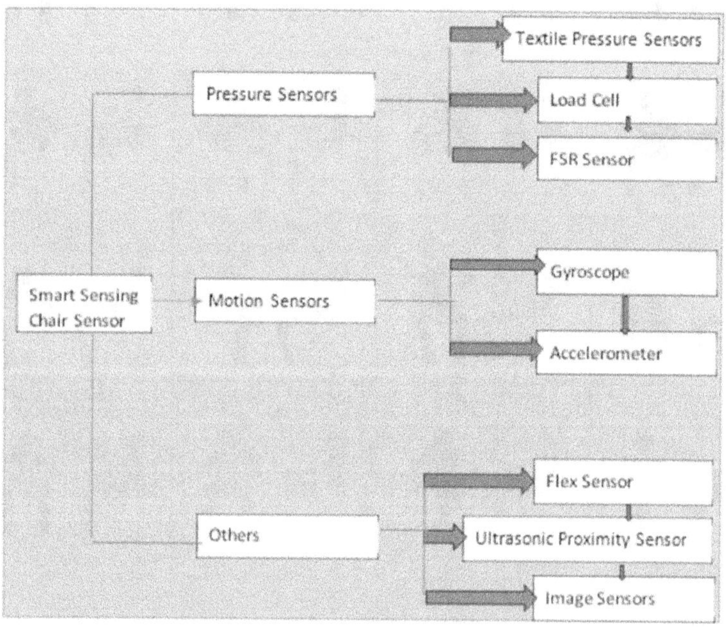

Figure 15.6. *Sensors used in smart sensing chairs*

Sr. No	Sensor	Description
		Pressure sensor
1.	Textile pressure sensor	Composed of soft fabric-based material consisting of a conductive thread pattern placed over a dielectric material serving as a substrate between threads.
2.	Load cell	This is another type of force sensor. It works by altering applied power-driven forces into a measurable digital signal which can be read by microcontrollers. Examples of load cells include strain gauges, piezoelectric, hydraulic and capacitive load cells.
3.	Force sensing/sensitive sensor (FSR)	Commonly used to measure the forces and physical pressure applied to its surface. The sensor varies its output resistance in response to the pressure executed on it. It is composed of a conductive polymer-based material which is integrated between two metal electrodes.
		Motion sensors
1.	Accelerometer	Described in detail in Chapter 14
2.	Gyroscope	
		Other
1.	Flex sensor	Also known as a bend sensor, it measures the degree of displacement resulting from the bending action being applied over it. It is composed of a flexible composite material which has a conductive ink material that changes in resistance as the sensor bends.
2.	Ultrasonic proximity sensor	Described in detail in Chapter 14.
3.	Image sensor	These are cameras and 3D image sensors, often cohesive with computer vision algorithms which identify visual elements from images and videos, thus helping to classify posture easily.

Table 15.1. *Sensors and their description*

15.1.5.1. *Dense sensor configuration*

A number of sensors are organized in a network, which affect the ability to detect a signal source, the distance between the sensor and the signal source, and the signal-to-noise ratio. Various sensors are organized as follows, as evidenced by various authors.

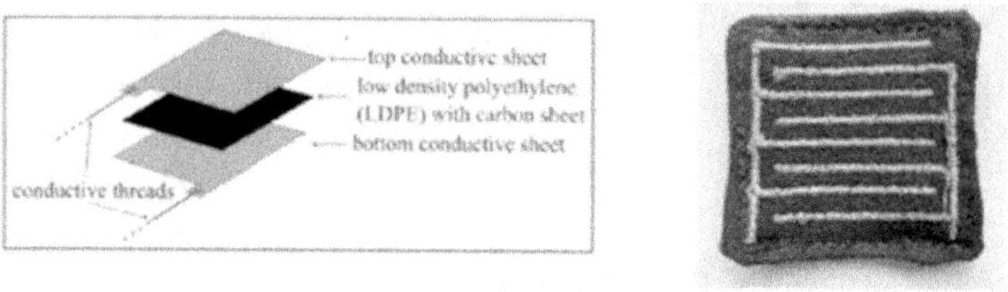

Figure 15.7. *Textile pressure sensors*

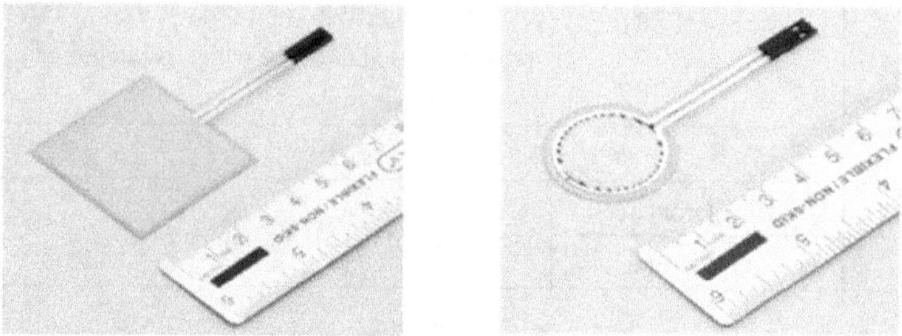

Figure 15.8. *Force sensitive sensors (FSRs)*

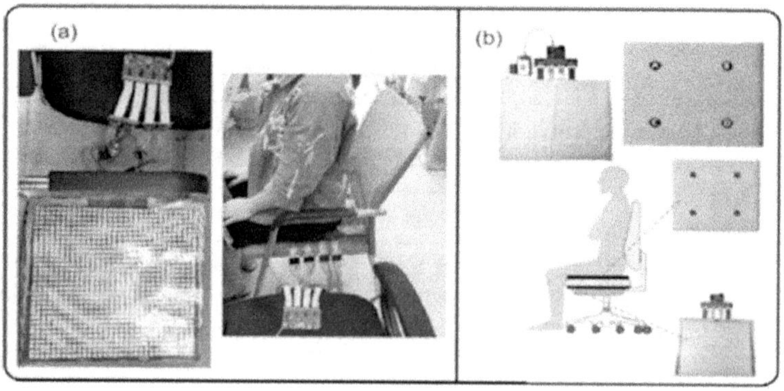

Figure 15.9. *Example of a dense sensor configuration*

The authors also demonstrated sensors used in various published papers (Odesola et al. 2024).

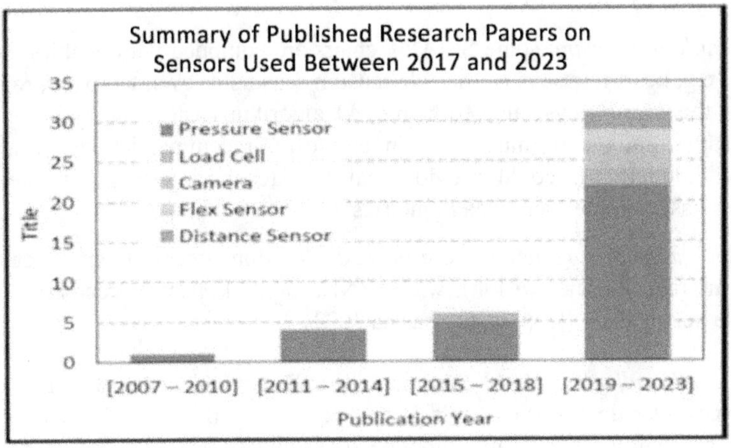

Figure 15.10. *Summary of published research papers*

The analysis indicated that force sensing resistors (FSRs) are the primary sensors used for posture detection, while CNNs and artificial neural networks (ANNs) are the predominant ML models for posture classification (Odesola et al. 2024).

15.1.6. *AI integration in SMART chairs*

SMART chairs can be considered a part of AI, depending on the features and technology integrated into them. SMART chairs typically refer to chairs that incorporate advanced technologies such as sensors, actuators and AI algorithms to provide enhanced functionality beyond a traditional chair. These features may include posture detection, ergonomic adjustments, health monitoring and personalized experiences.

How AI is used in SMART chairs

– *Posture and ergonomics*: many SMART chairs use AI and sensor technology to monitor the user's posture. AI can analyze data from built-in sensors (e.g. pressure sensors, accelerometers or cameras) to detect if the user is sitting correctly, whether their posture needs adjustment and even remind them to change positions to prevent discomfort or long-term health issues (Reddy and Mishra 2022). AI can also adjust the chair's angle, height or lumbar support automatically based on the user's posture, making it more ergonomic and comfortable.

- *AI algorithm*: CNNs analyze visual data to detect posture, while SVMs classify correct and incorrect postures. Reinforcement learning (RL) optimizes ergonomic adjustments based on user movements.

– *Health monitoring*: some SMART chairs are equipped with health-monitoring sensors (e.g. heart rate sensors, muscle tension sensors or even temperature sensors) to track the user's physical well-being. AI algorithms can analyze these data to detect stress, fatigue or muscle strain and offer recommendations or alerts. In advanced models, AI could provide real-time feedback or suggest breaks and exercises based on the user's health metrics.

- *AI algorithm*: random forests and decision trees classify health data (e.g. heart rate, muscle tension), while RNNs or LSTM track trends and provide real-time health alerts (Gupta and Sharma 2023).

– *Adaptive adjustments*: AI can enable the SMART chair to adapt to the user's preferences over time. For example, it may learn from the user's body movements, posture and sitting habits to automatically adjust for optimal comfort. These adjustments can include altering the chair's tilt, lumbar support or even offering a massage based on the user's needs, all determined by AI analysis of the user's patterns.

- *AI algorithm*: RL learns user preferences to adjust the chair's settings (e.g. tilt, lumbar support). K-means clusters sitting habits for personalized adjustments (Patel and Joshi 2023).

– *Integration with other smart devices*: some SMART chairs can connect with other devices in a smart home ecosystem (e.g. voice assistants such as Alexa or Google Assistant), using AI to interact with the user. For example, the chair could change settings based on voice commands or integrate with a user's calendar to prompt them to adjust posture after sitting for long periods.

- *AI algorithm*: NLP models (e.g. transformers) enable voice command understanding, while rule-based AI syncs with other smart devices for schedule-based adjustments (Kumar and Verma 2024).

– *AI-powered user feedback*: the SMART chair could also feature AI-driven feedback systems, such as using a companion app that provides the user with insights into their sitting habits and health. The AI could suggest personalized sitting schedules, recommend ergonomic improvements or track long-term sitting behavior to prevent injury.

- *AI algorithm*: recommender systems (e.g. collaborative filtering) suggest personalized sitting schedules, and supervised learning analyzes behavior for corrective feedback (Sharma and Tiwari 2023).

15.1.7. *Literature review*

Research conducted by Piñero-Fuentes et al. (2021) studied the effectiveness of a posture detection system for preventing telework-related musculoskeletal disorders. The system is prepared to be placed on a computer screen, where the image acquisition device, a webcam, has to be pointing at the person so that it can capture the posture from head to arms.

Initially, the device is connected to the embedded system where the video is processed in two steps:

– The user's joint positions are estimated using a neural network (Piñero-Fuentes et al. 2021) classifier.

– The user's whole posture is processed and validated to provide recommendation or detect change (Piñero-Fuentes et al. 2021).

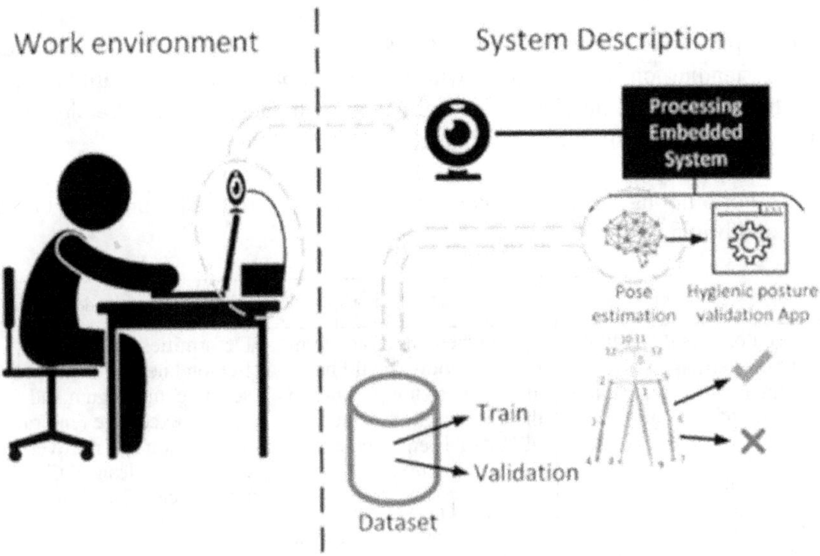

Figure 15.11. *Posture detection system*

Odesola et al. (2024) conducted a comprehensive review on smart sensing chairs for sitting posture detection, classification and monitoring. The authors studied in detail various other studies using smart sensing chairs for sitting posture.

15.2. Force plates

15.2.1. *Introduction*

Like the force plates introduced in the previous chapter, the main function of the biomechanical system lies in the detection of pressure under the foot while standing on it. It analyzes the pressure under the foot by various mechanical laws, one vital which is ground reaction force (GRF) (Nascimento et al. 2020).

The device is frequently used in many therapeutic labs and has shown significant results. The conclusion is that it is a beneficial device to be used for various neurological, orthopedic and geriatric conditions. The device can also be used to enhance sports-related or athletic skills (Kędziorek and Błażkiewicz 2020).

15.2.2. *Uses of force plates*

Force plates are generally used to assess the center of pressure on individuals who are standing on it. Individuals with balance impairment or disequilibrium are prone to have unequal distribution of weight on the bilateral lower extremity, which may increase the risk of falling (Kędziorek and Błażkiewicz 2020).

It is worth noting the conditions where individuals suffer with impaired balance or disequilibrium.

Osteoarthritis	Spinal conditions	Lower extremity deformities
A musculoskeletal condition where the cartilage starts wearing and tearing, leading to erosion of bony articular surfaces. If the condition worsens, then the bones get locked in such planes and angles, causing imbalance because of a shift in the line of gravity.	Spine when observed from posterior view should be aligned in a straight line, so that the weight distribution of H-A-T is equal on either lower extremity. Abnormal curvatures of the spine such as concavity and convexity. Shortening of structures or gathering on concave side and over-stretching or loosening or weakening of structures on the convex side. Thus, there is a change in LOG.	Deformities such as fixed flexion knee deformity, severe genu valgum and varum, excessive femoral anteversion and retroversion or CTEV, calcaneal tilt imbalance in standing position. D/t unequal weight bearing on bilateral lower extremity. Deviation of LOG. Risk of falling.

Table 15.2. *Uses of force plates*

Parkinson's	Sensory ataxia	Stroke
A condition where the cerebellum and medulla oblongata are affected, thus decreasing the production of dopamine. This causes resting pill-rolling tremors of the fingers and difficulty initiating any dynamic movements. Impaired dynamic balance. Increased risk of falling.	A condition where the cerebellum is dysfunctional Decrease in coordination and postural instability because of decreased brain sensory input Easy loss of balance. Increased risk of falling.	Also known as a brain attack, where there is decreased blood supply to the area of brain. Symptoms vary with respect to the area affected. For example, frontal lesion and impaired motor activity. Extension or flexion synergy. Loss of dynamic as well as static balance (if trunk is involved).

Table 15.3. *Neurological conditions*

15.2.3. *Working principle*

Human movement is complex when compared to robotic movement. Analysis of human motion is a technical task, where both quantitative and qualitative assessments are given equal importance. Historically, walking, running, and stepping were assessed manually by therapists who observed and evaluated human movement. For example, walking was assessed by eye and the different phases of the gait cycle were noted down. The therapist would also use a metronome to count the number of steps/temporal parameters and use the ink dip method to calculate spatial gait variables.

Nowadays, assessing complex human movement has become easy with the help of advanced bio-engineering and various sensors.

Force plates are commonly used in motion capture experiments. The plates collect the forces between the ground and foot. Force plates have sensors at each corner of the square which collect the current arising from small displacement of human body weight bearing on the force plates. The amount of current can determine the amount of force exerted by the external environment on the human body.

Force platforms are a great source of movement analysis. They are platforms with embedded force sensors and flexi force adaptors. The force plate is meant to measure the GRF which is the external force acting on our body when we weight-bear in any activity in an opposite direction. Force platforms measure the amount of GRF acting on the human body. We can assess activities such as walking, drop landing, stepping, vertical jumping, etc.

These sensors are important for measuring limb to limb asymmetries and the amount of force going through each lower extremity while performing any weight bearing action. Depending on the quantitative analysis of the force in newtons exerted by each lower extremity, we can manipulate the quantity of force (Kędziorek and Błażkiewicz 2020).

This quantitative analysis helps us with biomechanical findings and rehabilitation. The quantitative movements can be visualized in the form of graph on a laptop/computer which is connected to the force plate.

A force–time curve is an important feature of force plate movement analysis. Measuring any human weight bearing activity on a force plate means the plate can measure the amount of force produced during the activity. Graphical representation of this force production against time taken to perform the activity produces a force–time curve. It is thus recognized as an effective and significant method for studying athletic movements along with weight bearing activities by giving an insightful thought for proper movement analysis (Sundararajan and Sharma 2021).

Movement recorded by the force plate = $r \times F + T$ = Co-ordinates of r and T are r_x, r_y, r_z and T_x, T_y, and T_z.

T_x	T_z
r_x =	$T_x = 0$
r_y =	$T_y = 0$
$r_z = 0$	T_z =

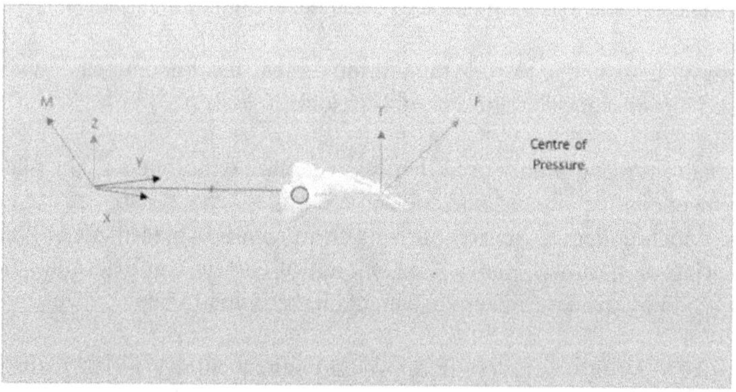

Figure 15.12. *Human weight bearing activity*

It is important to consider the foot's center of pressure where the equivalent single force is applied on the foot by the ground.

The device is known to measure the GRF during static and dynamic tests. It is a bio-mechatronic device, in-built with numerous pressure sensors and those which obtain plantar center of pressure displacement. The device measures the pressure between the foot and the platform, through the voltage change matrix obtained by a large number of in-built pressure sensors. The computer is coded to express different pressure values with different color codes, thus institutively displaying the information of pressure distribution throughout the sole (Chen et al. 2021).

15.2.4. *AI integration in force plates*

A force plate itself is not inherently a part of AI. Rather, it is a specialized tool used to measure forces and moments, typically during activities such as walking, running or standing. Force plates are equipped with sensors that detect the distribution and magnitude of forces applied to the plate's surface, such as GRF during movement.

However, AI can be integrated with force plates to enhance capabilities in certain applications, particularly in data analysis, pattern recognition and predictive modeling.

AI can be incorporated in force plate systems by:

– *Data analysis and interpretation*: force plates generate large volumes of data, particularly in dynamic activities where forces change rapidly. AI, especially ML algorithms, can be used to process and analyze these data efficiently (Gupta 2021). For example, AI can:

- identify abnormal patterns in a person's gait or balance that could indicate injury or risk of falls;

- recognize movement abnormalities and classify different types of movements, such as walking, running or jumping, by analyzing force plate data.

– *Predictive modeling*: AI can be used to predict outcomes based on force plate measurements. For instance, AI models could analyze how forces exerted during certain movements correlate with the likelihood of developing musculoskeletal injuries or predict recovery progress in patients undergoing rehabilitation. AI-driven models could suggest customized interventions or exercises based on individual force patterns (Singh 2022).

– *Biomechanical assessment*: in biomechanics, AI can be used alongside data from force plates to provide deeper insights into the forces involved in human movement (Patel 2023) AI algorithms could combine force plate data with other sensor inputs (e.g. motion capture, electromyography, etc.) to build a more comprehensive understanding of a person's biomechanics and help in designing more effective training or rehabilitation programs.

– *Real-time feedback*: AI-powered systems could provide real-time feedback to users based on data from force plates (Chaudhary 2022). For example, in sports training or rehabilitation, AI could analyze the force distribution during an athlete's jump or a patient's gait and provide immediate adjustments or suggestions to optimize performance or prevent injury.

15.2.5. *Literature review*

A study conducted by Tyler J. Collings, Yuri Lopes Lima, Benjamin Dutaillis and Matthew N. Bourne at Griffith University, Gold Coast Campus, Australia, in 2024 concluded that the device proves to be a highly accurate method of assessing vertical GRF and the center of pressure during a range of movement, strength and balance tests. The device VALD ForceDecks is a valid and reliable alternative to laboratory force plates when strict standardized testing and data analysis procedures are followed (Collings et al. 2024).

The subjects were told to perform the same sets of movements such as:

1. quiet standing;
2. single leg range of stability;
3. squat assessments;
4. countermovement jumps;
5. squat jumps;
6. drop jumps;
7. hop tests;
8. single leg hop tests;
9. shoulder ISO-Is;
10. shoulder ISO-Ys;
11. shoulder ISO-Ts;
12. isometric mid-thigh pulls.

All tests were performed in accordance with published protocols. The movements were performed on both VALD Force Decks and the laboratory force plates. The study thus concluded that the device demonstrated good or excellent validity (< 10% difference) and 79% had good or excellent reliability (intraclass correlation coefficient > 0.75).

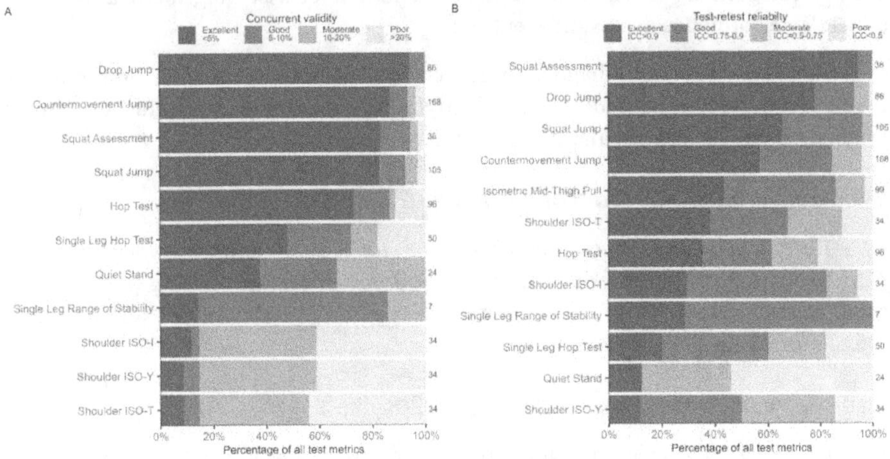

Figure 15.13. *Laboratory force plates (source: Collings et al. 2024)*

The study on 112 Division I athletes examined the reliability of force–time metrics in isometric mid-thigh pulls and countermovement jump tests. It found that most metrics, except for the rate of force development, were reliable, with some discrepancies in the isometric mid-thigh pull measures. The results showed that countermovement jumps and isometric mid-thigh pulls were generally reliable for performance monitoring, with some variations in eccentric measures such as deceleration and lower limb stiffness.

The authors emphasized the usefulness of integrating impulse-momentum and impulse-displacement techniques in countermovement jumps, as force plates reduce errors in measurements. The study also highlighted the potential for using these tests to assess neuromuscular function and fatigue during eccentric phases, suggesting their value for monitoring long-term neuromuscular adaptations (Merrigan et al. 2021).

The types of algorithms that were used in the device and described in the study likely included a combination of the following.

ML algorithms (Gorib and Bougoulias 2021)

– *Regression analysis*: to predict force–time metrics, velocity, displacement and other biomechanical parameters based on sensor data from force plates.

– *Support vector machines (SVM)*: to classify and analyze performance data, such as categorizing levels of neuromuscular fatigue or assessing the quality of the athlete's performance across different tests.

– *Random forests*: to handle large sets of biomechanical data and improve prediction accuracy, particularly in identifying the most relevant features for force production and neuromuscular performance.

Time-series analysis (Jadid and Sadeghi 2020)

– *Recurrent neural networks (RNN) or long short-term memory networks (LSTM)*: to analyze time-series data from the force plates, identifying patterns over time, such as rate of force development or eccentric phase duration, and detecting fatigue or changes in performance.

Deep learning algorithms (Zhao et al. 2020)

– *Convolutional neural networks (CNN)*: for analyzing video or motion capture data, recognizing movement patterns and identifying abnormal motion or postural deviations that indicate biomaterial failure or performance issues.

– *Autoencoders*: to detect anomalies in the collected data, such as small deviations (e.g. 2–3 milliseconds or 0.9 centimeters), which may signal discrepancies in performance or measurement accuracy.

Predictive modeling algorithms (Zhang et al. 2021a)

– *Gradient boosting machines (GBM)*: for predicting future changes in performance or identifying patterns of neuromuscular fatigue by analyzing current and past performance data.

Clustering algorithms (Zhang et al. 2021a)

– *K-means or DBSCAN (density-based spatial clustering of applications with noise)*: to group athletes into categories based on their force–time metrics, helping identify performance outliers or track individual progression.

These algorithms worked together to analyze and interpret data from the force plates, wearable sensors and motion tracking systems, providing insights into an athlete's performance, neuromuscular function and recovery. They helped the system identify issues such as biomaterial degradation, fatigue or abnormal movement patterns, facilitating personalized training and intervention.

Another study by Lee et al. (2020) examined the effects of plyometric versus ankle stability training on lower limb biomechanics in taekwondo. The study found that plyometric training improved ankle stability and shock absorption, reducing injuries during taekwondo activities, as measured by a force plate system. Fourteen participants were divided into two groups: one for plyometric activities and the other for ankle stability exercises, completing sessions twice a week for eight weeks. Both groups showed improvements in dynamic balance and postural sway, but the plyometric group exhibited better results by compensating for ankle instability with controlled landing techniques involving the knee and hip joints.

In this study, the AI algorithms used to analyze the impact of plyometric versus ankle stability training on lower limb biomechanics differed in the following ways.

Biomechanical data analysis

– *Plyometric training*: AI algorithms, such as convolutional neural networks (CNNs), were used to analyze landing techniques through motion capture or video analysis, focusing on joint angles and movement patterns at the knee, hip and ankle. Recurrent neural networks (RNNs) or long short-term memory networks (LSTMs) tracked the dynamic progression of movement over time to assess how plyometric training influences stability and shock absorption during landings (Liu and Zhang 2020).

– *Ankle stability training*: AI algorithms such as support vector machines (SVM) or random forests were applied to classify and compare postural sway and dynamic balance data from force plates and the Y-balance test, identifying changes in stability between groups (Ding et al. 2019). These models distinguished between dynamic balance improvements driven by plyometric exercises and those from ankle stability training.

Kinematic and kinetic data processing

Both groups generated kinematic and kinetic data, which were processed using multilayer perceptrons (MLPs) or gradient boosting machines (GBM) to evaluate forces, joint movements and balance performance (Liu et al. 2018). These algorithms help to identify patterns of improvement or compensation in the participants' movements.

Predictive modeling (Zhang et al. 2021b)

– *Plyometric group*: predictive models could be developed to forecast future stability improvements based on earlier training data, tracking of how plyometric training improves the control of knee and hip joints during landings, and how this compensates for ankle instability.

– *Ankle stability group*: predictive algorithms could track the progression of ankle joint stability over time and correlate it with improvements in dynamic balance and postural sway.

Anomaly detection (Schlegel et al. 2020)

Anomaly detection algorithms (e.g. autoencoders) could be used to identify outliers or abnormal movement patterns in both groups, helping to spot any inefficiencies or improper landing techniques that may hinder progress or lead to injury.

Performance comparison

Cluster analysis (e.g. K-means) could be used to group participants based on their performance improvements, allowing AI to compare the effectiveness of the two training methods by clustering participants with similar biomechanics and stability outcomes (Patel and Gupta 2019).

A study by Iwasa et al. (2023) at Kio University, Japan, examined the impact of physical therapy on postural control in an individual with spinocerebellar ataxia type (Liu and Zhang 2020), which causes severe postural balance disturbances. The rehabilitation focused on standing position and dynamic coordination exercises. Outcome measures included the center of gravity sway (using a BASYS CoP sway meter) and electromyography (EMG) to assess lower limb muscle activity. The CoP sway meter recorded parameters such as 95% confidence ellipse area (sway area), mean velocity (sway velocity) and power spectrum analysis (classified into low, medium and high frequencies). Muscle recruitment strategies during posture and daily activities were studied using the muscle BIT surface EMG system, with Bluetooth transmission to collect muscle data.

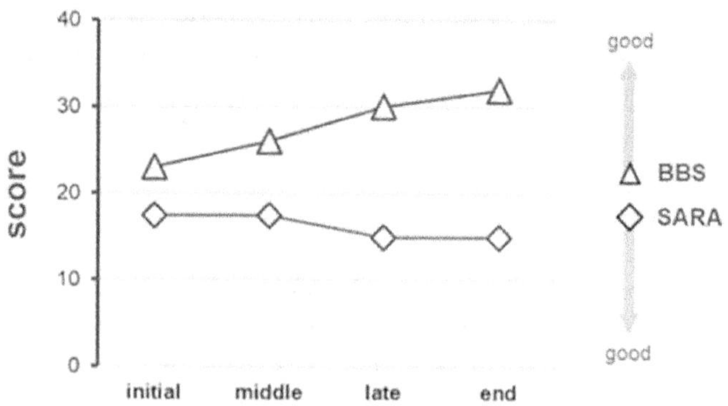

Figure 15.14. *ADL progress given by the authors*

The authors compared the ADL improvement and gait abilities of subjects using the Berg Balance Scale (BBS) and Scale for the Assessment and Rating of Ataxia (SARA) scores at four time points from the initial to the end stage of the intervention:

1. Initial, six weeks after fracture.
2. Middle, seven weeks after fracture.
3. Late, eight weeks after fracture.
4. Final, nine weeks after fracture.

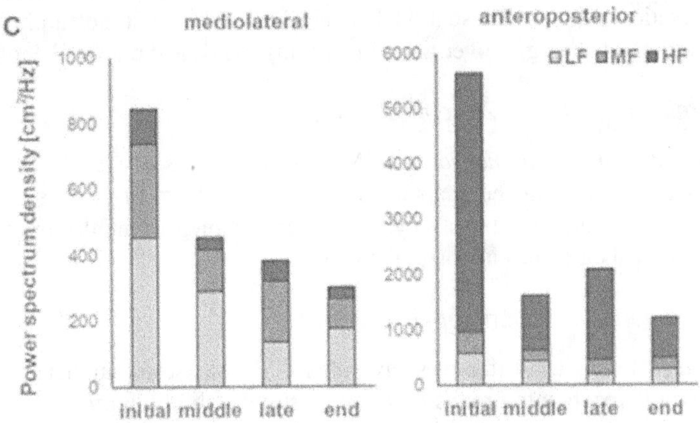

Figure 15.15. *Progress of standing COG sway values*

The authors concluded that the exercises improved standing postural balance and daily activities which contributed to a decrease in postural sway. This was graphically detected by the sway meter which followed the principle of center of pressure changes (similar to the force plate) (Iwasa et al. 2023).

In this study on spinocerebellar ataxia type (Liu and Zhang 2020), AI algorithms could be used to analyze postural control and muscle activity data collected through the CoP sway meter and electromyography (EMG). The following AI algorithms could be used.

ML algorithms (Gorib and Bougoulias 2021)

– *Support vector machines (SVM)*: to classify different levels of postural stability based on CoP data (e.g. detecting patterns in sway velocity or sway area and correlating them with balance dysfunction).

– *Random forests*: to feature select and classify the postural data into various stability categories, helping identify improvements in posture or muscular control over time.

– *Gradient boosting machines (GBM)*: to predict postural instability or muscle recruitment patterns based on historical data, taking into account the dynamic changes in balance and muscle activity during the rehabilitation process.

Time-series analysis (Jadid and Sadeghi 2020)

– *Recurrent neural networks (RNN) or long short-term memory (LSTM) networks*: to analyze the time-dependent nature of sway data, joint movements and muscle activation patterns. These models could track changes in postural control or muscle recruitment strategies over time, identifying trends and potential anomalies.

Deep learning algorithms (Zhao et al. 2020)

– *Convolutional neural networks (CNN)*: while CNNs are typically used in image analysis, they could be adapted to analyze data from the CoP sway meter, detecting intricate patterns in sway dynamics across time, potentially linking these patterns to clinical outcomes in postural balance.

Clustering algorithms (Zhang et al. 2021a)

– *K-means*: to cluster different sway patterns or muscle recruitment strategies, classifying participants into groups based on postural stability improvements during rehabilitation.

Key differences from previous studies

Data focus

The first two studies focus more on biomechanical performance (athletic rehabilitation and performance tracking) and dynamic stability, while the third focuses on neurological conditions with a focus on postural stability and muscle activity.

AI techniques

The first two studies apply ML and deep learning for biomechanical data and movement analysis, while the third study uses time-series analysis and postural sway detection for neurological rehabilitation.

Rehabilitation context

The first two studies are more athletic performance-oriented, while the third study deals with medical conditions and the challenges in rehabilitating neurological patients with ataxia.

Impact on physiotherapy diagnosis and assessment

– *Early detection*: AI enhances the ability to detect subtle changes in muscle activity and movement patterns, aiding physiotherapists in early diagnosis of issues such as muscle fatigue, joint instability or degenerative conditions such as ataxia.

– *Personalized rehabilitation*: AI enables customized treatment plans, offering real-time analysis to help adjust exercises based on individual performance and progress, ensuring that rehabilitation is always aligned with the patient's evolving needs.

– *Increased precision*: by analyzing biomechanical and neurological data more accurately, AI can improve the precision of diagnosis and monitoring, ensuring more effective interventions that minimize the risk of further injury or progressive damage.

– *Long-term monitoring*: AI provides tools for long-term tracking of rehabilitation, offering insights that improve not just short-term recovery but also long-term outcomes by identifying pattern shifts that could indicate future issues.

15.3. Conclusion

The convergence of physiotherapy principles, sensor technology and AI marks a paradigm shift in how posture and movement disorders are understood, detected and managed. Beginning with the foundational McKenzie method, the chapter outlines how early classification of postural disorders such as postural syndrome, dysfunction syndrome and derangement syndrome enables targeted and effective interventions. As poor posture continues to be a pervasive issue, the evolution of real-time posture detectors and smart chairs demonstrates the value of integrated AI systems in addressing ergonomic and clinical challenges. These devices not only offer continuous monitoring and feedback but also facilitate behavioral correction through user-friendly, noninvasive interventions. Force plates, another advanced biomechanical assessment tool discussed in this chapter, enhance clinical decision-making by providing precise kinetic data regarding GRF and balance. The synergy of ML algorithms and force plate analytics enables clinicians to detect subtle abnormalities in gait and stability, which is particularly crucial for patients with neurological conditions such as Parkinson's disease or spinocerebellar ataxia. AI models such as CNNs, RNNs and SVMs are shown to enhance the predictive and diagnostic potential of these tools, making rehabilitation more efficient and personalized. In summary, the integration of AI into posture detection and biomechanical assessment devices fosters an evidence-based, preventive and precision-focused approach in modern physiotherapy and rehabilitation sciences. It not only improves diagnostic accuracy and therapy outcomes but also contributes to long-term musculoskeletal wellness, injury prevention and enhanced quality of life.

15.4. References

Chakraborty, M. and Saha, S. (2023). AI-driven posture detection systems for healthcare: A review. *Journal of Artificial Intelligence and Vision*, 10(2), 45–61.

Chan, M., Dyck, M., Thevasagayam, G., Yap, R., Ouellet, J., Robbins, S.M. (2021). Inter-rater reliability of the McKenzie method of mechanical diagnosis and therapy for the provisional classification of low back pain in adolescents and young adults. *Journal of Manual & Manipulative Therapy*, 29(4), 255–261.

Chaudhary, R. (2022). Real-time feedback using AI in sports and rehabilitation: Applications of force plate data. *Journal of Sports Science and Technology*, 40(3), 155–168.

Chen, B., Liu, P., Xiao, F., Liu, Z., Wang, Y. (2021). Review of the upright balance assessment based on the force plate. *International Journal of Environmental Research and Public Health*, 18(5), 2696.

Collings, T.J., Lima, Y.L., Dutaillis B, Bourne, M.N. (2024). Concurrent validity and test–retest reliability of VALD ForceDecks' strength, balance, and movement assessment tests. *Journal of Science and Medicine in Sport*, 7(8), 572–580.

Conroy, V.M., Murray Jr, B.N., Alexopulos, Q.T., McCreary, J. (2022). *Kendall's Muscles: Testing and Function with Posture and Pain*. Lippincott Williams & Wilkins, Philadelphia, PA.

Ding, Z., Zhang, X., Chen, H. (2019). Machine learning techniques for postural sway analysis in human biomechanics: Applications in rehabilitation and sports. *Journal of Biomechanics*, 93, 116–123.

Gorib, S. and Bougoulias, D. (2021). Application of machine learning techniques for biomechanical data analysis: A review. *Journal of Biomechanics*, 116, 110140.

Gupta, R. (2021). Machine learning applications in force plate data analysis for dynamic activities. *Journal of Biomechanics*, 54(2), 45–59.

Gupta, A. and Sharma, N. (2023). AI-enabled smart chairs for health monitoring: Applications in stress detection and ergonomics. *Journal of Intelligent Systems in Healthcare*, 16(2), 103–118.

Huang, X., Xue, Y., Ren, S., Wang, F. (2023). Sensor-based wearable systems for monitoring human motion and posture: A review. *Sensors*, 23(22), 9047.

Iwasa, S., Akaguchi, R., Okuno, H., Nakanishi, K., Ueta, K., Morioka, S. (2023). Changes in standing postural control ability in a case of spinocerebellar ataxia type 31 with physical therapy focusing on the center of gravity sway variables and lower leg muscle activity. *Cureus*, 15(12), e51033. doi: 10.7759/cureus.51033.

Jadid, M. and Sadeghi, H. (2020). Application of recurrent neural networks for time-series analysis in biomechanics: Modeling and prediction of force-time profiles. *Computer Methods in Biomechanics and Biomedical Engineering*, 23(14), 1072–1082.

Kędziorek, J. and Błażkiewicz, M. (2020). Nonlinear measures to evaluate upright postural stability: A systematic review. *Entropy*, 22(12), 1357.

Kumar, A. and Verma, S. (2024). AI integration in smart chairs: Enhancing comfort and interaction through smart ecosystems. *Journal of Smart Technologies and Automation*, 22(1), 56–72.

Lee, H.M., Oh, S., Kwon, J.W. (2020). Effect of plyometric versus ankle stability exercises on lower limb biomechanics in taekwondo demonstration athletes with functional ankle instability. *International Journal of Environmental Research and Public Health*, 7(10), 3665. doi: 10.3390/ijerph17103665.

Liu, H. and Zhang, L. (2020). Application of deep learning in biomechanics: A review of convolutional neural networks and recurrent neural networks in human motion analysis. *Journal of Biomechanics*, 109, 109934.

Liu, W., Zhang, X., Li, L. (2018). Application of machine learning in biomechanics: A review on the use of MLPs and gradient boosting models for kinematic and kinetic data analysis. *Computers in Biology and Medicine*, 104, 107–115.

Merrigan, J.J., Stone, J.D., Hornsby, W.G., Hagen, J.A. (2021). Identifying reliable and relatable force–time metrics in athletes – Considerations for the isometric mid-thigh pull and countermovement jump. *Sports*, 9(1), 4.

Nascimento, L.M., Bonfati, L.V., Freitas, M.L., Mendes Junior, J.J., Siqueira, H.V., Stevan Jr, S.L. (2020). Sensors and systems for physical rehabilitation and health monitoring – A review. *Sensors*, 20(15), 4063.

Odesola, D.F., Kulon J, Verghese S, Partlow A, Gibson, C. (2024). Smart sensing chairs for sitting posture detection, classification, and monitoring: A comprehensive review. *Sensors*, 24(9), 2940.

Patel, D. (2023). Integrating AI with force plate and sensor data for comprehensive biomechanical assessments. *Journal of Biomechanical Engineering*, 45(7), 210–222.

Patel, S. and Gupta, A. (2019). Performance comparison of training interventions using cluster analysis: A study on biomechanical and stability outcomes in sports rehabilitation. *Journal of Sports Science and Medicine*, 18(4), 687–695.

Patel, R. and Joshi, V. (2023). AI-driven adaptive adjustments in smart chairs: Learning user preferences for optimal comfort. *Journal of Artificial Intelligence in Smart Systems*, 19(1), 25–40.

Patel, S. and Kumar, P. (2021). Wearable sensors and AI for posture detection and physical activity tracking. *International Journal of Wearable Technology and Health Monitoring*, 7(1), 35–50.

Piñero-Fuentes, E., Canas-Moreno, S., Rios-Navarro, A., Domínguez-Morales, M., Sevillano, J.L., Linares-Barranco, A. (2021). A deep-learning based posture detection system for preventing telework-related musculoskeletal disorders. *Sensors*, 21(15), 5236. doi: 10.3390/s21155236.

Reddy, S. and Mishra, R. (2022). AI-driven smart chairs for ergonomic posture monitoring and adjustment. *Journal of Artificial Intelligence in Design and Technology*, 11(3), 74–88.

Rybski, M.F. and Juckett, L. (2024). Posture. In *Kinesiology for Occupational Therapy*, Rybski, M.F. (ed.). Routledge, Oxford.

Schlegel, C., Muench, F., Krueger, B. (2020). Anomaly detection in biomechanics: Leveraging autoencoders for identifying abnormal movement patterns in sports and rehabilitation. *IEEE Transactions on Biomedical Engineering*, 67(3), 861–868.

Sharma, K. and Tiwari, P. (2023). AI-driven user feedback systems in smart chairs: Personalized health and ergonomics monitoring. *Journal of Wearable Technology and Human-Computer Interaction*, 14(3), 145–160.

Singh, M. (2022). Predictive modeling in biomechanics: AI-driven insights from force plate data. *Journal of Applied Biomechanics*, 38(5), 123–134.

Singh, A. and Kapoor, R. (2022). Real-time posture detection and monitoring using AI: Applications in ergonomics, fitness, and behavioral analysis. *Journal of AI and Human-Computer Interaction*, 15(4), 98–110.

Sundararajan, V. and Sharma, S. (2021). Machine learning in posture detection: Applications and advancements. *International Journal of Artificial Intelligence and Machine Learning*, 12(3), 215–229.

Zhang, Z. and Liu, Y. (2020). Pose estimation and human body analysis using deep learning: Tools and techniques. *Journal of Computer Vision and AI Research*, 18(2), 112–126.

Zhang, H., Zhang, Y., Wang, X. (2021a). Machine learning approaches in predictive modeling and clustering for biomechanics and sports performance. *Journal of Sports Science and Medicine*, 20(4), 707–717.

Zhang, Z., Li, Y., Liu, H. (2021b). Predictive modeling of biomechanical performance: Using machine learning techniques for tracking joint stability and dynamic balance improvements. *Journal of Biomechanical Engineering*, 143(7), 071009.

Zhao, Y., Liu, Q., Liu, Y. (2020). Deep learning for biomechanical data analysis: Applications in movement recognition and anomaly detection. *IEEE Access*, 8, 132527–132537.

16

Key Technologies of Artificial Intelligence: Robotics, Wearables and Big Data

Artificial intelligence (AI) is proving to be a revolutionary technology in almost every field of life including healthcare, sports, the fashion industry, military, etc. The intervention of AI in day-to-day life improves the standard of living and transforms industries into automation, making life easy and luxurious. AI systems and subset technologies such as machine learning (ML), artificial neural networks and deep learning (DL) empower machines to solve complex problems and make optimized decisions from a number of options. While AI improves the standard of living, it also impacts on privacy and data security. Therefore, AI must be handled and implemented responsibly and thus some prospects and challenges accompanying AI must be considered in order to shape a better future.

16.1. Introduction

The rapid advancement of the digital age has led to widespread automation, replacing human labor faster than expected. Some scientists believe this trend will significantly impact human civilization, ultimately leading to extensive job automation. Ahmed (2019) argues that AI and automation can weaken even well-established businesses, and they emphasize that labor unions have a moral duty to protect both workers' rights and broader ethical standards.

Chapter written by Rohini BAWA and Surekha RANI.
For a color version of all of the figures in this chapter, see www.iste.co.uk/kumar/ai-driven1.zip.

The Covid-19 pandemic further disrupted businesses, causing supply-chain issues and reduced productivity (Allam et al. 2019; Awotunde et al. 2021). It also accelerated the shift towards automation and remote work as companies sought alternatives to traditional labor. Studies indicate that these changes had significant negative effects globally. Therefore, it is crucial to examine how technological advancements have influenced economic growth and why recent innovations are particularly transformative (Bragazzi et al. 2020).

The AI literature provides an extensive study of three key issues such as:

– relevant education for automation and education;

– apprehension of the young and older people towards new technologies;

– whether bosses interested in computerization leads to decreased labor.

Regarding the current study, the second issue is pertinent (Bharadiya et al. 2023). Understanding the underlying reasons behind the growing apprehension about automation is necessary for a thorough examination of how technology developments affect the labor market (Benke et al. 2018). The causes of this have been extensively studied by researchers.

16.2. Intervention of artificial intelligence (AI) in robotics and automation

Artificial intelligence (AI) has emerged as a major technical development with considerable potential to completely transform the area of robotics. In addition to technological developments, ethical issues surrounding AI's application in robots have gained a lot of attention (Deshpande et al. 2018; Chopra et al. 2023). Safety, accountability, openness and prejudice in decision-making are examples of ethical dilemmas that must be discussed to guarantee the moral and responsible use of AI-powered robots across a range of fields. The basic boundaries of AI algorithms in managing some unpredictable and active environments, problems with safety, toughness and interpretability still exist despite the rapid advancements in AI for robotics (Dhana et al. 2022). It is important to carefully consider the societal and economic effects of the widespread use of AI in robots, including the possible effects on employment as well as social standards as revealed in Figure 16.1.

Basic applications – healthcare

The integration of AI and IoMT in customer health apps has the potential to transform healthcare. It empowers people by improving the care provided by healthcare authorities (Dong et al. 2020; Grischke et al. 2020). These apps provide more personalized feedback, with guidance and support for maintaining good health.

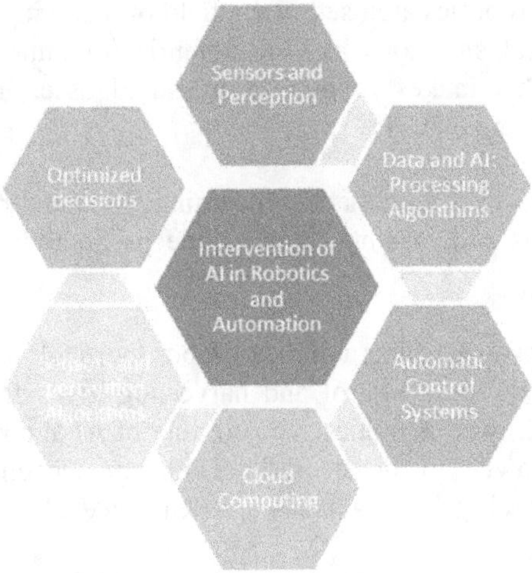

Figure 16.1. *Intervention of AI and robotics*

As an example, there is increased accuracy in the detection of cancer and early-stage detection with the use of AI. Reviewing and translating mammograms can now be done up to 30 times faster with precision up to 99%. This will not only speed up diagnosis but also lead to needless biopsies being dropped (Gaur et al. 2021). AI supports healthcares to take a more comprehensive approach to disease management, helping them coordinate care plans and achieve better management for fulfilling long-term treatment plans (Gao et al. 2021).

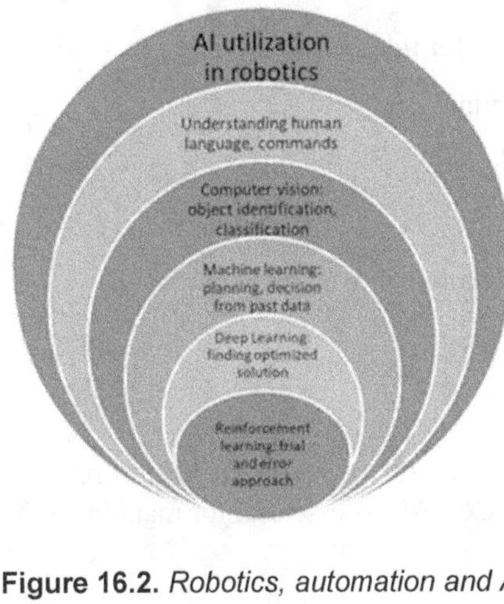

Figure 16.2. *Robotics, automation and AI*

In addition to AI, robotics are used in the field of medicine, and the combination of robotics in medical situations has the potential to improve the efficiency of patient outcomes. This makes robotics a treasured asset in modern healthcare (Jin et al. 2020; Haick et al. 2021).

The incorporation of AI with machine learning (ML) and robotics in agriculture creates powerful tools and insights to boost the productivity of farms (Jin et al. 2020).

The use of robotics in farming automates labor thorough tasks such as irrigation with dispersal of seeds, pest control and harvesting gives farmers' more time to concentrate on more fruitful activities. The mixture of AI and robotics in agriculture brings about the power to drive positive change, leading to the complete development of the global agricultural landscape (Khogali et al. 2023; Kuru et al. 2024).

In storage and warehouse settings, the use of AI has significantly improved safety by helping machines understand their surroundings better, especially with the help of thermal and haptic sensors. These sensors act as the "eyes and touch" for robots, guiding their decisions and allowing them to operate smoothly and safely. Automated guided vehicles (AGVs) and carts (AGCs) are commonly used to move stock around, enabling 24/7 operations while keeping costs consistent.

Automation robotics plays a dynamic part in the automotive industry. The rewards of robotics in motorcars are diverse including true vision for positioning and situating required particulars, enabling tasks such as fitting door buffers, panels and other factors (Lee et al. 2019; Lutz et al. 2019).

AI supported by all robotic and automation units can be seen in Figure 16.2.

Some algorithms used for robotics are:

– reinforcement learning;

– supervised learning;

– computer vision;

– simultaneous localization and mapping (SLAM);

– evolutionary algorithms;

– deep learning (DL).

Each of the several algorithms used in AI for robotics has advantages and disadvantages of its own. SLAM is essential for mapping and navigation, supervised learning is useful for tasks involving labeled data, DL is strong for perception and control, reinforcement learning is excellent for adaptive decision-making and computer vision offers visual perception capabilities (Miller et al. 2020). Task requirements such as data availability with computational resources and desired degree of robot autonomy all influence which method is best.

Chopra et al. (2023) give an overview on the exciting connection of AI and robotics, reshaping both economic and organizational dynamics. Together, AI and robotics have unlocked many new opportunities in areas of human–robot interaction, social robotics and perceptive robotics. Recent highlights of AI in robotics are innovative research trends with some technical tactics, and real-world applications which also raises some ethical, social and economic questions in regard to this rapidly growing field. AI in robotics sheds light on the prospects of basic tasks connected with this technology.

As employers adopt automation technologies developed by AI, workers may fear job displacement, creating challenges for both parties (Naha et al. 2022). The study "Automation Fears: Drivers and Solutions" surveyed 502 people in Bulgaria, revealing that individuals prioritize their own solutions over profitable or social ones due to rising automation concerns. This basic survey found that fears of job automation are influenced by personal beliefs and demographics (Nadikattu 2021). Key factors driving these fears include peer pressure, the likelihood of a job being automated, concerns about technology dehumanizing work and an individual's self-perception of professionalism.

The main objective of this chapter is to explore the long-term societal impacts of current progress in AI technology and computerization. The findings are shaped by various subtopics addressed throughout the research (Nam et al. 2021). Each section begins with an explanation of the chapter's structure. The study expresses common features and establishes a theoretical framework. It also examines the drivers of automation and AI, as well as how society perceives and accepts these technologies in general (Leary et al. 2013).

Hisham et al. (2023) explored social-impact assessment theory, as described by Dietz et al. (2024). It involves identifying, evaluating and measuring how an event affects society.

Similarly, the societal effects of AI should be carefully analyzed, just as the impact of scientific research is assessed. To track AI's influence amid rapid technological growth, a hybrid approach can be used. This approach combines narrative evaluation, which identifies relevant studies on a specific issue, and thematic pattern analysis, which detects recurring themes, trends and concepts within texts such as transcripts.

Definitions that have been adopted

The "impact of AI on jobs" suggests that AI in the workplace can either create and enhance new job opportunities or lead to significant job losses.

The "impact of AI on workers' well-being" suggests that while automation and AI may increase productivity and wages for some employees, they can also negatively affect job security and overall well-being.

The "impact of AI on organizational dehumanization" describes employees' perceptions of being mistreated by their organizations, feeling undervalued and being treated more such as machines than humans due to increased AI integration.

"Fears from the job automation" refers to the concern that widespread automation may lead individuals to worry about job security due to the increasing reliance on automated systems.

"AV worries" relates to concerns in autonomous vehicle (AV) engineering, where safety validation goes beyond traditional methods. Ensuring the functional performance of these new vehicle types presents a unique challenge in guaranteeing their safety (Obschonka et al. 2020).

Basic research problems

The link between AI advancements and possible job losses underscores the significant risks associated with the common adoption of AI, despite its many benefits. This research aims to conduct a detailed analysis of how automation and AI influence various long-term societal factors. The primary objective is to determine the global impact of automation and AI on businesses and human civilization, ultimately assessing whether these technologies benefit or harm society (Panesar et al. 2019).

Because of the rapidly changing global shifts in AI technology and innovation, the current study presents the following core questions (Ramasamy et al. 2022):

– How are society and businesses affected by the impending rise in the use of AI?

– What social problems are the latest developments in AI technology causing?

AI powered literature reviews can provide support for issues that require complex social, business and ethical information. Social science methods are required for technological societal and ethical suggestions because they deliver more flexibility when considering composite data.

There are some theories related to AI's societal and ethical impacts.

Societal impact theories

AI presents risks such as job displacement, increased poverty, instability, unemployment and serious privacy concerns. To harness its benefits safely, regulations and stronger security measures are essential (Raj et al. 2019). While technology can improve various aspects of life such as health monitoring, social connections, information access, communication and convenience, it also reduces privacy and introduces new challenges. Despite enabling instant communication, technology can contribute to loneliness and expose individuals to new forms of manipulation and intimidation. According to basic research, AI is constructed for adopting social environments. The AI system can be trained to carry out social behaviors and launch social networks (Shi et al. 2020).

In order to assess social impact theories, it is important to ascertain who is most likely to be impacted, figure out how to identify those who will be impacted, assess potential social ramifications, put management strategies in place to minimize harmful effects and maximize benefits, and enable regular monitoring and the pursuit of goals.

Principled impact theories

This theory has assisted in the sorting of appropriate data on the ethical implications of AI. Ethical viewpoints include several subcategories, such as utilitarianism, deontology, rights and virtues. Ethical behavior plays a crucial role in improving factors such as employee performance, job satisfaction, organizational commitment, trust and organizational citizenship behaviors.

Ethics, as a set of guiding principles, varies across societies, with cultural beliefs and standards shaping ethical norms. Since cultural environments influence ethical perspectives, AI is expected to have a wide range of social impacts depending on societal values (Siau et al. 2018).

Some of the biggest societal effects of AI

In regard to the financial impact of AI, there are basic growing concerns that AI could replace human labor, leading to job displacement and economic instability (Soori et al. 2023). The gap of wealth between the wealthiest and poorest members of society will grow because only those who can afford and have access to the requisite services and knowledge to use AI systems for financial gain will do so.

In regard to public health and security, AI and robotics are rapidly transforming the healthcare sector by playing an progressively vital role in clinical analysis and treatment. Robots are already being used for patient diagnosis, improving efficiency and accuracy (Tussyadiah et al. 2020). However, as their presence grows, so does the likelihood of harm; mostly with assistive robots and drones, which must make critical decisions that straightly affect human safety and well-being. This raises important ethical and regulatory concerns regarding their responsible use. As society uses information technology, it may be basically altered by the use of AI, mainly in regard to how personal information will be connected and how cybercriminals get access to private information. Facial recognition powered with AI can help to secure locations but there is a risk of misuse by cybercriminals who may hypothetically compromise the systems and exploit them maliciously (Wisskirchen et al. 2017).

Some negative ideals connected with AI

– *Inequality*: one of the worst consequences of AI technology is the widening of the wealth gap. While the use of this knowledge reduces the numbers of human employees in numerous organizations, AI-driven enterprises will be the only ones benefiting from it.

– *Privacy*: AI is often disapproved for invading people's privacy, but it also holds the potential to help solve important social challenges when used responsibly.

– *Ecological impact*: AI can be used to sort waste in order to decrease pollution. It can do so with the deployment of AVs to decrease greenhouse-gas discharges and traffic congestion. In addition, DL methodologies can be used to expand biodiversity and global conservation.

Here are some basic results of the investigation with a few important features of long-term AI social impacts.

AI is widely used in society, but those who do not engage with it may perceive it negatively.

The rise of AI and automation naturally brings about concern and anxiety, especially when it comes to job security and broader changes in society. While physical robots can replace certain manual jobs and contribute to job losses, computers and AI often have a different impact. Rather than directly replacing workers, they tend to change how jobs are done, such as automating tasks, improving efficiency and sometimes even creating new kinds of roles.

AI leads to higher-throughput levels in job-role specialization with human capabilities, problem-solving including quantitative aids and impactful work.

16.3. Future fabric: smart clothing with AI

Smart fabric or smart textile is based on the integration of fabric material with integrated circuits or chips which enable clothes to monitor health as well as look trendy (Zheng et al. 2021). Smart fabric, also called e-fabric, has digital components and biosensors which collect data from biomarkers and process them. These data are then transmitted to remote hospital units for diagnostic purposes or stored for future records depicted in Figure 16.3.

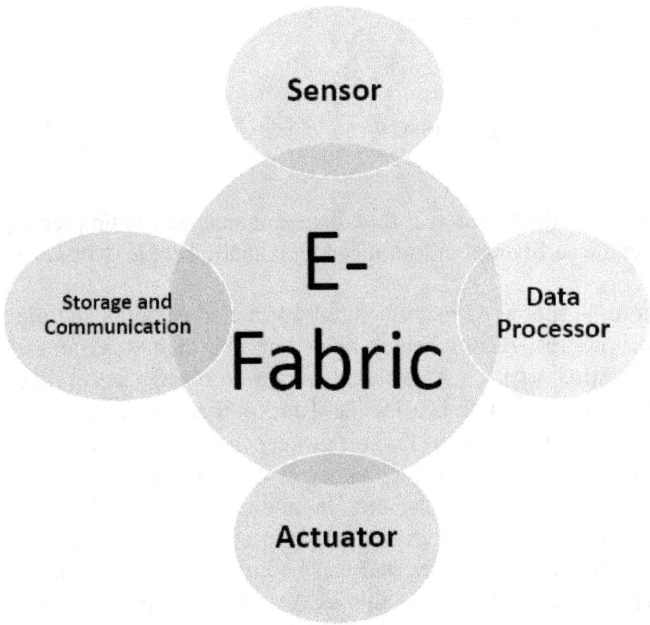

Figure 16.3. *E-fabric technology*

E-fabric equipped with sensors enables it to interact with the environment and adapt according to stimuli (Zhao et al. 2022). By blending AI with textiles, smart fabric can give a personalized experience to the wearer. Smart fabric equipped with AI sensors can track heart rate, body temperature, respiratory rate and provide alerts on potential issues. One such example is smart socks made by Senoria. They are equipped with a biosensor which monitors foot pressure, steps and walking, as shown in Figure 16.4.

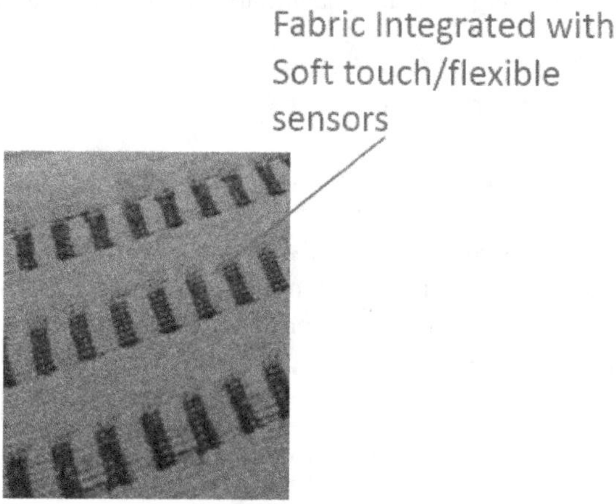

Figure 16.4. *E-fabric with sensor ICs*

AI algorithms collect these real-time data and analyze running techniques. They can also detect the early onset of foot ulcers in diabetic people (Kuru et al. 2024).

Furthermore, in sports, AI manufactured smart clothes optimize training routines and keep their progress records which helps improve their weak areas and prevent injuries. In addition to this, AI-embedded fabrics change according to weather conditions. One such example is a Jacquard Jacket that was launched by Google in collaboration with Levi et al. (2021). This jacket senses the wearer's movements and gestures and enables the wearer to control their digital gadgets such as smartwatches, check notifications, etc. In addition to this, the jacket provides important data on ultraviolet exposure, pollution levels and the air quality index. It also prioritizes wearer preferences and can be customized accordingly to give the user a perfect experience. Similar to this jacket, another company made a Hexoskin smart shirt and Xenoma e-Skin supported by an AI algorithm to monitor the wearer's physiological data, to help them explore the environmental effect on the body (Lee et al. 2019).

In the developing field of e-fabrics, one famous name is MIT Media laboratory, which is working to develop fabric that can vary color according to temperature. They are also working to produce electricity from the wearer's movement.

Therefore, smart fabric or e-fabric has many applications in different fields which can be summarized in the following sections and can be presented with the help of Figure 16.5.

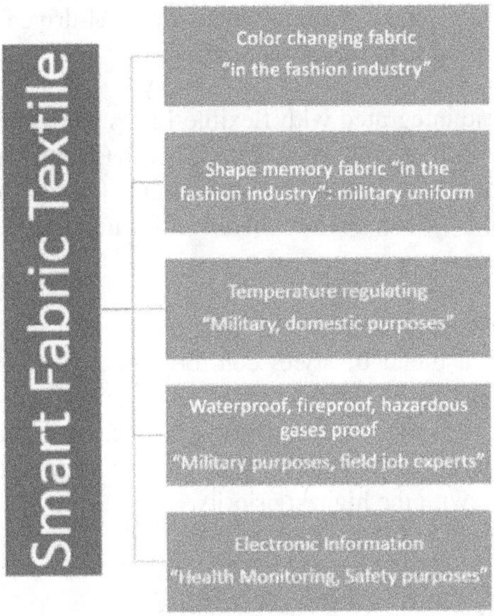

Figure 16.5. *AI-equipped e-fabric for energy harvesting*

To make e-fabric more advanced and self-sufficient, scientists at the University of California, Berkeley, are working in the field to produce energy from body temperature and movement. This energy will be used to operate the low power sensors that are embedded in e-fabric, avoiding the need for bulky batteries. This research enables e-fabric to work continuously without the need for an external supply (Jin et al. 2020).

16.4. AI wearables for the military

In the field of defense, e-textiles can be a potential candidate for soldier safety. Army uniform equipped with AI sensors can record environmental conditions and also provide awareness of surroundings at the base station (Gao et al. 2021). One such uniform is currently a research topic for the U.S. Army Research Laboratory, which tracks heart rate, body temperature, water level in the body, exposure to UV rays and hazardous gases to achieve better decision-making in the battle grounds.

16.5. AI wearables in the fashion industry

Smart fabric embedded with AI sensors has proven to be a revolutionary approach in the fashion and consumer industries. AI-driven e-fabric can change color or texture according to seasonal requirements.

Furthermore, fabric integrated with flexible LEDs can give brighter and magical effects to clothes which seems very attractive. AI-controlled LED enables the clothes to change color and design pattern within a fraction of seconds. One such prototype is designed by Cutecuircuit which is a fashion brand known for its unique and innovative designs.

Therefore, as discussed above, we can get a clear picture of AI-equipped wearables presenting a blend of style, comfort, protection and health monitoring (Muller et al. 2020). But as AI sensors record all of a user's personal data, they must be handled ethically with the utmost privacy. Hence, the bottom line is that while developing AI wearables, responsibility for maintaining privacy and handling data ethically will be taken with the highest priority.

16.6. AI technology for handling big data

Big data refers to very dynamic, large and complex datasets unable to be handled with traditional data processing technologies. For handling and processing such vast data, AI intervention in terms of either DL or ML is required. One particular example of big data is the data coming from the Internet of Things (IoT) world or weather forecasting which includes real-time data and data with a large variety.

Therefore, to store and handle such large data, scientists need large storage spaces and efficient algorithms that can process data within a fraction of seconds (Gao et al. 2021). Big data not only comes from the IoT but as wireless technology advances. People are using more social media to share and process lots of videos, increasing the amount of data. In addition to this, various sensory markets which exchange a lot of real-time data also require smart AI algorithm for processing.

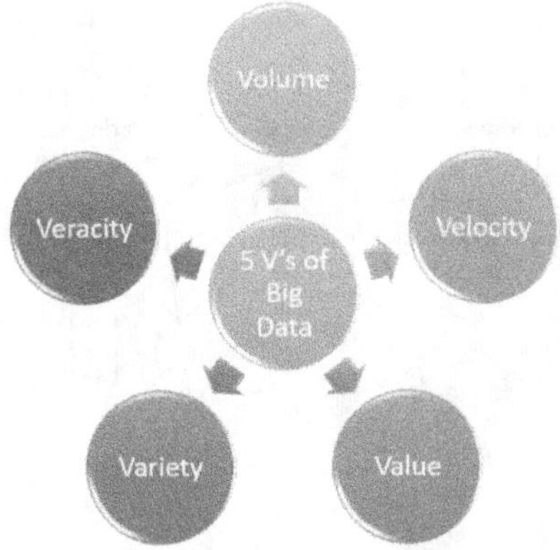

Figure 16.6. *5 Vs of big data*

Furthermore, to support AI wearables or smart homes discussed in the section above, big data need to be processed which can only be possible with the help of either ML or DL, part of AI (Ahmed et al. 2019). Big data is characterized by the 5 Vs: volume, velocity, value, veracity and variety as depicted in Figure 16.6.

– Volume: the main volume of data comes from real-time monitoring, including data from mobiles, videos, transaction history, sensex data and social sites, etc.

– Velocity: for example, when data speed is less than one second and it has to be processed within less than one second to avoid data traffic and congestion.

– Varity: when data is very unstructured and random, such as data from wireless signal noise which is unable to be fitted in any classification. To process such data, algorithms are based on statics and have lots of feedback loops making them complex.

– Value: data include lots of privacy and security features such as transactions, dealing with money in banking, data collected in business operations, etc. (Allam et al. 2019).

– Veracity: ensuring the truthfulness, correctness and integrity of large data. Veracity is very important when dealing with data related to banks or any commercial companies.

With the advancement in wireless technology, data are coming constantly from different sources. These data need to be sorted and processed according to user requirement. To do so, the AI algorithm needs to be developed according to user requirements. One such algorithm has been developed by Google called "MapReduce algorithm" (Benke et al. 2018) as elaborated in Figure 16.7.

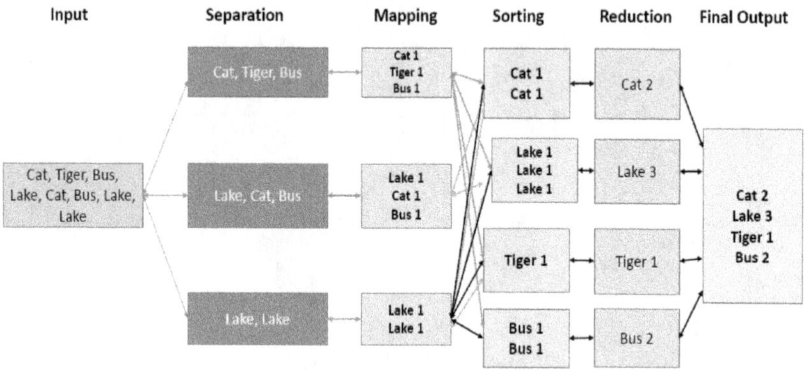

Figure 16.7. *MapReduce algorithm for sorting big data*

16.7. Use of AI-handled big data

To support and process big data, ML or DL has been used. ML is a subset of artificial learning whose focus is to train the machine so that it can sort the complex data and change the algorithm as per demand (Dong et al. 2020).

Companies are expecting more advancements in this operation. Thus, it has become an emerging technology in today's world.

In the following section, some important points are enlisted, mentioning the advantages of AI-handled data in various day-to-day fields.

Business

Big data handled with efficient algorithms store the customer's habits, interests and purchase records, and provides suggestions after analyzing previous data records. It gives the customer a personalized feel and increased customer satisfaction.

Healthcare

To monitor the patient's vitals, AI-equipped sensors have proven to be a potential candidate. Data stored by these sensors can be processed and transmitted to doctors in remote areas to help them with diagnoses, tracking health records and reactions to any medicine or therapy. This increases the provision of remote healthcare (Chopra et al. 2023).

Automobiles and transportation

Big data processed by AI can create a smart automobile, able to automatically adapt speed or even route on the basis of data such as terrain topography, weather data, destination point distance, traffic jams or traffic conditions in particular areas, etc. The big data can also lead to a smart car which can be fully automatic, meaning driven with no physical driver (Dhana et al. 2022).

Key AI technologies such as big data, robotics, automation and wearables are at the forefront in the advancement of various industries. Due to a large number of advantages, the demand for AI tools is increasing day by day, thus increasing the business opportunities in the field of AI. Figure 16.8 shows the increased demand for AI and average capital which will be spent on AI tools by different companies in the coming years.

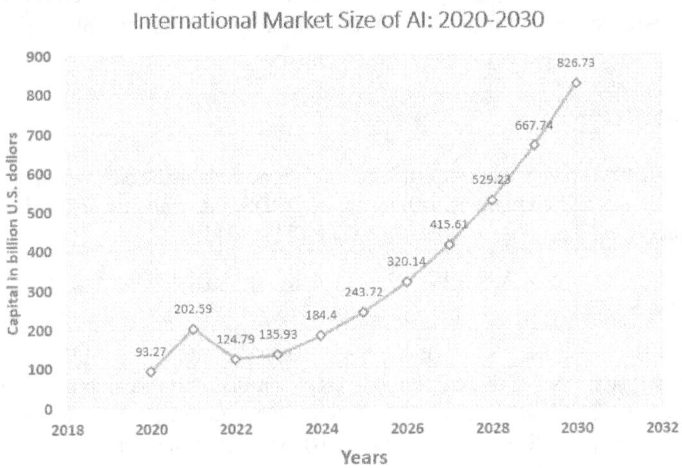

Figure 16.8. *AI tool demand in the coming years*

AI-supported robotics and automation can solve complex problems and help make optimized decisions which can mimic human behavior. Furthermore, AI-equipped wearable sensors enable people to track their physical records to help with diagnosis or other disease treatments. Wearable sensors also make remote health monitoring services easy and accurate. To support all of these facilities, real-time monitoring is required, which generates big and complex data (Soori et al. 2023). To store, process, handle and transport such big data, again AI proves a potential solution. The subset of AI technology such as DL and ML train machines to process big data and get a useful output from it. Thus, AI is a boon to human society when implemented ethically and responsibly.

16.8. Conclusion and future scope

Robotics, big data and wearables equipped with AI technologies are fundamentally restructuring the way industries operate and how we interact with machines. Robotics supported by AI algorithms leads to human-like machines and optimized decision-making from a number of options that can improve productivity and ensure human safety. Biosensors and wearables integrated with AI technologies can monitor human health, regulate body temperature and generate electricity from human temperature, making it useful and self-efficient without the need for external voltage. Furthermore, all of the complex data generated from the technologies discussed above are again handled and processed by AI. This unlocks vast opportunities for smart future predictions on the basis of previous records and optimized decision-making which can avoid natural or humanmade disasters. Thus, the bottom line is that AI has proven to be a boon to human society, but privacy must be considered, and it must be handled ethically and responsibly.

16.9. References

Ahmed, G. (2019). Management of artificial intelligence enabled smart wearable devices for early diagnosis and continuous monitoring of CVDS. *International Journal of Innovative Technology and Exploring Engineering*, 9(1), 1211–1215.

Allam, Z. and Dhunny, Z.A. (2019). On big data, artificial intelligence and smart cities. *Cities*, 89, 80–91.

Awotunde, J.B., Folorunso, S.O., Bhoi, A.K., Adebayo, P.O., Ijaz, M.F. (2021). Disease diagnosis system for IoT-based wearable body sensors with machine learning algorithm. In *Hybrid Artificial Intelligence and IoT in Healthcare*, Kumar Bhoi, A., Kumar Mallick, P., Mohanty, M.N., de Albuquerque, V.H.C. (eds). Springer, Singapore.

Benke, K. and Benke, G. (2018). Artificial intelligence and big data in public health. *International Journal of Environmental Research and Public Health*, 15(12), 2796.

Bharadiya, J.P. (2023). A comparative study of business intelligence and artificial intelligence with big data analytics. *American Journal of Artificial Intelligence*, 7(1), 24–30.

Bragazzi, N.L., Dai, H., Damiani, G., Behzadifar, M., Martini, M., Wu, J. (2020). How big data and artificial intelligence can help better manage the Covid-19 pandemic. *International Journal of Environmental Research and Public Health*, 17(9), 3176.

Chopra, R. (2023). Artificial intelligence in robotics. *International Journal for Research in Applied Science & Engineering Technology (IJRASE)*, 11(4), 2345–2349. doi: 10.22214/ijraset.2023.50635.

Chopra, B., Singha, A., Fariha, A., Gulwani, S., Parnin, C., Tiwari, A., Henley, A.Z. (2023). Conversational challenges in AI-powered data science: Obstacles, needs, and design opportunities. *arXiv:2310.16164*. doi: 10.48550/arXiv.2310.16164.

Deshpande, A. and Kumar, M. (2018). *Artificial Intelligence for Big Data: Complete Guide to Automating Big Data Solutions using Artificial Intelligence Techniques*. Packt Publishing Ltd, Birmingham.

Dhana, K., Franco, O.H., Ritz, E.M., Ford, C.N., Desai, P., Krueger, K.R., Holland, T.M., Dhana, A., Liu, X., Aggarwal, N.T. et al. (2022). Healthy lifestyle and life expectancy with and without Alzheimer's dementia: Population based cohort study. *BMJ*, 377, e068390. doi: 10.1136/bmj-2021-068390.

Dhanabalan, T. and Sathish, A. (2018). Transforming Indian industries through artificial intelligence and robotics in industry 4.0. *International Journal of Mechanical Engineering and Technology*, 9(10), 835–845.

Dietz, T., Dotzauer, M. (2024). The rise of sustainability provisions in international investment agreements. *Global Environmental Politics*, 24(4), 10–36.

Dong, K., Peng, X., Wang, Z.L. (2020). Fiber/fabric-based piezoelectric and triboelectric nanogenerators for flexible/stretchable and wearable electronics and artificial intelligence. *Advanced Materials*, 32(5), 1902549.

Gao, P., Li, J., Liu, S. (2021). An introduction to key technology in artificial intelligence and big data driven e-learning and e-education. *Mobile Networks and Applications*, 26(5), 2123–2126.

Gaur, L., Afaq, A., Singh, G., Dwivedi, Y.K. (2021). Role of artificial intelligence and robotics to foster the touchless travel during a pandemic: A review and research agenda. *International Journal of Contemporary Hospitality Management*, 33(11), 4079–4098.

Grischke, J., Johannsmeier, L., Eich, L., Griga, L., Haddadin, S. (2020). Dentronics: Towards robotics and artificial intelligence in dentistry. *Dental Materials*, 36(6), 765–778.

Haick, H. and Tang, N. (2021). Artificial intelligence in medical sensors for clinical decisions. *ACS Nano*, 15(3), 3557–3567.

Hisham, I., Khalil, M.I., Abbas, H. (2023). Multi-label ophthalmological disease classification using vision transformers. In *2023 5th Novel Intelligent and Leading Emerging Sciences Conference*. IEEE, Giza.

Jin, X., Liu, C., Xu, T., Su, L., Zhang, X. (2020). Artificial intelligence biosensors: Challenges and prospects. *Biosensors and Bioelectronics*, 165, 112412. doi: 10.1016/j.bios.2020.112412.

Khogali, H.O. and Mekid, S. (2023). The blended future of automation and AI: Examining some long-term societal and ethical impact features. *Technology in Society*, 73, 102232. doi: 10.1016/j.techsoc.2023.102232.

Kuru, K. (2024). Use of wearable miniaturised medical devices with artificial intelligence (AI) in enhancing physical medicine. In *World Congress on Physical Medicine and Rehabilitation*, Tokyo.

Leary, T.G., Green, R., Denson, K., Schoenfeld, G., Henley, T., Langford, H. (2013). The relationship among dysfunctional leadership dispositions, employee engagement, job satisfaction, and burnout. *The Psychologist-Manager Journal*, 16(2), 112–122.

Lee, S., Shi, Q., Lee, C. (2019). From flexible electronics technology in the era of IoT and artificial intelligence toward future implanted body sensor networks. *APL Materials*, 7(3), 031302.

Levi, R., Carli, F., Arévalo, A.R., Altinel, Y., Stein, D.J., Naldini, M.M., Celi, L.A. (2021). Artificial intelligence-based prediction of transfusion in the intensive care unit in patients with gastrointestinal bleeding. *BMJ Health & Care Informatics*, 28(1), e100245.

Lutz, C. (2019). Digital inequalities in the age of artificial intelligence and big data. *Human Behavior and Emerging Technologies*, 1(2), 141–148.

Miller, E.D. (2020). The COVID-19 pandemic crisis: The loss and trauma event of our time. *Journal of Loss and Trauma*, 25(6–7), 560–572.

Müller, V.C. (2020). Ethics of artificial intelligence and robotics. In *The Stanford Encyclopedia of Philosophy, Summer 2020 Edition*, Zalta, E.N. and Nodelman, U. (eds). The Metaphysics Research Lab – Stanford University, Stanford [Online] Available at: https://plato.stanford.edu/archives/sum2020/entries/ethics-ai/.

Nadikattu, A.K.R. (2021). Influence of artificial intelligence on robotics industry. *International Journal of Creative Research Thoughts (IJCRT)*, 9(1), 2320–2882.

Naha, A., Teixeira, A., Ahlén, A., Dey, S. (2022). Sequential detection of replay attacks. *IEEE Transactions on Automatic Control*, 68(3), 1941–1948.

Nahavandi, D., Alizadehsani, R., Khosravi, A., Acharya, U.R. (2022). Application of artificial intelligence in wearable devices: Opportunities and challenges. *Computer Methods and Programs in Biomedicine*, 213, 106541.

Nam, K., Dutt, C.S., Chathoth, P., Daghfous, A., Khan, M.S. (2021). The adoption of artificial intelligence and robotics in the hotel industry: Prospects and challenges. *Electronic Markets*, 31, 553–574.

Obschonka, M. and Audretsch, D.B. (2020). Artificial intelligence and big data in entrepreneurship: A new era has begun. *Small Business Economics*, 55, 529–539.

O'Leary, D.E. (2013). Artificial intelligence and big data. *IEEE Intelligent Systems*, 28(2), 96–99.

Panesar, S., Cagle, Y., Chander, D., Morey, J., Fernandez-Miranda, J., Kliot, M. (2019). Artificial intelligence and the future of surgical robotics. *Annals of Surgery*, 270(2), 223–226.

Raj, M. and Seamans, R. (2019). Primer on artificial intelligence and robotics. *Journal of Organization Design*, 8(1), 11.

Ramasamy, L.K., Khan, F., Shah, M., Prasad, B.V.V.S., Iwendi, C., Biamba, C. (2022). Secure smart wearable computing through artificial intelligence-enabled internet of things and cyber-physical systems for health monitoring. *Sensors*, 22(3), 1076.

Shi, Q., Dong, B., He, T., Sun, Z., Zhu, J., Zhang, Z., Lee, C. (2020). Progress in wearable electronics/photonics – Moving toward the era of artificial intelligence and internet of things. *InfoMat*, 2(6), 1131–1162.

Siau, K. and Wang, W. (2018). Building trust in artificial intelligence, machine learning, and robotics. *Cutter Business Technology Journal*, 31(2), 47.

Soori, M., Arezoo, B., Dastres, R. (2023). Artificial intelligence, machine learning and deep learning in advanced robotics, a review. *Cognitive Robotics*, 3, 54–70.

Tussyadiah, I. (2020). A review of research into automation in tourism: Launching the annals of tourism research curated collection on artificial intelligence and robotics in tourism. *Annals of Tourism Research*, 81, 102883.

Wisskirchen, G., Biacabe, B.T., Bormann, U., Muntz, A., Niehaus, G., Soler, G.J., von Brauchitsch, B. (2017). Artificial intelligence and robotics and their impact on the workplace. *IBA Global Employment Institute*, 11(5), 49–67.

Zhao, H., Su, R., Teng, L., Tian, Q., Han, F., Li, H., Cao, Z., Xie, R., Li, G., Liu, X. (2022). Recent advances in flexible and wearable sensors for monitoring chemical molecules. *Nanoscale*, 14(5), 1653–1669.

Zheng, Y., Tang, N., Omar, R., Hu, Z., Duong, T., Wang, J., Wu, W., Haick, H. (2021). Smart materials enabled with artificial intelligence for healthcare wearables. *Advanced Functional Materials*, 31(51), 2105482.

17

Enhancing Health Data Security and Intelligence with Blockchain and Artificial Intelligence

This chapter goes into great detail about the interesting things that blockchain and artificial intelligence (AI) can do for the healthcare business, especially when it comes to safely sharing health data. It shows how important it is to share health information safely in today's healthcare systems and that old methods do not always work. It also discusses how AI may help with handling healthcare data, which makes it easier to analyze and understand data, and thus make decisions. The chapter uses real-life examples to show how blockchain and AI can work together. For example, it shows how AI can improve sharing health data on the blockchain by adding features such as dynamic consent management and predictive analytics. It also talks about important topics such as ethics, permission, data security, scalability, interoperability and new industry standards. The chapter also looks ahead and imagines how blockchain and AI could make it easier to share healthcare data by making it easier to work with other systems, providing AI-driven clinical help, managing identities in a decentralized way and having strong rules. Finally, it stresses how important it is to involve all stakeholders in this changing path towards a data-driven future in healthcare, while also sharing how safe health data exchange could change the way healthcare works and the entire industry.

17.1. Introduction

In today's rapidly changing healthcare system, sharing and exchanging patient data is very important for providing care that is effective, efficient and patient-focused. Blockchain and artificial intelligence (AI) are two new technologies that make it easier to send health data safely. The rise of this new technology makes these changes easier (Tagde et al. 2021). In this digital age, healthcare is no longer limited

Chapter written by Charu CHHABRA, Sohrab Ahmad KHAN, Harsirjan KAUR, Kusum YADAV and Mohd. MAZHAR.

to just one hospital or clinic. A lot of different hospitals, professionals and providers treat patients, which creates a network of healthcare data. There are a number of reasons why approved stakeholders must share sensitive information in a safe way. Secure health data exchange lets doctors and nurses see all of a patient's medical history, which helps them make better and more personalized treatment decisions. In the end, this makes therapy better and helps patients more. Quick access to data makes it easier for healthcare professionals to work together on patients with complicated conditions. This cuts down on unnecessary tests, stops medications from interacting with each other and makes it easier for patients to switch between different types of treatment. Combining health data from different sources also means medical research and innovation is better, which changes how health problems are treated and prevented and how treatment is provided. Quick data exchange is needed for keeping an eye on epidemiological trends, dealing with outbreaks and handling public health emergencies (Lee et al. 2022). Sharing healthcare data in the past has caused a number of problems and issues, such as relying on old paper records and computer systems that do not work together. It is easy to break the law and gain access to, change or mess with paper records. But if electronic systems do not have enough protection, they could be hacked and data could be stolen (Seh et al. 2020). Different healthcare professionals may use electronic health record (EHR) systems that do not work well together, which makes problems worse. Data fragmentation makes it very hard to give patients consistent therapy. It is harder to protect patient privacy and get consent when information is shared by many groups. Managing rights in the traditional way may not work, which could lead to mistakes and wasted time. This makes it harder for healthcare professionals to talk to each other, especially when it comes to making sure that data are accurate and that interoperability standards are met (Ozair et al. 2015).

This chapter will look at how blockchain and AI could help to reduce the risks and problems that come with sharing healthcare data. We want to show how blockchain can protect data, stop changes and build trust in healthcare settings, as well as how AI can make blockchain better by helping healthcare professionals analyze, interpret and make decisions about data more effectively. We will look at real-world examples of how blockchain and AI are used, finishing the chapter with suggestions for further reading.

17.2. The need for secure health data sharing

As healthcare changes quickly, it is more important than ever to share health data safely. The amount of healthcare data being created is growing very quickly. The growth could be due to better medical technology, putting patient data online, the increase in wearable devices and the ability to combine information from different healthcare sources. The large amount of data coming in, such as patient histories, genetic information, medical imaging and clinical records, shows how important it is to have safe ways to communicate (Negro-Calduch et al. 2021). Figure 17.1 shows some of the benefits of transferring data correctly. The smooth exchange of healthcare data has many benefits, which shows how important it is in modern medicine.

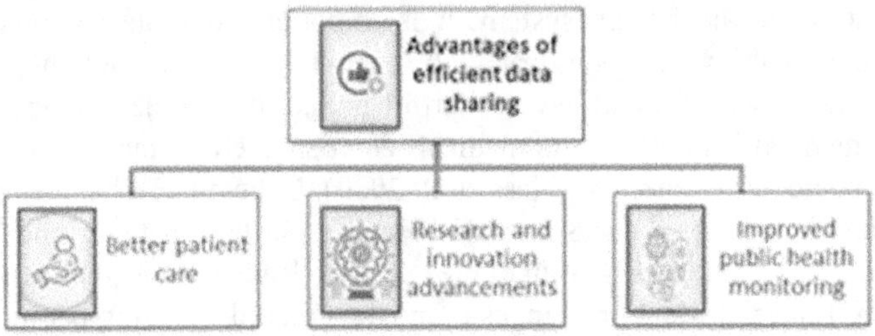

Figure 17.1. *Advantages of efficient data sharing*

17.2.1. *Better patient care*

When healthcare professionals have access to all relevant information, they can get a full picture of a patient's medical history and current health. With this information, they may be able to make better, more individualized treatment choices, which will lead to better outcomes for patients. Because it is easy to share data, the quality of care stays the same whether a patient sees their primary care doctor, a specialist or receives treatment at another facility (Adane et al. 2019).

17.2.2. *Research and innovation advancements*

Combining health data from different sources is very important for making progress in medical research and new ideas. Researchers can use large databases to find patterns in diseases, see how well treatments work and come up with new ways to treat them. This study helps not only the patients who take part, but it also adds to the body of medical knowledge and makes healthcare better (Dash et al. 2019).

17.2.3. *Improved public health monitoring*

In order to keep an eye on and fix public health problems, it is very important to share complete and up-to-date health data. This lets administrators quickly take steps to stop the spread of illness, keep an eye on how it spreads and find outbreaks. It plays a big role in managing chronic diseases by making sure that patients get the right treatment for their health (Edelstein et al. 2018).

17.3. Privacy concerns and security constraints

It may be very hard to share health information safely and securely. There are a number of significant problems in this area. Healthcare data often contain private information about a person's medical history, illnesses and treatments. To maintain patient trust in the healthcare system, it is important to protect patient privacy. People who gain access to this information without permission could hugely violate people's privacy, which could have a hurtful impact on people, making them less trusting of medical institutions. As medical records have become digital, protecting data has become very important (Nass et al. 2009). Cyberattacks happen more often on EHRs and healthcare databases. Data breaches put both patients and healthcare professionals at a huge risk since they could lead to the theft of personal information, financial fraud or changes to medical records. Getting permission from patients to share data can be very hard, especially when the data have to pass through a lot of different systems and companies. Patients need to be able to control who can see their health information and why they can see it (Yeo et al. 2022). To do this, there needs to be strong permission-management protocols in place. The existing problems and the greater need for sharing health data show that new ideas are needed. Blockchain and AI are two new technologies that could change how medical data are managed by making sure they are private and can be shared quickly. The next sections will look at how AI and blockchain technology could help to manage the problems safely and quickly, with a focus on the patient sending healthcare data (Tagde et al. 2021).

17.4. Blockchain technology

Blockchain technology could change a lot of industries, including healthcare, in a big way. Blockchain is basically a ledger system that is not controlled by one person. It is unique because it is decentralized and cannot be changed (Gad et al. 2022). Blockchain is different from traditional centralized systems because it works on a network of decentralized computers called nodes. A centralized system stores all of its data on one server and is run by a central authority. Blockchain uses a peer-to-peer network, which means that each node can help keep an eye on and confirm transactions. There are blocks in this system, and each block contains a piece of data. The fact that blockchain is decentralized is what makes it special. When something is decentralized, it means that there is not one person or group in charge of the whole network. On the other hand, data are spread out over many nodes and checked, which makes the data more durable and harder to change. Because no single group can change the system, the lack of a central authority encourages openness and trust among members. One more important thing about blockchain is that it cannot be changed. Adding data to a block makes it very hard to change or delete, which creates a chain of information that cannot be changed. There is a reference (or hash) to the block before it in each block, which makes the chain unbreakable. The strong cryptographic methods used means that changing one block requires all blocks that come after it to be changed, making it very difficult (Yaga et al. 2019).

In the field of healthcare data, blockchain technology has a number of important benefits. It protects the accuracy of healthcare data by making sure it does not change. Blockchain technology makes patient histories, test results and EHRs safe and unchangeable. This means it is much less likely that people will commit fraud or make changes without permission. This information is stored safely on many nodes and is encrypted, which keeps it safe. It is very hard to obtain unauthorized access or carry out data breaches because an attacker has to break into multiple nodes at once, which is hard to do because the system is decentralized and encrypted. Blockchain could make it easier for different healthcare professionals and systems to talk to each other. By providing a standardized and safe way for data to be shared, it reduces the fragmentation that is common in traditional healthcare information systems. This interoperability makes it easier for different healthcare organizations to share data and talk to each other (Shi et al. 2020).

Figure 17.2 shows some real-world examples of how blockchain is used in healthcare systems.

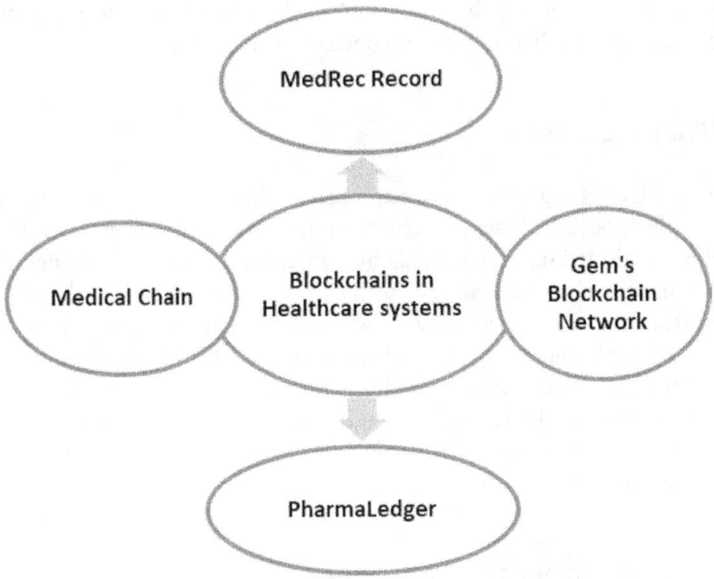

Figure 17.2. *Various real-world blockchains in the healthcare system*

17.4.1. *MedRec record*

The MedRec project, created by MIT researchers, looks into how blockchain technology can be used to create a patient-centered, decentralized EHR system. MedRec allows patients to maintain the highest levels of privacy and security, while still allowing healthcare to take place (Ekblaw et al. 2016).

17.4.2. *Medicalchain*

Medicalchain is a blockchain network that lets people safely store their medical records and gives healthcare professionals access to them when required. It strengthens the connection between patients and providers by making sure that data are correct, making telemedicine consultations easier, and making medical billing easier (Capece and Lorenzi 2020).

17.4.3. *Gem blockchain network*

Gem, a well-known name in the blockchain field, has created a network specifically for the healthcare industry. This network helps with a number of healthcare problems, such as managing the supply chain, processing claims and checking doctors' credentials. Gem uses blockchain technology to make healthcare processes safer, more reliable and more open (Ghosh et al. 2023).

17.4.4. *PharmaLedger*

PharmaLedger is a group of drug companies that use blockchain technology to make the pharmaceutical supply chain more open and easier to follow. This blockchain-based method makes medicines safer and means real medicines can be used, lowering the risks that come with fake medicines (Zakari et al. 2022). These examples from the real world show how blockchain technology can be used and adapted in the healthcare field. Blockchain could take the healthcare industry into a new era of trust, efficiency and creativity by making data safer, making it easier for different systems to work together and putting the patient first in terms of treatment. This includes everything from EHRs to managing the supply chain and making sure drugs are safe (Shi et al. 2020).

17.5. Blockchain in health data sharing

Many people are interested in using blockchain technology to securely share health data across different parts of the healthcare system (Fiore et al. 2023).

Table 17.1 lists a number of the main ways that blockchain can ensure that health data are shared safely.

A.	EHRs can be developed
B.	A standardized and safe platform for communication
C.	Automated and enforced consent management using smart contracts
D.	Improves traceability and transparency
E.	Safe data exchange between institutions and researchers
F.	Auditability and traceability of transactions

Table 17.1. *Benefits of blockchain*

17.5.1. *Patient-centered health records*

Blockchain means EHRs that put the patient's needs first can be created. This technology lets people control their medical information and lets doctors and nurses look at it when they need to. This patient-centered approach ensures that people can safely share and actively manage their health information, which helps them be more involved in making decisions about their treatment (Dubovitskaya et al. 2020).

17.5.2. *Interoperability*

Blockchain is a safe and standardized way for different healthcare providers and systems to talk to each other. It makes it easier for a lot of groups to share important information about patients (Schmeelk et al. 2022).

17.5.3. *Consent management*

Blockchain makes it possible to automate and enforce consent management through the use of smart contracts. Patients have the right to decide who can see their information and when they can see it. This makes it easier to get permission and makes sure that sharing data is legal and what patients want (Alhajri et al. 2022).

17.5.4. *Supply chain management*

Blockchain makes it easier to see and follow things in the supply chains of drugs and medical equipment. It checks items to make sure they are real. This keeps fake drugs from being sold and keeps patients safer (Gomasta et al. 2023).

17.5.5. *Research and clinical trials*

Businesses and schools can safely share data with blockchain. It is easier to access research data, which speeds up medical discoveries and clinical trials (Maslove et al. 2018). Blockchain fixes big issues in terms of who owns data, how access is controlled and how permission can be obtained. Patients can clearly state and enforce their consent choices with smart contracts based on blockchain technology. They can choose who can see their data, what information can be seen and how long that permission lasts. Automated consent management makes it easier to ensure that patients' wishes about sharing their data are followed. Blockchain gives patients control over their health data, which protects their rights and freedom over their information. Patients can let doctors or researchers see their data while still keeping it. Blockchain's access control systems make sure that only people or groups who are allowed to see certain health data can do so. This makes privacy and security better because it lowers the chance of unauthorized access (Khan et al. 2021).

When it comes to keeping an eye on and checking health data transactions, blockchain has a lot of good points. Data stored on the blockchain cannot be changed, and it will stay that way. This unchangeability keeps health data accurate because it stops people from making changes or deletions that are not allowed. The blockchain keeps track of all transactions in a clear and chronological order. This openness makes it possible to check the record, which makes people more responsible for how they obtain and share data. Every record has timestamps, which are important for keeping track of who has looked at the data and making sure that data sharing laws and agreements are followed. Blockchain uses cryptography to keep these audit trails from being changed. The user would not be able to change any of the following blocks, so trying to change records would make the audit less reliable (Jafri et al. 2022). Blockchain technology makes it very likely that patients, providers and other interested parties can safely share health data. This makes it easier for different systems to work together, respects patients' requests for permission, gives them control over their data and keeps a full record of who has access to and shares the data. The benefits of using blockchain in healthcare include making data safer, more open and more focused on what patients need (Ali et al. 2023).

17.6. Leveraging AI in health data sharing

AI is changing the way healthcare data are handled and the way decisions are made (Bajwa et al. 2021). Deep learning, machine learning and natural language processing are all elements of AI. Because it can analyze and understand big, complicated datasets, AI is very helpful for working with healthcare data. This is very important in a field where there are a lot of data that need to be used for patient care. AI's main strength is that it can find patterns, make useful conclusions and help people make smart choices in a wide range of healthcare settings. It can do more than just find out what is wrong with someone; it can also suggest treatments, help care for patients and make health systems better (Bajwa et al. 2021). AI helps people share healthcare data by using systems for data analysis, interpretation and decision support. It is great for data analysis because it can quickly look at big datasets and find patterns, outliers and connections that people would miss. This knowledge is very important for academics and healthcare workers who want to use different datasets to help them make decisions based on facts. AI systems can look at different types of medical data, such as genetic data, clinical records and medical images. Radiologists may use AI-powered algorithms to look for problems in MRIs or X-rays. This makes it easier to spot problems early and fix them quickly. AI-powered decision support systems help healthcare professionals right away by putting together patient data, medical knowledge and best practices to help with diagnosis, treatment planning and giving medication. These systems give advice that is in line with best practices, which improves the care (Jungwirth and Haluza 2023). AI-driven insights are going to make healthcare better. AI may be able to help create personalized treatment plans by analyzing patient data. It may suggest drugs that are more likely to work by looking at a patient's genetic profile, medical history and how they reacted to past treatments. This plan cuts down on the trial-and-error process and harmful effects. AI models can tell when people will get sick, when they will go to the hospital and when they will get better. This ability to see the future helps healthcare organizations make the best use of their resources, schedule patient appointments in the best way and take steps to avoid serious outcomes. AI also helps healthcare workers find groups of patients who are at high risk so they can give them the right drugs. AI can look at patient data to find people who are most likely to benefit from chronic illness management programs or preventive treatments (Davenport and Kalakota 2019). AI can help in ways such as setting up appointments, sending out bills and processing claims more smoothly. This efficiency lowers the costs of running the business and gives healthcare workers more time to care for patients. AI speeds up the process of making new drugs by looking at big datasets to find possible treatments and guess how well they will work. This is good for patients and helps medical research move faster by making new drugs and treatments faster (Shen et al. 2019; Robinson 2023). AI makes it easier for the healthcare industry to analyze, understand and make decisions about data by adding it to data exchange. AI-generated insights could change how healthcare is delivered and managed, making it more efficient and better for patients. AI in healthcare opens up endless possibilities for new ideas and improvements, which is good news for both patients and healthcare workers (Bajwa et al. 2021).

17.7. Combining blockchain and AI for enhanced security

Using blockchain technology and AI together to manage healthcare data is an example of how working together can help solve the difficult issues of privacy, security, data analysis and making decisions. Blockchain and AI work together to improve healthcare. Because blockchain is decentralized and cannot be changed, it is a safe and reliable way to share health data. It does away with the need for middlemen and promotes open, honest and patient-centered management of health records. Blockchain is mostly a record book, but it does not automatically look at or understand data. AI can handle and make sense of a lot of complicated data because it can learn and improve over time. AI can look at a lot of different sources, such as EHR, medical imaging, genetic sequences and patient histories, to find strange things, complicated patterns and connections in healthcare data (Tagde et al. 2021). AI could make it easier to share health data on the blockchain by using smart contracts, which are agreements that automatically enforce certain terms and conditions. These contracts can change as needed because they have AI in them. For example, if a patient's health changes, AI may automatically change the consent settings so that the patient's new wishes are followed when it comes to data access. This dynamic consent management system gives patients a lot more privacy and control. Using AI-powered predictive analytics with blockchain-based healthcare systems is also very useful. AI can look at old patient records very closely to find out what could cause diseases, what complications could occur and what can be done to stop them. For example, if an AI algorithm shows that a patient may be at a higher risk for a certain illness, the blockchain system could automatically limit access to some sensitive information or suggest more testing (Tagde et al. 2021). Medicalchain is a great example of how AI and blockchain can work together to share health data. This new technology uses AI to look at health data as it happens and blockchain to protect the data and limit who can see it. Patients still have control over their own data, even though AI algorithms may help doctors figure out what is wrong with them, what treatments to use and how to take their medications (Vyas et al. 2022). Another example is Covantis. Even though it was made for the agricultural industry, it shows how blockchain and AI can work together to manage a supply chain. It uses AI and machine learning to keep an eye on how safe and good the products are. Data are written down on the blockchain so that they are clear and easy to find. In the healthcare industry, this could also be used in pharmaceutical and medical supply chains (Dedeturk et al. 2021). Nebula Genomics is a genomics platform that uses AI and blockchain technology to look at and make sense of data. Users can safely share their genomic data on the blockchain. Then, AI algorithms can help them learn about their genetic risks, family history and any health problems they could have in the future. Genetic data are more useful for medical research and personalized therapy because blockchain is secure and AI can analyze data (Mazlan et al. 2020).

17.8. Challenges and concerns

Figure 17.3 shows some of the problems that need to be carefully thought about when using blockchain and AI to share healthcare data, even if there are many benefits to doing so.

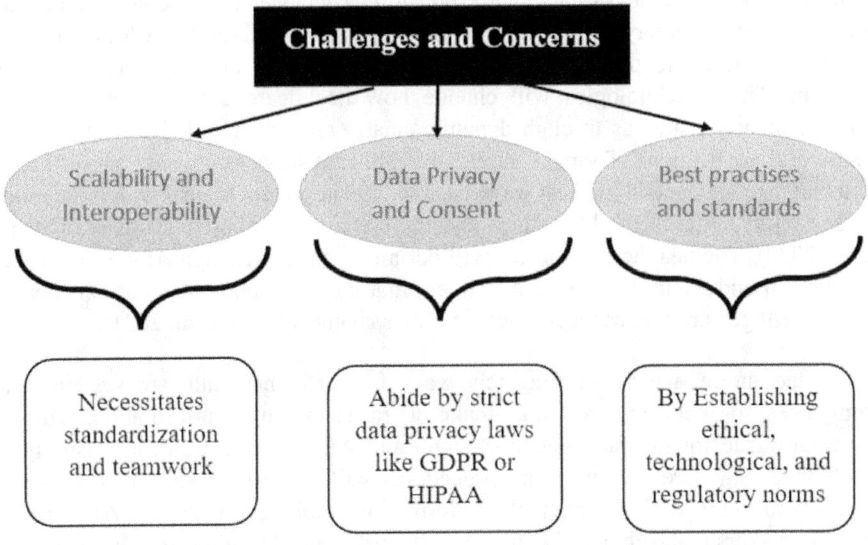

Figure 17.3. *Challenges and concerns of using AI and blockchain. For a color version of this figure, see: www.iste.co.uk/kumar/ai-driven1.zip*

17.8.1. *Scalability and interoperability*

When there are a lot of healthcare transactions, blockchain networks could have trouble growing. This is a big problem that everyone needs to help solve so that different blockchain systems and healthcare IT infrastructures can work together without any issues (Shuaib et al. 2021).

17.8.2. *Data privacy and consent*

AI systems that use blockchain must follow strict rules to keep data safe, such as HIPAA and GDPR. It is harder to keep track of consent when AI changes who can access things based on new information. To ease these worries about privacy, it is important to make the rules clear, have strong permission processes and let patients know (Upadrista et al. 2023).

17.8.3. *Best practices and standards*

It is important to set new industry standards and best practices as blockchain and AI become more common in healthcare. Healthcare providers, tech companies, regulators and patient advocacy groups all need to work together to make sure that healthcare data are shared safely and responsibly. They need to come up with ethical, technical and regulatory frameworks to do this (Tagde et al. 2021). Blockchain and AI work together to make sharing healthcare data safer, more private and more efficient. These technologies will change how healthcare data are handled in a number of ways, such as through dynamic smart contracts, predictive analytics and real-world applications. To make sure that healthcare services are safe and effective and that they are used in the best way possible, it is important to deal with issues such as scalability, interoperability, ethics and the creation of industry standards (Mehta et al. 2020). The healthcare industry will definitely have been significantly impacted blockchain and AI in the future. It is likely that the connection between blockchain and AI will get stronger as technology and rules change (Tagde et al. 2021).

In the future, we might find new ways for different healthcare systems and companies to work together that make it easier to share information. This is especially true for the fact that blockchain and AI are being used increasingly in healthcare. Standardized rules and procedures will make it easier for doctors and patients to share and obtain health information (Bohr et al. 2020). AI's job of helping doctors make choices is going to get harder. AI algorithms will be able to do more than just help with diagnosis and treatment. They will help doctors figure out when diseases are likely to break out, make the best use of resources and make treatment plans that are right for each patient. This constant change will make healthcare work better and faster (Javed et al. 2021). Decentralized identifiers (DIDs) are a way to keep track of digital identities using blockchain. We should see more of them in the future. This will give people more power over their health information and identities, which will in turn make healthcare interactions safer and more private (Taherdoost 2023). Regulatory bodies will set rules for how to use AI and blockchain in healthcare that are fair and ethical. Even if rules and morals are stricter, protecting patient privacy and data will still be very important (Taherdoost 2023).

17.9. Conclusion

Sharing health data safely could have a big impact on the health of patients and the healthcare industry as a whole. Blockchain and AI are becoming more common in healthcare. This could lead to better patient care, more research and new ideas, easier administrative tasks and better security. Making healthcare more patient-centered means giving people more control over their health data. Healthcare workers, IT developers, regulators and patients all need to work together to make this vision come true. This change needs to take into account moral issues, people's right to privacy and the rules. As blockchain and AI become more common for sharing health data, the main goal will be to make sure that all solutions put safety, effectiveness and patient needs first. We can picture a time when technology will help doctors to do a better job of caring for patients, speed up medical research and make health outcomes better. These technologies could improve the world for everyone by making people healthier, more connected and able to use their data.

17.10. References

Adane, K., Gizachew, M., Kendie, S. (2019). The role of medical data in efficient patient care delivery: A review. *Risk Management and Healthcare Policy*, 12, 67–73.

Alhajri, M., Rudolph, C., Shahraki, A.S. (2022). A blockchain-based consent mechanism for access to fitness data in the healthcare context. *IEEE Access*, 10, 22960–22979.

Ali, A., Al-Rimy, B.A.S., Almazroi, A.A., Alsubaei, F.S., Almazroi, A.A., Saeed, F. (2023). Securing secrets in cyber-physical systems: A cutting-edge privacy approach with consortium blockchain. *Sensors*, 23(16), 7162.

Bajwa, J., Munir, U., Nori, A., Williams, B. (2021). Artificial intelligence in healthcare: Transforming the practice of medicine. *Future Healthcare Journal*, 8(2), e188.

Bohr, A. and Memarzadeh, K. (eds) (2020). The rise of artificial intelligence in healthcare applications. In *Artificial Intelligence in Healthcare*. Academic Press, Cambridge, MA.

Capece, G. and Lorenzi, F. (2020). Blockchain and healthcare: Opportunities and prospects for the EHR. *Sustainability*, 12(22), 9693.

Dash, S., Shakyawar, S.K., Sharma, M., Kaushik, S. (2019). Big data in healthcare: Management, analysis and future prospects. *Journal of Big Data*, 6(1), 1–25.

Davenport, T. and Kalakota, R. (2019). The potential for artificial intelligence in healthcare. *Future Healthcare Journal*, 6(2), 94.

Dedeturk, B.A., Soran, A., Bakir-Gungor, B. (2021). Blockchain for genomics and healthcare: A literature review, current status, classification and open issues. *PeerJ*, 9, e12130.

Dubovitskaya, A., Baig, F., Xu, Z., Shukla, R., Zambani, P.S., Swaminathan, A., Wang, F. (2020). ACTION-EHR: Patient-centric blockchain-based electronic health record data management for cancer care. *Journal of Medical Internet Research*, 22(8), e13598.

Edelstein, M., Lee, L.M., Herten-Crabb, A., Heymann, D.L., Harper, D.R. (2018). Strengthening global public health surveillance through data and benefit sharing. *Emerging Infectious Diseases*, 24(7), 1324.

Ekblaw, A., Azaria, A., Halamka, J.D., Lippman, A. (2016). MedRec: A case study for blockchain in healthcare. White Paper, MIT Digital Currency Initiative Media Lab, Cambridge, MA [Online]. Available at: https://www.media.mit.edu/publications/medrec-whitepaper/.

Fiore, M., Capodici, A., Rucci, P., Bianconi, A., Longo, G., Ricci, M., Golinelli, D. (2023). Blockchain for the healthcare supply chain: A systematic literature review. *Applied Sciences*, 13(2), 686.

Gad, A.G., Mosa, D.T., Abualigah, L., Abohany, A.A. (2022). Emerging trends in blockchain technology and applications: A review and outlook. *Journal of King Saud University – Computer and Information Sciences*, 34(9), 6719–6742.

Ghosh, P.K., Chakraborty, A., Hasan, M., Rashid, K., Siddique, A.H. (2023). Blockchain application in healthcare systems: A review. *Systems*, 11(1), 38.

Gomasta, S.S., Dhali, A., Tahlil, T., Anwar, M.M., Ali, A.M.S. (2023). PharmaChain: Blockchain-based drug supply chain provenance verification system. *Heliyon*, 9(7), e17957.

Jafri, R. and Singh, S. (2022). Blockchain applications for the healthcare sector: Uses beyond Bitcoin. In *Blockchain Applications for Healthcare Informatics*, Tanwar, S. (ed.). Academic Press, Cambridge, MA.

Javed, I.T., Alharbi, F., Bellaj, B., Margaria, T., Crespi, N., Qureshi, K.N. (2021). Health-ID: A blockchain-based decentralized identity management for remote healthcare. *Healthcare*, 9(6), 712.

Jungwirth, D. and Haluza, D. (2023). Artificial intelligence and public health: An exploratory study. *International Journal of Environmental Research and Public Health*, 20(5), 4541.

Khan, S.N., Loukil, F., Ghedira-Guegan, C., Benkhelifa, E., Bani-Hani, A. (2021). Blockchain smart contracts: Applications, challenges, and future trends. *Peer-to-Peer Networking and Applications*, 14, 2901–2925.

Lee, S., Kim, J., Kwon, Y., Kim, T., Cho, S. (2022). Privacy preservation in patient information exchange systems based on blockchain: System design study. *Journal of Medical Internet Research*, 24(3), e29108.

Maslove, D.M., Klein, J., Brohman, K., Martin, P. (2018). Using blockchain technology to manage clinical trials data: A proof-of-concept study. *JMIR Medical Informatics*, 6(4), e11949.

Mazlan, A.A., Daud, S.M., Sam, S.M., Abas, H., Rasid, S.Z.A., Yusof, M.F. (2020). Scalability challenges in healthcare blockchain system – A systematic review. *IEEE Access*, 8, 23663–23673.

Mehta, S., Grant, K., Ackery, A. (2020). Future of blockchain in healthcare: Potential to improve the accessibility, security and interoperability of electronic health records. *BMJ Health & Care Informatics*, 27(3), e100217.

Nass, S.J., Levit, L.A., Gostin, L.O. (eds) (2009). The value and importance of health information privacy. In *Beyond the HIPAA Privacy Rule: Enhancing Privacy, Improving Health Through Research.* National Academies Press, Washington, D.C.

Negro-Calduch, E., Azzopardi-Muscat, N., Krishnamurthy, R.S., Novillo-Ortiz, D. (2021). Technological progress in electronic health record system optimization: Systematic review of systematic literature reviews. *International Journal of Medical Informatics*, 152, 104507.

Ozair, F.F., Jamshed, N., Sharma, A., Aggarwal, P. (2015). Ethical issues in electronic health records: A general overview. *Perspectives in Clinical Research*, 6(2), 73–76.

Robinson, C. (2023). Revolutionizing healthcare: The impact of AI, wearables, and robotics. LinkedIn [Online]. Available at: https://www.linkedin.com/pulse/revolutionizing-healthcare-impact-ai-wearables-charlie-robinson/.

Schmeelk, S., Kanabar, M., Peterson, K., Pathak, J. (2022). Electronic health records and blockchain interoperability requirements: A scoping review. *JAMIA Open*, 5(3), ooac068.

Seh, A.H., Zarour, M., Alenezi, M., Sarkar, A.K., Agrawal, A., Kumar, R., Khan, R.A. (2020). Healthcare data breaches: Insights and implications. *Healthcare*, 8(2), 133.

Shen, B., Guo, J., Yang, Y. (2019). MedChain: Efficient healthcare data sharing via blockchain. *Applied Sciences*, 9(6), 1207.

Shi, S., He, D., Li, L., Kumar, N., Khan, M.K., Choo, K.K.R. (2020). Applications of blockchain in ensuring the security and privacy of electronic health record systems: A survey. *Computers & Security*, 97, 101966.

Shuaib, M., Alam, S., Alam, M.S., Nasir, M.S. (2021). Compliance with HIPAA and GDPR in blockchain-based electronic health record (withdrawn). *Materials Today Proceedings*. doi: 10.1016/j.matpr.2021.03.059.

Tagde, P., Tagde, S., Bhattacharya, T., Tagde, P., Chopra, H., Akter, R., Kaushik, D., Rahman, M.H. (2021). Blockchain and artificial intelligence technology in e-Health. *Environmental Science and Pollution Research*, 28(38), 52810–52831.

Taherdoost, H. (2023). The role of blockchain in medical data sharing. *Cryptography*, 7(3), 36. doi: 10.3390/cryptography7030036.

Upadrista, V., Nazir, S., Tianfield, H. (2023). Secure data sharing with blockchain for remote health monitoring applications: A review. *Journal of Reliable Intelligent Environments*, 9(3), 349–368.

Vyas, S., Shabaz, M., Pandit, P., Parvathy, L.R., Ofori, I. (2022). Integration of artificial intelligence and blockchain technology in healthcare and agriculture (retracted). *Journal of Food Quality*, 1–11. doi: 10.1155/2024/9853093.

Yaga, D., Mell, P., Roby, N., Scarfone, K. (2019). Blockchain technology overview. *arXiv:1906.11078*. doi: 10.48550/arXiv.1906.11078.

Yeo, L.H. and Banfield, J. (2022). Human factors in electronic health records cybersecurity breach: An exploratory analysis. *Perspectives in Health Information Management*, 19(2), 1i.

Zakari, N., Al-Razgan, M., Alsaadi, A., Alshareef, H., Alashaikh, L., Alharbi, M., Alotaibi, S. (2022). Blockchain technology in the pharmaceutical industry: A systematic review. *PeerJ Computer Science*, 8, e840. doi: 10.7717/peerj-cs.840.

Zarrin, J., Wen Phang, H., Babu Saheer, L., Zarrin, B. (2021). Blockchain for decentralization of Internet: Prospects, trends, and challenges. *Cluster Computing*, 24(4), 2841–2866.

List of Authors

Sachin AHUJA
Department of Engineering
Chandigarh University
Mohali
India

Mahendra ALATE
Krishna Vishwa Vidyapeeth
(Deemed To Be University)
Karad
India

S. ANANDH
Krishna College of Physiotherapy
Krishna Vishwa Vidyapeeth
(Deemed To Be University)
Karad
India

Priya BATTA
Amity School of Engineering and Technology
Amity University Punjab
Mohali
India

Rohini BAWA
Department of Electronics and Communication Engineering
Chandigarh University
Mohali
India

Suresh J. BHOSALE
Department of Surgery
KIMS
Krishna Vishwa Vidyapeeth
(Deemed To Be University)
Karad
India

Nitin K. CHAUDHARY
Bhagwati Engineering College
Nagpur
India

Charu CHHABRA
Department of Physiotherapy
Jamia Hamdard University
New Delhi
India

Radhika CHINTAMANI
Department of Manual Therapy
Krishna College of Physiotherapy
Krishna Vishwa Vidyapeeth
(Deemed To Be University)
Karad
India

Vivek DESHPANDE
Department of Mechanical
Engineering
VIIT
Pune
India

Ankita DURGAWALE
Krishna College of Physiotherapy
Krishna Vishwa Vidyapeeth
(Deemed To Be University)
Karad
India

Anand GUDUR
Department of Oncology
KIMS
Krishna Vishwa Vidyapeeth
(Deemed To Be University)
Karad
India

Rashmi GUDUR
Department of Oncology
KIMS
Krishna Vishwa Vidyapeeth
(Deemed To Be University)
Karad
India

G. HIMASHREE
Department of Physiology
Krishna Vishwa Vidyapeeth
(Deemed To Be University)
Karad
India

Namrata KADAM
Krishna College of Physiotherapy
Krishna Vishwa Vidyapeeth
(Deemed To Be University)
Karad
India

Harsirjan KAUR
Department of Physiotherapy
SGT University
Gurugram
India

Sohrab Ahmad KHAN
Department of Physiotherapy
Jamia Hamdard University
New Delhi
India

Abhishek KUMAR
Chandigarh University
Mohali
India

Kalpana MALPE
HOD Department of Computer
Science
Sarvodaya Institute of Engineering
Nagpur
India

Dhiraj Kumar MANE
Krishna Vishwa Vidyapeeth
(Deemed To Be University)
Karad
India

Mohd. MAZHAR
KR Mangalam University
Gurugram
India

Vivek PARHATE
Suryodaya College of Engineering and Technology
Nagpur
India

Chandrakant PATIL
Krishna College of Physiotherapy
Krishna Vishwa Vidyapeeth (Deemed To Be University)
Karad
India

Surekha RANI
Department of Electronics and Communication Engineering
Chandigarh University
Mohali
India

Pramod Singh RATHORE
Manipal University Jaipur
India

Omkar SOMADE
Krishna College of Physiotherapy
Krishna Vishwa Vidyapeeth (Deemed To Be University)
Karad
India

Manoj VAIRALKAR
Govind Rao College of Engineering and Technology
Nagpur
India

G. VARADHARAJULU
Krishna College of Physiotherapy
Krishna Vishwa Vidyapeeth (Deemed To Be University)
Karad
India

Vibha VYAS
Electronics and Telecommunication Department
COEP Technological University (COEP Tech)
Pune
India

Kusum YADAV
Department of Physiotherapy
Sanskaram University
Haryana
India

Trupti YADAV
Krishna College of Physiotherapy
Krishna Vishwa Vidyapeeth (Deemed To Be University)
Karad
India

Index

A, B

C, D

F, H

I, M

N, O

P, Q

R, S

T, V, W

Summary of Volume 2

Chapter 7. Future Directions in Oncology: Emerging Technologies and Artificial Intelligence in Physiotherapy

Chandrakant PATIL, Dhiraj Kumar MANE, Anand GUDUR and Kalpana MALPE

Chapter 8. Deep Learning in Oncology and Physiotherapy: Enhancing Diagnosis and Recovery

Poonam PATIL, Neeraja ASWALE, Dhiraj Kumar MANE and Anand GUDUR

Chapter 10. Physiotherapy: A Holistic Approach to Rehabilitation and Functional Wellness

G. HIMASHREE, Radhika CHINTAMANI and G. VARADHARAJULU

Chapter 11. AI and Wearable Technology in Physiotherapy for Oncology Patients

Trupti YADAV, Anand GUDUR and Vibha VYAS

Chapter 13. Recurrent Neural Networks for Predictive Modeling in Cancer Time Series Data

Mayiri BURUNGALE, Suresh J. BHOSALE, Rashmi GUDUR and Shamla MANTRI

Chapter 16. Role of Physiotherapy in Cancer Recovery: A Review and Recommendation

Sougata PANDA

Chapter 17. The Role of Artificial Intelligence in Modern Healthcare: Transforming Diagnosis, Treatment and Rehabilitation

Mamta

Other titles from

ISTE

in

Computer Engineering

2025

CHIKH Azeddine
Systemic Approach to Categorizing and Modeling Requirements

TANEJA Ashu, KUMAR Abhishek, LIMKAR Suresh Vishnudas, OUAISSA Mariya, OUAISSA Mariyam
Driving Innovation through AI and Digital Twin for 6G Powered Sustainable Ultra Smart Cities

TANEJA Ashu, KUMAR Abhishek, LIMKAR Suresh Vishnudas, OUAISSA Mariya, OUAISSA Mariyam
6G Urban Innovation: AI and Digital Twin for Next-Gen Sustainable Cities

2024

CHAKRABORTY Rajdeep, GHOSH Anupam, KUMAR MANDAL Jyotsna, CHOUDHURY Tanupriya, CHATTERJEE Prasenjit
Smart Edge Computing: An Operation Research Perspective (International Perspectives in Decision Analytics and Operations Research Set – Volume 2)

HOMÈS Bernard
Fundamentals of Software Testing (2nd edition, revised and updated)

JERBI Abdessalem
Discrete-Event Simulation: Concepts and Production in Arena

PRADHAN Devasis, KUMAR SAHU Prasanna, MYO TUN Hla, CHATTERJEE Prasenjit
Artificial and Cognitive Computing for Sustainable Healthcare Systems in Smart Cities
(International Perspectives in Decision Analytics and Operations Research Set – Volume 3)

2023

CHOUDHURY Tanupriya, KHANNA Abhirup, CHATTERJEE Prasenjit, UM Jung-Sup, BHATTACHARYA Abhishek
Blockchain Applications in Healthcare: Innovations and Practices
(International Perspectives in Decision Analytics and Operations Research Set – Volume 1)

PAI G A Vijayalakshmi
A Textbook of Data Structures and Algorithms 1: Mastering Linear Data Structures
A Textbook of Data Structures and Algorithms 2: Mastering Nonlinear Data Structures
A Textbook of Data Structures and Algorithms 3: Mastering Advanced Data Structures and Algorithm Design Strategies

2022

HOMÈS Bernard
Advanced Testing of Systems-of-Systems 1: Theoretical Aspects
Advanced Testing of Systems-of-Systems 2: Practical Aspects

MEHTA Shikha, TIWARI Sanju, SIARRY Patrick, JABBAR M.A.
Tools, Languages, Methodologies for Representing Semantics on the Web of Things

SIDHOM Sahbi, KADDOUR Amira
Systems and Uses of Digital Sciences for Knowledge Organization
(Digital Tools and Uses Set – Volume 9)

ZAIDOUN Ameur Salem
Computer Science Security: Concepts and Tools

2021

DELHAYE Jean-Loïc
Inside the World of Computing: Technologies, Uses, Challenges

HARDIN Thérèse, JAUME Mathieu, PESSAUX François,
VIGUIÉ DONZEAU-GOUGE Véronique
Concepts and Semantics of Programming Languages 1: A Semantical Approach with OCaml and Python
Concepts and Semantics of Programming Languages 2: Modular and Object-oriented Constructs with OCaml, Python, C++, Ada and Java

MKADMI Abderrazak
Archives in The Digital Age: Preservation and the Right to be Forgotten (Digital Tools and Uses Set – Volume 8)

TOKLU Yusuf Cengiz, BEKDAŞ Gebrail, NIGDELI Sinan Melih
Metaheuristics for Structural Design and Analysis
(Optimization Heuristics Set – Volume 3)

2020

DARCHE Philippe
Microprocessor 1: Prolegomenes – Calculation and Storage Functions – Models of Computation and Computer Architecture
Microprocessor 2: Core Concepts – Communication in a Digital System
Microprocessor 3: Core Concepts – Hardware Aspects
Microprocessor 4: Core Concepts – Software Aspects
Microprocessor 5: Software and Hardware Aspects of Development, Debugging and Testing – The Microcomputer

DUVAUT Patrick, DALLOZ Xavier, MENGA David, KOEHL François,
CHRIQUI Vidal, BRILL Joerg
Internet of Augmented Me, I.AM: Empowering Innovation for a New Sustainable Future

LAFFLY Dominique
TORUS 1 – Toward an Open Resource Using Services: Cloud Computing for Environmental Data
TORUS 2 – Toward an Open Resource Using Services: Cloud Computing for Environmental Data
TORUS 3 – Toward an Open Resource Using Services: Cloud Computing for Environmental Data

LAURENT Anne, LAURENT Dominique, MADERA Cédrine
Data Lakes
(Databases and Big Data Set – Volume 2)

OULHADJ Hamouche, DAACHI Boubaker, MENASRI Riad
Metaheuristics for Robotics
(Optimization Heuristics Set – Volume 2)

SADIQUI Ali
Computer Network Security

VENTRE Daniel
Artificial Intelligence, Cybersecurity and Cyber Defense

2019

BESBES Walid, DHOUIB Diala, WASSAN Niaz, MARREKCHI Emna
Solving Transport Problems: Towards Green Logistics

CLERC Maurice
Iterative Optimizers: Difficulty Measures and Benchmarks

GHLALA Riadh
Analytic SQL in SQL Server 2014/2016

TOUNSI Wiem
Cyber-Vigilance and Digital Trust: Cyber Security in the Era of Cloud Computing and IoT

2018

ANDRO Mathieu
Digital Libraries and Crowdsourcing
(Digital Tools and Uses Set – Volume 5)

ARNALDI Bruno, GUITTON Pascal, MOREAU Guillaume
Virtual Reality and Augmented Reality: Myths and Realities

CARDON Alain
Beyond Artificial Intelligence: From Human Consciousness to Artificial Consciousness

HOMAYOUNI S. Mahdi, FONTES Dalila B.M.M.
Metaheuristics for Maritime Operations
(Optimization Heuristics Set – Volume 1)

JEANSOULIN Robert
JavaScript and Open Data

PIVERT Olivier
NoSQL Data Models: Trends and Challenges
(Databases and Big Data Set – Volume 1)

SEDKAOUI Soraya
Data Analytics and Big Data

SALEH Imad, AMMI Mehdi, SZONIECKY Samuel
Challenges of the Internet of Things: Technology, Use, Ethics
(Digital Tools and Uses Set – Volume 7)

SZONIECKY Samuel
Ecosystems Knowledge: Modeling and Analysis Method for Information and Communication
(Digital Tools and Uses Set – Volume 6)

2017

BENMAMMAR Badr
Concurrent, Real-Time and Distributed Programming in Java: Threads, RTSJ and RMI

BOUHAÏ Nasreddine, SALEH Imad
Internet of Things: Evolutions and Innovations
(Digital Tools and Uses Set – Volume 4)

HÉLIODORE Frédéric, NAKIB Amir, ISMAIL Boussaad, OUCHRAA Salma, SCHMITT Laurent
Metaheuristics for Intelligent Electrical Networks
(Metaheuristics Set – Volume 10)

MA Haiping, SIMON Dan
Evolutionary Computation with Biogeography-based Optimization
(Metaheuristics Set – Volume 8)

PÉTROWSKI Alain, BEN-HAMIDA Sana
Evolutionary Algorithms
(Metaheuristics Set – Volume 9)

PAI G A Vijayalakshmi
Metaheuristics for Portfolio Optimization: An Introduction using MATLAB®
(Metaheuristics Set – Volume 11)

REYES-GARCIA Everardo
The Image-Interface: Graphical Supports for Visual Information
(Digital Tools and Uses Set – Volume 3)

REYES-GARCIA Everardo, BOUHAÏ Nasreddine
Designing Interactive Hypermedia Systems
(Digital Tools and Uses Set – Volume 2)

SZONIECKY Samuel, BOUHAÏ Nasreddine
Collective Intelligence and Digital Archives: Towards Knowledge Ecosystems
(Digital Tools and Uses Set – Volume 1)

2016

BLUM Christian, FESTA Paola
Metaheuristics for String Problems in Bio-informatics
(Metaheuristics Set – Volume 6)

DEROUSSI Laurent
Metaheuristics for Logistics
(Metaheuristics Set – Volume 4)

DHAENENS Clarisse, JOURDAN Laetitia
Metaheuristics for Big Data
(Metaheuristics Set – Volume 5)

LABADIE Nacima, PRINS Christian, PRODHON Caroline
Metaheuristics for Vehicle Routing Problems
(Metaheuristics Set – Volume 3)

LEROY Laure
Eyestrain Reduction in Stereoscopy

LUTTON Evelyne, PERROT Nathalie, TONDA Alberto
Evolutionary Algorithms for Food Science and Technology
(Metaheuristics Set – Volume 7)

MAGOULÈS Frédéric, ZHAO Hai-Xiang
Data Mining and Machine Learning in Building Energy Analysis

RIGO Michel
Advanced Graph Theory and Combinatorics

ZELM Martin, DOUMEINGTS Guy, MENDONÇA Joao Pedro
Enterprise Interoperability: in the Digitized and Networked Factory of the Future

2015

BARBIER Franck, RECOUSSINE Jean-Luc
COBOL Software Modernization: From Principles to Implementation with the BLU AGE® Method

CHEN Ken
Performance Evaluation by Simulation and Analysis with Applications to Computer Networks

CLERC Maurice
Guided Randomness in Optimization
(Metaheuristics Set – Volume 1)

DURAND Nicolas, GIANAZZA David, GOTTELAND Jean-Baptiste, ALLIOT Jean-Marc
Metaheuristics for Air Traffic Management
(Metaheuristics Set – Volume 2)

MAGOULÈS Frédéric, ROUX François-Xavier, HOUZEAUX Guillaume
Parallel Scientific Computing

MUNEESAWANG Paisarn, YAMMEN Suchart
Visual Inspection Technology in the Hard Disk Drive Industry

2014

BOULANGER Jean-Louis
Formal Methods Applied to Industrial Complex Systems
Formal Methods Applied to Complex Systems: Implementation of the B Method

EXPOSITO Ernesto, DIOP Codé
Smart SOA Platforms in Cloud Computing Architectures

GARDI Frédéric, BENOIST Thierry, DARLAY Julien, ESTELLON Bertrand, MEGEL Romain
Mathematical Programming Solver based on Local Search

KRICHEN Saoussen, CHAOUACHI Jouhaina
Graph-related Optimization and Decision Support Systems

LARRIEU Nicolas, VARET Antoine
Rapid Prototyping of Software for Avionics Systems: Model-oriented Approaches for Complex Systems Certification

LAURAS Matthieu, ZELM Martin, ARCHIMÈDE Bernard, BÉNABEN Frédérick, DOUMEINGTS Guy
Enterprise Interoperability: Interoperability for Agility, Resilience and Plasticity of Collaborations

OUSSALAH Mourad Chabane
Software Architecture 1
Software Architecture 2

QUESNEL Flavien
Scheduling of Large-scale Virtualized Infrastructures: Toward Cooperative Management

RIGO Michel
Formal Languages, Automata and Numeration Systems 1: Introduction to Combinatorics on Words
Formal Languages, Automata and Numeration Systems 2: Applications to Recognizability and Decidability

SAINT-DIZIER Patrick
Musical Rhetoric: Foundations and Annotation Schemes

TOUATI Sid, DUPONT DE DINECHIN Benoit
Advanced Backend Code Optimization

2013

ANDRÉ Etienne, SOULAT Romain
The Inverse Method: Parametric Verification of Real-time Embedded Systems

BOULANGER Jean-Louis
Safety Management for Software-based Equipment

DELAHAYE Daniel, PUECHMOREL Stéphane
Modeling and Optimization of Air Traffic

FRANCOPOULO Gil
LMF — Lexical Markup Framework

GHEDIRA Khaled
Constraint Satisfaction Problems: CSP Formalisms and Techniques

GOVAERT Gérard, NADIF Mohamed
Co-Clustering: Models, Algorithms and Applications

LANDRAGIN Frédérick
Man–Machine Dialogue: Design and Challenges

LUCAS Laurent, LOSCOS Céline, REMION Yannick
3D Video: Models, From Capture to Diffusion

ROCHANGE Christine, UHRIG Sascha, SAINRAT Pascal
Time-Predictable Architectures

WAHBI Mohamed
Algorithms and Ordering Heuristics for Distributed Constraint Satisfaction Problems

ZELM Martin, VAN SINDEREN Marten, FERRAIRA PIRES Luis, DOUMEINGTS Guy
Enterprise Interoperability: Research and Applications in the Service-oriented Ecosystem

2012

ARBOLEDA Hugo, ROYER Jean-Claude
Model-Driven and Software Product Line Engineering

BLANCHET Gérard, DUPOUY Bertrand
Computer Architecture

BOULANGER Jean-Louis
Industrial Use of Formal Methods: Formal Verification
Formal Methods: Industrial Use from Model to the Code

CALVARY Gaëlle, DELOT Thierry, SÈDES Florence, TIGLI Jean-Yves
Computer Science and Ambient Intelligence

MAHOUT Vincent
Assembly Language Programming: ARM Cortex-M3 2.0: Organization, Innovation and Territory

MARLET Renaud
Program Specialization

SOTO Maria, SEVAUX Marc, ROSSI André, LAURENT Johann
Memory Allocation Problems in Embedded Systems: Optimization Methods

2011

BICHOT Charles-Edmond, SIARRY Patrick
Graph Partitioning

BOULANGER Jean-Louis
Static Analysis of Software: The Abstract Interpretation

CAFERRA Ricardo
Logic for Computer Science and Artificial Intelligence

KORDON Fabrice, HADDAD Serge, PAUTET Laurent, PETRUCCI Laure
Distributed Systems: Design and Algorithms
Models and Analysis in Distributed Systems

LORCA Xavier
Tree-based Graph Partitioning Constraint

RIDHA MAHJOUB A.
Progress in Combinatorial Optimization: ISCO2010

TRUCHET Charlotte, ASSAYAG Gerard
Constraint Programming in Music

VICAT-BLANC Pascale, GOGLIN Brice, GUILLIER Romaric,
SOUDAN Sébastien
Computing Networks: From Cluster to Cloud Computing

2010

AUBRUN Christophe, SIMON Daniel, SONG Ye-Qiong
Co-design Approaches for Dependable Networked Control Systems

AUDIBERT Pierre
Mathematics for Informatics and Computer Science

BABAU Jean-Philippe, BLAY-FORNARINO Mireille, CHAMPEAU Joël,
ROBERT Sylvain, SABETTA Antonino
Model Driven Engineering for Distributed Real-Time Systems: MARTE modeling, model transformations and their usages

BOULANGER Jean-Louis
Safety of Computer Architectures

SIGAUD Olivier, BUFFET Olivier
Markov Decision Processes in Artificial Intelligence

MONMARCHÉ Nicolas, GUINAND Frédérick, SIARRY Patrick
Artificial Ants

SOLNON Christine
Ant Colony Optimization and Constraint Programming

2009

FOURNIER Jean-Claude
Graph Theory and Applications: with exercises and problems

GUÉDON Jeanpierre
The Mojette Transform: theory and applications

JARD Claude, ROUX Olivier H.
Communicating Embedded Systems: Software and Design

LECOUTRE Christophe
Constraint Networks: Techniques and Algorithms

2008

BANÂTRE Michel, MARRON Pedro Jose, OLLERO Anibal, WOLITZ Adam
Cooperating Embedded Systems and Wireless Sensor Networks

MERZ Stephan, NAVET Nicolas
Modeling and Verification of Real-time Systems: Formalisms and Software Tools

PASCHOS Vangelis Th.
Combinatorial Optimization and Theoretical Computer Science: Interfaces and Perspectives

WALDNER Jean-Baptiste
Nanocomputers and Swarm Intelligence

2007

BENHAMOU Frédéric, JUSSIEN Narendra, O'SULLIVAN Barry
Trends in Constraint Programming

JUSSIEN Narendra
A to Z of Sudoku

2006

BABAU Jean-Philippe, CHAMPEAU Joël, GÉRARD Sébastien
From MDD Concepts to Experiments and Illustrations – DRES 2006

HABRIAS Henri, FRAPPIER Marc
Software Specification Methods: An Overview Using a Case Study

MURAT Cécile, PASCHOS Vangelis Th.
Probabilistic Combinatorial Optimization on Graphs

PANETTO Hervé, BOUDJLIDA Nacer
Interoperability for Enterprise Software and Applications: Proceedings of the Workshops and the Doctorial Symposium of the Second IFAC/IFIP I-ESA International Conference: EI2N, WSI, IS-TSPQ 2006

2005

GÉRARD Sébastien, BABAU Jean-Philippe, CHAMPEAU Joël
Model Driven Engineering for Distributed Real Time Embedded Systems

PANETTO Hervé
Interoperability of Enterprise Software and Applications: Workshops of the INTEROP-ESA 2005 International Conference (EI2N, WSI, ISIDI and IEHENA 2005)

www.ingramcontent.com/pod-product-compliance
Lightning Source LLC
LaVergne TN
LVHW022314291125
826349LV00003B/91
9781836690474